Advances in Computer-Aided Technology

Advances in Computer-Aided Technology

Editors

Marek Kočiško
Martin Pollák

MDPI • Basel • Beijing • Wuhan • Barcelona • Belgrade • Manchester • Tokyo • Cluj • Tianjin

Editors
Marek Kočiško
Technical University of Kosice
Slovakia

Martin Pollák
Technical University of Kosice
Slovakia

Editorial Office
MDPI
St. Alban-Anlage 66
4052 Basel, Switzerland

This is a reprint of articles from the Special Issue published online in the open access journal *Machines* (ISSN 2075-1702) (available at: https://www.mdpi.com/journal/machines/special_issues/computer_aided).

For citation purposes, cite each article independently as indicated on the article page online and as indicated below:

LastName, A.A.; LastName, B.B.; LastName, C.C. Article Title. *Journal Name* **Year**, *Volume Number*, Page Range.

ISBN 978-3-0365-6746-4 (Hbk)
ISBN 978-3-0365-6747-1 (PDF)

Contents

About the Editors

Marek Kočiško

Prof. Dr. Marek Kočiško graduated in 2001 from the Faculty of Manufacturing Technologies of the Slovak Technical University of Kosice and completed his PhD. in 2004 in the field of engineering technologies and materials. He currently works as a professor (2022) and the head of the Department of Computer Aided Manufacturing Technologies at the Faculty of Manufacturing Technologies with a seat in Presov, TUKE. His work focuses on computer-aided production processes, CAD/CAM/CAE systems, simulation of production systems, computer networks, virtual (augmented) reality, and reverse- and simultaneous engineering. His strengths include knowledge of PC hardware, computer networks, notification technology, excellent knowledge of operating systems, office systems, computer simulation systems, and CAD/CAM/CAE systems (Creo, Unigraphics NX, Solid Edge, etc.). He is an expert member of many national and international associations and a reviewer member of many journals registered in the Current Content, Web of Science and Scopus databases.

Martin Pollák

Dr. Martin Pollák graduated in 2014 from the Faculty of Manufacturing Technologies of the Slovak Technical University of Kosice and completed his PhD. in 2017 in the field of engineering technologies and materials. He currently works as an assistant professor (2017) at the Faculty of Manufacturing Technologies in Presov, Department of Computer Aided Manufacturing Technologies. His work focuses on CA systems, computer simulation, machine vision using UI, automation and manufacturing technologies with a focus on the use of additive technology in robotics. His strengths include knowledge of working in CAD/CAM/CAE systems and simulation systems for the design of robotic workplaces, programming of industrial, collaborative and educational robots, diagnostics of production equipment and work with programming languages. He has extensive experience in solving scientific research and design tasks, which he actively applies in the process of design and programming. He is a reviewer member of many journals registered in the Current Content, Web of Science and Scopus databases.

machines

MDPI

Editorial

Editorial for Special Issue "Advances in Computer-Aided Technology"

Martin Pollák * and Marek Kočiško

Faculty of Manufacturing Technologies with a Seat in Presov, Technical University of Kosice, Bayerova 1, 080 01 Presov, Slovakia
* Correspondence: martin.pollak@tuke.sk

The determination of this Special Issue topic, in the field of Advances in Computer-Aided Technology belonging to the section "Mechatronic and Intelligent Machines", was well received by the community of scientists and researchers. Computer-aided technologies (CAx) encompass the use of computer technology to aid in the design, analysis, and manufacturing of products. Significant progress has also been made in this area, due to the rapid expansion of science and technology. Advanced CAx tools combine many different aspects of product lifecycle management (PLM), including design, finite element analysis (FEA), manufacturing, production planning, and product. In terms of the transition to the Industry 4.0 concept, the idea of the digital twin comes to the fore, and existing CAx systems must adapt to this trend. This book presents 13 selected papers that are related to industrial applications in several major topics of the scientific main areas such as New trends in CAx systems, Digital manufacturing, Internet of Things in manufacturing, Simulation of production systems and processes, Systems for advanced finite element analysis, Material engineering, Digitization, and 3D scanning.

A multi-case load spectrum-compiling method is proposed in the study of authors Wang et al., for the dynamic modeling of a hydraulic excavator stick to simplify and accelerate the fatigue bench test. Fatigue tests were performed using a program load spectrum and using a random spectrum [1]. The main goal of the paper of authors Skrivanek et al. is to present the effect of a change in the drive structure of a small-diameter knitting machine. The novelty of the work was a comprehensive comparison of the existing and modified drive structure with a description of the possible benefits in terms of dynamic behavior, overall simplification of the machine concept and expected savings in terms of electricity consumption [2]. The article by Mitalova et al. describes the initial study and development in the field of machining one kind of composite material with natural reinforcement WPC with AWJ. The influence of selected technological factors on the topography of resulting surfaces has been studied in particular [3]. The article of the authors Lishchenko et al. describes research on how to develop information signals in time and frequency domains containing the surface quality diagnostic features to monitor the 3D-printing technological system state. The article shows how it is possible to increase the efficiency of the online monitoring of the quality of the 3D printing technological process using an optical contactless high-performance measuring instrument. The research results are applicable to 3D-printing machines [4]. The study of the authors Ivanov et al. describes the production of flexible fixtures with modern machining technologies capable of machining similar shapes and sizes of parts with complex geometries [5]. The paper from the authors Slapak et al. describes a new measurement approach to obtain the necessary transmission parameters together with its mathematical model. A mathematical model was created to calculate the nonlinear friction of the gearbox in all four quadrants with a detailed description of the simulation of the servo drive with the modeled gearbox. The results are compared with the real measurement of the system [6]. The paper from the authors Komarek et al. deals with the modification of the existing mechanical system of the needle bar. The purpose of the

Citation: Pollák, M.; Kočiško, M. Editorial for Special Issue "Advances in Computer-Aided Technology". *Machines* **2023**, *11*, 227. https://doi.org/10.3390/machines11020227

Received: 1 February 2023
Accepted: 2 February 2023
Published: 3 February 2023

published research was to design a completely new principle for controlling the grip and release of the needle by means of an electromagnet located in the body of the needle bar. This innovative solution will eliminate the noise generated by the release of the needle on the original system [7]. The usage area of digital impressions, which is the primary stage of digital dentistry, is increasing worldwide; there is a lack of performance data on intraoral scanners, especially in the field of implantology, which is dealt with in more detail in the study by the authors Albayrak et al. [8]. The aim of the research by the authors Pollák et al. was to determine the deep-drilling process input conditions. The focus of the analysis was on how the monitored technological and physical impacts translate into achieving the required gun-drill life and the quality and dimensional accuracy of deep holes, as well as their overall impact on tool life [9]. The study by the authors Ryumin et al. describes the development in the field of design of ship structures, focusing on the detailed description of an integrated system capable of performing both the design and modeling of the structure of merchant ships [10]. The paper by the authors Grozav et al. proposes a study on the feasibility of implementing Deep Neural Networks for predicting the dimensional accuracy and the mechanical characteristics of components obtained through the FDM method using empirical data acquired by high-precision metrology [11]. The paper by the authors Solfronk et al. presents an investigation of the computational strategy used in FEA on the springback prediction of a thin sandwich material made of micro-alloyed steel. The Vegter yield criterion combined both with isotropic and kinematic hardening law was used for their own calculation [12]. The article by the authors Stejskal et al. presents a direct analytical expression covering the rank of two tensors in a plane. The method was applied to the simplest type of tensor, which represented an advantage for the analytical approach of the method. Numerical methods for tensor rendering can be replaced with an accurate analytical method [13].

The editorial team thanks all authors of the published articles as well as authors of articles who were interested in publishing in the Special Issue. We especially thank the reviewers, who examined all the submitted articles in detail and, through their scientific experience, expressed their opinion on the solved issue and helped with valuable advice in improving the presented results of the authors' studies.

Conflicts of Interest: The authors declare no conflict of interest.

References

1. Wang, P.; Xiang, Q.; Królczyk, G.; Lu, P.; Wang, B.; Li, Z. Dynamic Modeling of a Hydraulic Excavator Stick by Introducing Multi-Case Synthesized Load Spectrum for Bench Fatigue Test. *Machines* **2022**, *10*, 741. [CrossRef]
2. Skrivanek, J.; Bilek, M.; Batka, O. Modified Drive Structure of a Knitting Machine. *Machines* **2022**, *10*, 615. [CrossRef]
3. Mital'ová, Z.; Botko, F.; Vandžura, R.; Litecká, J.; Mital', D.; Simkulet, V. Machining of Wood Plastic Machining of Wood Plastic Composite Using AWJ Technology with Controlled Output Quality. *Machines* **2022**, *10*, 566.
4. Lishchenko, N.; Pitel', J.; Larshin, V. Online Monitoring of Surface Quality for Diagnostic Features in 3D Printing. *Machines* **2022**, *10*, 541. [CrossRef]
5. Ivanov, V.; Botko, F.; Dehtiarov, I.; Kočiško, M.; Evtuhov, A.; Pavlenko, I.; Trojanowska, J. Development of Flexible Fixtures with Incomplete Locating: Connecting Rods Machining Case Study. *Machines* **2022**, *10*, 493. [CrossRef]
6. Šlapák, V.; Ivan, J.; Kyslan, K.; Hric, M.; Ďurovský, F.; Paulišin, D.; Kočiško, M. Measurement and Modelling of a Cycloidal Gearbox in Actuator with Permanent Magnet Synchronous Machine. *Machines* **2022**, *10*, 344. [CrossRef]
7. Komárek, J.; Klogner, V. Design of Electromagnetic Control of the Needle Gripping Mechanism. *Machines* **2022**, *10*, 309. [CrossRef]
8. Albayrak, B.; Korkmaz, I.H.; Wee, A.G.; Sukotjo, C.; Bayındır, F. Assessing the Effect of Interimplant Distance and Angle on Different Impression Techniques. *Machines* **2022**, *10*, 293. [CrossRef]
9. Pollák, M.; Kočiško, M.; Petrus, J.; Grozav, S.D.; Ceclan, V. Research into the Impact of Spindle Speed and Feed Rate Changes on the Life of a Deep-Drilling Technology Tool. *Machines* **2022**, *10*, 268. [CrossRef]
10. Ryumin, S.; Tryaskin, V. Computer-Aided System for Parametric Design of Ship Hull Structures—CADS-Hull. *Machines* **2022**, *10*, 262. [CrossRef]
11. Grozav, S.D.; Sterca, A.D.; Kočiško, M.; Pollák, M.; Ceclan, V. Feasibility of Predictive Models for the Quality of Additive Manufactured Components Based on Artificial Neural Networks. *Machines* **2022**, *10*, 128. [CrossRef]

12. Solfronk, P.; Sobotka, J.; Koreček, D. Effect of the Computational Model and Mesh Strategy on the Springback Prediction of the Sandwich Material. *Machines* **2022**, *10*, 114. [CrossRef]
13. Stejskal, T.; Svetlík, J.; Dobránsky, J. An Analytical Method for Tensor Visualization in a Plane. *Machines* **2022**, *10*, 89. [CrossRef]

machines

Article

Dynamic Modeling of a Hydraulic Excavator Stick by Introducing Multi-Case Synthesized Load Spectrum for Bench Fatigue Test

Penghui Wang [1], Qingyi Xiang [1,*], Grzegorz Królczyk [2], Pengmin Lu [1], Binhua Wang [1] and Zhixiong Li [3,*]

1 Key Laboratory of Road Construction Technology and Equipment Ministry of Education, Chang'an University, Xi'an 710064, China
2 Faculty of Mechanical Engineering, Opole University of Technology, 45-758 Opole, Poland
3 Yonsei Frontier Lab, Yonsei University, Seoul 03722, Korea
* Correspondence: xiangqy@chd.edu.cn (Q.X.); zhixiong.li@yonsei.ac.kr (Z.L.)

Abstract: A multi-case load spectrum compiling method is proposed in this study for dynamic modeling of a hydraulic excavator stick to simplify and accelerate the fatigue bench test. This new method includes a simplified criterion of small-load-omitting threshold based on the principle of invariable fatigue damage, an extreme value inference criterion based on the overflow characteristics of the hydraulic system, and a synthetic extrapolation method under various working conditions. Firstly, a one-dimensional spectrum of a medium-sized excavator stick was compiled. Then, the program load spectrum for the bench fatigue test was obtained by modifying the one-dimensional spectrum based on the damage consistency criterion and the damage equivalent principle. Lastly, the fatigue tests were conducted using the program load spectrum, as well as using the random spectrum. The comparison results demonstrate that the damage location and fatigue life distribution of the stick using these two spectra are generally consistent, with a relative error smaller than 8.8%; however, the proposed program load spectrum can accelerate the test process with less time consuming than that of the random spectrum. As a result, the multi-case load spectrum is feasible and reliable for dynamic modeling of the hydraulic excavator stick in practice.

Keywords: dynamic modeling; load spectrum reconstruction; fatigue test; hydraulic excavator

Citation: Wang, P.; Xiang, Q.; Królczyk, G.; Lu, P.; Wang, B.; Li, Z. Dynamic Modeling of a Hydraulic Excavator Stick by Introducing Multi-Case Synthesized Load Spectrum for Bench Fatigue Test. *Machines* **2022**, *10*, 741. https://doi.org/10.3390/machines10090741

Academic Editors: Marek Kočiško and Martin Pollák

Received: 3 August 2022
Accepted: 24 August 2022
Published: 28 August 2022

1. Introduction

The fatigue life of the key structural components is a critical index that affects the safety and reliability of the hydraulic excavator. To evaluate the fatigue life, the bench fatigue test is widely adopted. One prerequisite of the bench fatigue test is that the load in the test must reflect the influence of the real load on the fatigue life of structural components. As the structural components of the excavator usually suffer random load in practical applications, the random load spectrum, which can consider the load sequences, is the optimal load choice. However, the random fatigue test has high requirements on the dynamic characteristics of the loading system. Moreover, the test process is complex and costly, and the test cycle is very long, so the random load spectrum is often converted into a program spectrum to perform the fatigue test.

Currently, there is a lack of a published standard to compile the fatigue test program spectrum for excavator hydraulic sticks, so the program spectrum can only be obtained from the measured data. In some studies, the measured load was used to compile the program spectrum for the bench fatigue test. For example, based on several actual engineering failure cases, Bošnjak, Srđan, and Arsić et al. [1–5] analyzed the causes of fatigue failure of key structural parts of bucket wheel excavators. However, in their research, the external load on the structure was not measured. Yin et al. [6] studied the external load test method for the electric drive BE-395B front shovel excavator, and obtained the hinge joint force, cable tension, and bucket tip force to make the program spectrum. Bae et al. [7] proposed a method to extract the fundamental fatigue load from the measured stress time histories collected by sensors arranged on the boom. The above studies did not carry out the fatigue life assessment of structural parts from the perspective of compiling the program spectrum to carry out the bench fatigue test. Taking the stress at the dangerous points of the stick and the boom as the intermediate quantities, Gao [8] and Shi [9] calculated the loading force of the bucket tip based on the stress equivalent principle. Then, the loading force was compiled into a program spectrum to conduct the bench fatigue test of an excavator with a fixed attitude. However, this method relies on the position selection of the dangerous points. Moreover, the method can only ensure that the stress states of the selected dangerous points are the same as the actual ones, while it cannot effectively evaluate the fatigue life of other dangerous points.

Two issues exist for applying the measured load to compile the program spectrum used for the bench fatigue test of the key structural parts of the excavator: (1) It is difficult to obtain the external load of the excavator working device through directly test, because the working environment of the excavator is harsh, the load conditions under different working media are quite different, and the magnitude and direction of the external load of the working device are constantly changing during operations. It is necessary to propose a special load identification method. (2) Directly compiling the measured load into the bench fatigue test program spectrum is also hard due to the contradiction between the fixed attitude of the whole machine fatigue test and the time-varying attitude of the actual operation. Hence, it is necessary to propose an equivalent method to convert the measured complex load to the unidirectional load required for the bench fatigue test to ensure that the bench fatigue test can reflect the actual stress state of the structure. Considering that the external load is difficult to test, a cross-section stress testing method based on strain testing was proposed [10,11], and a three-dimensional axis pin force sensor was developed [12]. These two methods can measure all the loads of the excavator working device including the side load and eccentric load. In view of the contradiction between the fixed attitude of the whole machine fatigue test and the time-varying attitude of the actual operation, a method of respectively arranging the load spectrum of the stick in their local coordinate systems and performing bench fatigue tests was put forward [13]. At the same time, based on the damage consistency criterion of key fatigue points, the multi-directional load of the measured hinge point was equivalent to the unidirectional load required for the bench fatigue test. This method overcomes the problem that the relationship between the forces cannot be reproduced when the load components of each hinge point of the stick are compiled into a load spectrum for the fatigue test, whilst ensuring that the fatigue test of the bench can reflect the actual stress state as much as possible.

Based on our previous work, this study further investigates the compilation of the load spectrum. A program spectrum compilation method for bench fatigue test of hydraulic excavators is proposed in this study, and the following problems will be solved: (1) How to consider the diversity of the actual working medium of the excavator in the compiled load spectrum; (2) How to determine the threshold value of cyclic elimination of small-value loads in spectral compilation since external loads vary greatly in different operating media; (3) What is the influence of the dynamic characteristics of the excavator hydraulic system and the overload protection device on the extrapolation of extreme values during spectrum compilation; (4) How to standardize the compiled program spectrum to improve its applicability since there are many excavator manufacturers of the same tonnage. All these problems have not been resolved yet. To bridge this research gap, this study proposed a multi-case load spectrum compiling method to solve these problems.

The remainder of this work is organized as follows. Section 2 introduces the proposed method. Analysis and results are presented in Section 3. Section 4 concludes the main findings.

2. Materials and Methods

The compilation process of the bench fatigue test of the excavator stick is summarized in Figure 1. This paper mainly focuses on the following key steps: the threshold value of small load omitting; the extreme value inference criterion of the mean and amplitude; the synthesis and extrapolation of various typical operating conditions; and the correction, acceleration, and normalization of the program load spectrum.

Figure 1. Compiling flow of program load spectrum of excavator hydraulic stick.

2.1. Determining Threshold of Small Amplitude Cyclic Load

Many load cycles will be obtained after peak-valley extraction of random load-time history, among which the load cycles with smaller amplitudes account for a large proportion, which will cause a large workload to the rain flow counting. However, these load cycles do not necessarily cause fatigue damage to the structure, or do not affect the fatigue life of the structure, and can be omitted in advance.

The criteria for determining the omitting threshold value of small amplitude cyclic load (hereinafter referred to as 'small loads') have not been finalized or widely used because the relevant parameters in the calculation model are difficult to obtain [14,15]. The common omitting criteria are divided into three categories: ① Omit according to the percentage of the maximum load amplitude or maximum cyclic load range [16,17]; ② Omit according to the percentage of the material fatigue limit, such as the load less than 50~70% of the fatigue limit is usually removed [18,19]; ③ If the load belongs to the normal distribution, the load less than 1.75σ will be omitted (σ is the standard deviation of the cyclic load) [20]. However, these three empirical methods cannot reflect the contribution of the small load to fatigue damage, which may cause a large deviation in the structural life evaluation.

Considering the engineering practicability, this paper proposes a simple and convenient method to determine the omitting threshold value of small load from the perspective of fatigue damage. In this method, the omitting threshold value is determined as the critical value on the premise of keeping the fatigue damage at the dangerous point unchanged before and after the small load is omitted from the full load-time history. In other words, the omitting threshold is determined according to whether fatigue damage is caused to the structure. The "fatigue damage" here is calculated according to Miner's rule and the S-N curve of the structural details at specific dangerous points. It is worth noting that this threshold is not the fatigue limit in smooth materials.

For large and complex welded structures such as excavator sticks, the sensitivity of fatigue damage at each dangerous point of the structure to the load is not consistent when external loads are applied. Therefore, the omitting threshold value of each dangerous point should be determined respectively, and the minimum value of them should be taken as the final threshold value. The calculation process is shown in Figure 2.

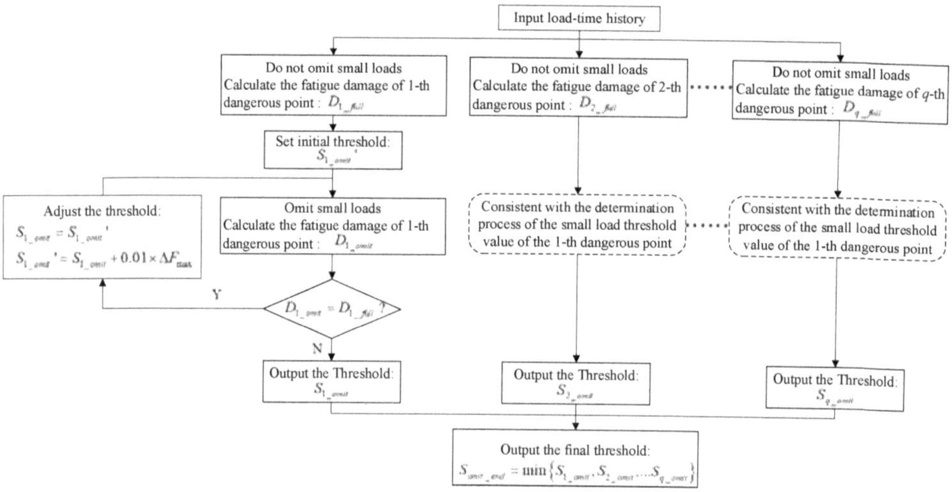

Figure 2. Simplified calculation process of the omitting threshold value of small load.

Table 1 compares the omitting threshold value calculated by the proposed method with the values of 10~15% of the maximum cyclic load range under four typical digging conditions of a medium-sized excavator's stick. It can be seen that the threshold value determined by this method, 19.185, is close to but not completely within the range of 10~15% of the maximum cyclic load range, indicating that it is reasonable but not completely reliable to use 10~15% of the maximum cyclic load range to determine the threshold value. In addition, the omitting threshold value can remain a constant under the different digging conditions.

Table 1. The small-load-omitting threshold value of the unidirectional equivalent stick load.

Threshold Value Calculated by the Proposed Method (kN)		19.185
10~15% of the maximum cyclic load range (kN)	Category I: Loose soil	15.962~23.943
	Category II: Original soil	20.764~31.145
	Category III: Clay with small stones	18.762~28.143
	Category IV: Heavy clay containing large stones	21.083~31.624

2.2. Load Extremum Extrapolation Criterion

In engineering, the load with an occurrence probability of 1×10^{-6} is generally regarded as the extreme load. Several commonly used distributions, such as normal distribution, log-normal distribution, two-parameter Weibull distribution, three-parameter

Weibull distribution, Gamma distribution, and extremal distribution are respectively used to conduct the Kolmogorov–Smirnov test (KS distribution test) on the load average and load amplitude of the medium-sized excavator's stick. Taking the Category IV digging object as an example, the distribution test results are shown in the Table 2.

Table 2. The different distributions' D value of the load amplitude of the stick using the KS distribution test method.

Critical D Value	Normal Distribution	Log-Normal Distribution	Two-Parameter Weibull Distribution	Three-Parameter Weibull Distribution	Gamma Distribution	Extremal Distribution
0.0355	0.1510	0.1726	0.1500	0.0303	0.0843	0.0380

It can be seen that the load amplitude value distribution is more in line with the three-parameter Weibull distribution (D value 0.0303 < critical D value 0.0355). So, the extreme load Y_{i_max} of amplitude value under the i-th digging object is:

$$Y_{i\_max} = c_i + a_i \sqrt[b_i]{-\ln P} \tag{1}$$

where a_i, b_i, and c_i are the scale, shape, and position parameter of the three-parameter Weibull distribution of the equivalent load amplitude under the i-th digging object; $P = 1 \times 10^{-6}$ is the probability of amplitude extremum.

The mean value obtained from the rain flow counting of the equivalent load-time history was fitted with log-normal distribution, so the extreme load X_{i_max} of the mean value under the i-th digging object is:

$$X_{i\_max} = e^{-U_P\sigma_i + \mu_i} \tag{2}$$

where μ_i and σ_i are the mean and standard deviation of the log-normal distribution of the equivalent load mean value under the i-th digging object; U_P can be obtained from the standard normal distribution.

Therefore, the extreme amplitude value Y_{max} after the synthesis of the four digging objects is:

$$Y_{max} = \max(Y_{i\_max}) \tag{3}$$

The extreme mean value X_{max} after the synthesis of the four digging objects is:

$$X_{max} = \max(X_{i\_max}) \tag{4}$$

With the above equations, the theoretical value of the extreme load of mean and amplitude are derived.

However, the hydraulic system of the excavator usually has an overload protection system. Therefore, limited by the set pressure of the main safety valve and the actuator relief valve in the hydraulic system, the theoretical extreme value may not be reached. If the main safety valve or the actuator relief valve overflows during the load spectrum test, it indicates that the load spectrum test has recorded the extreme load that may be generated in practice, and there is no need to infer the extreme value according to the Conover theory. Therefore, the cylinder pressure curves measured during the digging of the primary soil were analyzed, as shown in Figure 3.

It can be seen from Figure 3 that the pressure in the rodless chamber of the stick cylinder exceeded the preset pressure 316 bar, which is the main safety valve of the system and was obtained through the maximum digging force test. The overflow phenomenon also occurred: the measured peak pressure exceeded the set pressure of the system, then gradually attenuated and fluctuated, and finally stabilized at the rated pressure. This phenomenon is caused by the dynamic characteristics of the relief valve when the main safety valve suddenly changes from the closed state to the rated pressure state. This phenomenon occurred many times in the whole load spectrum test process, which indicates that the extreme load had been captured in the load spectrum test.

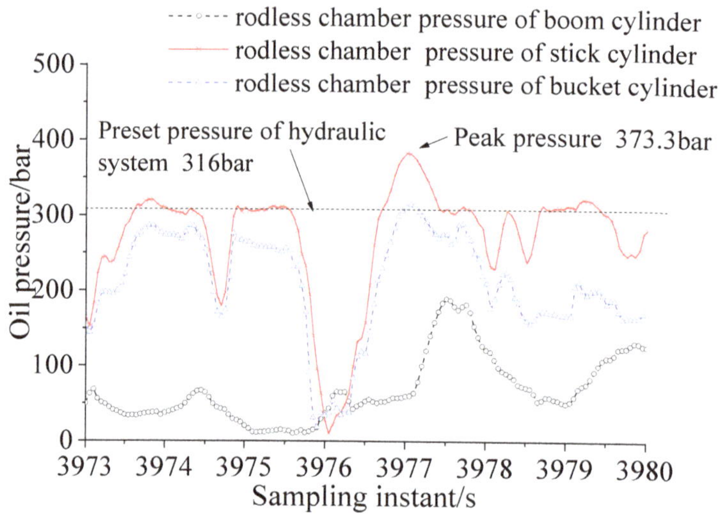

Figure 3. Pressure of the excavator.

The extreme amplitude values of the test prototype excavator's stick determined from three different perspectives, namely, the Conover theoretical extreme value, the extreme value measured in the static maximum digging force test, and the extreme values captured from the actual working conditions are compared in Table 3. It is seen that the extremum extrapolated according to the Conover theory was much larger than the extremum value measured in the actual digging process with overflow phenomenon, while the extremum value measured in the static maximum digging force test was much smaller than the extreme value measured in the digging process. Such a large difference existing between the extremum values obtained by different extrapolation criteria would inevitably cause great differences in the results of the load spectrum compilation. Therefore, during the compilation of the program load spectrum, the extreme load should be determined according to the actual digging process, instead of extrapolating according to the Conover theory.

Table 3. Extrapolation of extreme amplitude load of the test prototype of a medium-sized excavator.

① Conover Theoretical Extreme Value	② Extreme Value Measured in Static Maximum Digging Force Test	③ Extreme Value Measured in Actual Digging Process
212.2 kN	68.6 kN	105.4 kN

2.3. Extrapolation and Synthetic Method of Load Spectrum

In previous work [21], the common working conditions of hydraulic excavators were investigated, and four different digging objects and their proportions were determined. In order to reflect the actual load situation as much as possible, the excavator load spectrum should be measured while digging these four objects, respectively; hence, the data synthesis under different digging objects should be considered when compiling the program load spectrum. For this reason, a new method is proposed to simultaneously complete the data synthesis and frequency extrapolation of multiple digging objects in one step.

① According to the distribution parameters of the mean and amplitude in various digging objects, the corresponding joint probability density function $f(x,y)$ and joint probability distribution function $F(x,y)$ of the mean and amplitude could be obtained:

$$f(x,y) = \frac{1}{x\sigma\sqrt{2\pi}} \exp\left(\frac{-(\ln x-\mu)^2}{2\sigma^2}\right)\left(\frac{b}{a}\right)\left(\frac{y-c}{a}\right)^{b-1} \exp\left(-\left(\frac{(y-c)}{a}\right)^{b}\right) \tag{5}$$

$$F(x,y) = \int_{-\infty}^{y}\int_{0}^{x} f(x,y)dxdy \tag{6}$$

② Calculate the total frequency, N_i, of the mean and amplitude of the i-th digging object within the full sample length T:

$$N_i = n_i k_{1i} T/t \tag{7}$$

where T is the full sample length (number of digging cycles/bucket); k_{1i} is the proportion of the i-th digging object obtained from the investigation; t is the actual collection bucket number of the i-th digging object; n_i is the total number of cycles obtained from equivalent load rain flow counting of the i-th digging object; i = 1, 2, 3, … , q, and q is the total number of digging object types.

③ Calculate the extrapolation frequency, N_i', of the i-th digging object:

$$N_i' = k_{2i} \times 10^6 \tag{8}$$

where k_{2i} is the frequency extrapolation coefficient of the i-th digging object; $k_{2i} = N_i/(\sum N_i)$. i = 1, 2, 3, … , q, q is the total number of digging object types. It can be seen that the extrapolation frequency N_i' of the i-th digging object is independent of the full sample length T.

④ Synthesize the four digging objects to obtain the two-dimensional mean-amplitude program load spectrum. On the basis of the extreme mean value and extreme amplitude value determined above, the amplitude of various digging objects was divided into 8 levels according to the 1, 0.95, 0.85, 0.725, 0.575, 0.425, 0.275, and 0.125 times of the extreme amplitude value, and the mean value was divided into 8 levels with equal intervals. The frequency n_{iuv} on each mean-amplitude level of the i-th digging object is calculated by:

$$n_{iuv} = N_i' \int_{R_u}^{R_{u+1}}\int_{M_v}^{M_{v+1}} \frac{1}{x\sigma\sqrt{2\pi}} \exp\left(\frac{-(\ln x-\mu)^2}{2\sigma^2}\right)\left(\frac{b}{a}\right)\left(\frac{y-c}{a}\right)^{b-1} \exp\left(-\left(\frac{(y-c)}{a}\right)^{b}\right)dxdy \tag{9}$$

where M_v and M_{v+1} (v = 1, 2, 3, … , 8) are the lower and upper integral limits of the mean value of the v-th level under the i-th digging object; R_u and R_{u+1} (u = 1, 2, 3, … , 8) are the lower and upper integral limits of the amplitude value of the u-th level under the i-th digging object.

Then, the frequency n_{uv} corresponding to the v-th mean value and the u-th amplitude of the synthesized two-dimensional program load spectrum is:

$$n_{uv} = \sum_{i=1}^{q} n_{iuv} \tag{10}$$

Finally, a two-dimensional program load spectrum compiled according to the above steps of the stick of a medium-sized excavator can be obtained, and it is presented in Table 4.

Table 4. The mean-amplitude program load spectrum of a medium-sized excavator stick.

Level	Amplitude (kN)	Mean (kN)							
		−48.17	−23.94	0.29	24.52	48.75	72.98	97.20	121.43
1	13.18	7829	75,035	182,958	201,256	137,814	70,008	29,144	15,612
2	28.99	1838	17,819	44,317	49,747	34,609	17,769	7449	4015
3	44.80	671	6507	16,186	18,178	12,656	6503	2729	1474
4	60.61	247	2394	5950	6679	4650	2391	1004	543
5	76.42	87	841	2086	2340	1628	838	352	191
6	89.60	31	298	738	827	576	296	125	68
7	100.14	11	106	262	293	204	105	44	24
8	105.41	17	167	412	460	320	165	69	38

2.4. One-Dimensional Program Load Spectrum

In order to facilitate the test loading, the fluctuation central method was used to transform the two-dimensional program spectrum into one-dimensional to keep the amplitudes of each level in the two-dimensional program spectrum unchanged. Then, the eight-level mean values corresponding to the amplitude of each level were weighted and averaged. The mean value M_u corresponding to the u-th level amplitude is calculated as follows:

$$M_u = \sum_{v=1}^{8} M_v n_{uv} / \sum_{v=1}^{8} n_{uv} \tag{11}$$

The frequency n_u corresponding to the u-th level amplitude is:

$$n_u = \sum_{v=1}^{8} n_{uv} \tag{12}$$

Considering that the actual load is random, that is, the load does not appear in a specific order, in order to reduce the difference between the program spectrum loading and the actual load, the test should be repeated several times. The size of the subroutine block should be determined according to the preliminary life estimation results so that the whole test includes at least 10 subroutine blocks.

The fatigue life represented by the spectrum in Table 3 is denoted as l_P. Given that the full life of the structure is N and the total number of subroutine blocks is C, then the life represented by one subroutine block is: $l = N/C$; When the spectrum in Table 3 was extended to a subroutine block, the frequency corresponding to the u-th level amplitude is modified to $n_u{}'$:

$$n'_u = \beta n_u \tag{13}$$

where $\beta = l/l_p$ is the expansion factor.

According to industry experience, the full life target of excavator working device structure is generally about 8000~10,000 hours. In this paper, let a subroutine block represents 1000 h, and then the spectrum in Table 3 is further expanded in frequency according to Equation (13) to obtain the one-dimensional program load spectrum representing 1000 h.

Taking the test prototype of a medium-sized excavator as an example, the spectrum after frequency expansion is shown in Table 5.

2.5. Compilation of Program Load Spectrum

The program spectrum compiling process consists of many steps, and errors will inevitably occur in each step, eventually resulting in the error between the damage reproduced in the bench fatigue test and the damage in the actual service state. Therefore, the one-dimensional program load spectrum in Table 4 should be corrected before it is used in the bench fatigue test to reduce the error between the bench fatigue test and the actual service state.

Table 5. One-dimensional program load spectrum with variable mean for a medium-size excavator stick (1000 working hours).

Level	Amplitude (kN)	Mean (kN)	Frequency
1	13.18	26.91	1,120,745
2	28.99	27.67	276,525
3	44.80	27.70	101,077
4	60.61	27.70	37,155
5	76.42	27.69	13,024
6	89.60	27.70	4608
7	100.14	27.64	1634
8	105.41	27.63	2566

2.5.1. Fatigue Damage Consistency Correction of the Program Load Spectrum for Bench Fatigue Test

It is a very complicated problem to study the error of each step in detail. In this paper, the damage consistency criterion and the damage of the key fatigue measuring points on the structure were used to modify the load spectrum. The damage of the key fatigue measuring points on the structure caused by the program load spectrum should be close to the damage caused by the actual digging process on the premise of ensuring the damage of the key fatigue measuring points on the structure is not greater than or equal to the actual damage of each measuring point calculated by the measured stress spectrum. The specific steps are as follows:

Firstly, the stress time history of each fatigue key measuring point on the structure was compiled into a stress spectrum. Then, according to the S-N curve of welded joint and the Miner cumulative damage rule, the actual damage D_{rj} of each measuring point is calculated by:

$$D_{rj} = \sum_{i=1}^{n_s} \frac{n_{ij}(\sigma_{ij})^{m_j}}{C_j} \quad j = 1, 2, \ldots, h \tag{14}$$

where h is the number of key fatigue measuring points; m_j and C_j are the S-N curve constants of the j-th key fatigue measuring point, and n_s is the number of the stress spectrum levels; σ_{ij} and n_{ij} represent the i-th level amplitude value and corresponding frequency of the stress spectrum of the j-th key fatigue measuring point, respectively.

Secondly, the damage D_{pj} of the key fatigue measuring points using the one-dimensional program load spectrum listed in Table 4 is calculated by

$$D_{pj} = \sum_{k=1}^{n_p} \frac{n_k(P_k k_j)^{m_j}}{C_j} \quad j = 1, 2, \ldots, h \tag{15}$$

where n_p is the number of the one-dimensional program load spectrum levels, here, $n_p = 8$; P_k and n_k denote the k-th level amplitude value and its corresponding frequency of the one-dimensional program load spectrum, respectively. k_j is the load-stress proportional coefficient between P_k and the stress of the j-th key fatigue measuring point, which is only related to the location of the key fatigue measuring point on the structure.

Then, the correction factor γ is used to correct the one-dimensional program load spectrum, and the corrected amplitude of each level is $P_k' = \gamma P_k$. The damage D_{pj}' of the key fatigue measuring point caused by the modified one-dimensional program spectrum is:

$$D'_{pj} = \sum_{k=1}^{n_p} \frac{n_k(\gamma P_k k_j)^{m_j}}{C_j} \quad j = 1, 2, \ldots, h \tag{16}$$

The damage consistency correction optimization model is established as follows:

Objective function: $\min\left\{ \sum_{j=1}^{h} \left(D_{rj} - D'_{pj} \right)^2 \right\}$

Constraint condition: $D_{rj} \leq D'_{pj}$

Finally, Equations (14) and (16) were substituted into the optimization model to acquire the correction coefficient γ, and the program spectrum $P_k{}' = \gamma P_k$ after damage consistency correction that would be used as the bench fatigue test program spectrum is obtained. In the example of this article, the correction coefficient of the stick program load spectrum solved by the optimization model is 1.104.

2.5.2. Acceleration and Normalization of Program Load Spectrum for Bench Fatigue Test

In order to accelerate the test, the above program load spectrum needs to be concentrated in time main. The small load less than the fatigue cut-off line will be omitted, but the omission should have no influence on the damage of each fatigue key point on the structure.

In addition, to facilitate the test loading, the eighth level load with larger load values but fewer loading cycles was converted to the seventh level according to the principle of equal damage, and the new frequency $n_7{}'$ was calculated according to Equation (17); meanwhile, the second level load with smaller load values but more loading cycles was converted to the third level, and the new frequency $n_3{}'$ was calculated according to Equation (18).

$$n'_7 = n_7 + \frac{n_8 \sigma_8^{m_8} C_7}{\sigma_7^{m_7} C_8} \tag{17}$$

where σ_7 and n_7 represent the seventh load level and corresponding frequency in the above eight-level program load spectrum; m_7 and C_7 represent the constants of the S-N curve corresponding to the seventh load level; σ_8 and n_8 represent the eighth load level and corresponding frequency in the above eight-level program load spectrum; m_8 and C_8 represent the constants of the S-N curve corresponding to the eighth load level.

$$n'_3 = n_3 + \frac{n_2 \sigma_2^{m_2} C_3}{\sigma_3^{m_3} C_2} \tag{18}$$

It is known that the tonnage of excavators varies significantly, even for those belonging to the same tonnage category. For instance, the tonnage of the medium-sized excavator ranges from 12.5 to 25.5 tons, while the large-sized excavator ranges from 35.8 to 49.2 tons. Because different tonnage excavators suffer different external loads, the program load spectrum should be different for different tonnage excavators. However, making program load spectra according to each tonnage level is unrealistic, as this process is time consuming and expensive. Hence, the program load spectrum is normalized to make it applicable for various excavators with different tonnages. Considering the maximum bucket cylinder digging force is a characteristic parameter that indicates the working capacity of the excavator, the maximum bucket cylinder digging force of the medium-sized excavator test prototype, 138 kN, is taken as unit 1, and the relative values of the upper and lower load limits in the program load spectrum are given in Table 6. It is worth noting that this normalization is an approximation for engineering applications. The premise of this method is that the external load encountered in the actual work is basically proportional to the "maximum bucket cylinder digging force".

Table 6. Relative values of programmed load spectrum of different tonnage excavators.

Level	Upper Limit of the Amplitude (kN)	Lower Limit of the Amplitude (kN)	Frequency
1	0.56	−0.16	157,683
2	0.69	−0.28	37,155
3	0.81	−0.41	13,025
4	0.92	−0.52	4608
5	1.00	−0.60	4628

For an excavator with a different tonnage, its program spectrum can be obtained by multiplying the relative values in Table 5 by the maximum bucket cylinder digging force.

During the bench fatigue test, the subroutine of low-high-low sequence was repeated to load the tested object. For example, Figure 4 shows the program load spectrum loading sequence of the bench fatigue test of the medium-sized excavator's stick.

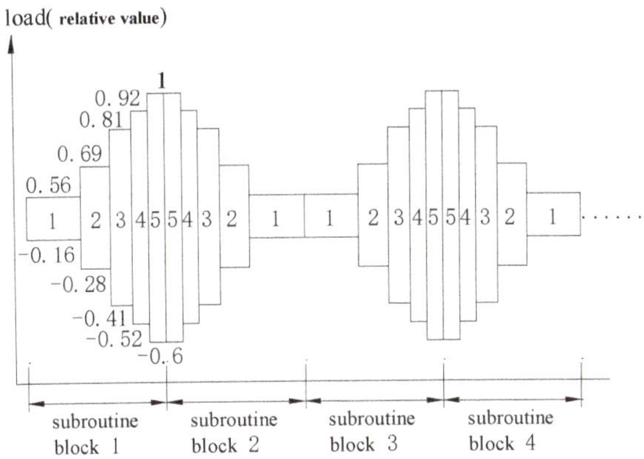

Figure 4. Loading sequence of program load spectrum for bench fatigue test.

3. Bench Fatigue Test

The fatigue life virtual prediction technology was used to verify the validity of the bench fatigue test program load spectrum compiled in this paper. The bench fatigue test program load spectrum and the random equivalent load-time history before compiled, which is hereinafter referred to as a random spectrum, were respectively used in the fatigue life tests, and the results using these two spectra are compared to evaluate the proposed program load spectrum compiling method.

3.1. Structural Stress Analysis under Unit Load

The random load at the hinge points of the excavator are low-frequency loads and does not include the natural frequency of the structure. Therefore, the stress results of static finite element analysis were used as the input of the fatigue analysis.

The finite element model of the stick was established in the finite element analysis software, ANSYS R15.0. The two kinds of load spectra to be applied were unidirectional load. Therefore, in the model, the 5 degrees of freedom of the hinge point of the stick and the boom and the 5 degrees of freedom of the fixed end of the stick cylinder rod were constrained, respectively. Then, the unit vertical force was applied at the hinge point of the stick and the bucket, as shown in Figure 5.

3.2. Fatigue Analysis and Discussion

The results of ANSYS were imported into the fatigue analysis software, nCode DesignLife, to predict the fatigue life. A fatigue analysis process that can withstand the above two load spectra is built. The random spectrum applied in the fatigue simulation process only represented the working time of 5.011 hours, while the representative working time of the applied bench fatigue test program load spectrum was 1000 hours. Therefore, the life of the structural details was the life value of the simulation results multiplied by the representative working time of the corresponding spectrum.

Figure 6 compares the overall life distribution cloud diagram of the stick structure under the action of the two load spectra. It can be seen that the life distribution of the stick structure was basically the same, and there were four areas with short life: ① The front end of the bucket cylinder support ear plate located on the upper wing plate of the stick; ② Near the reaming hole on the ear plate of the stick cylinder support; ③ Part of the weld near the reaming hole of the stick and the boom; ④ Intersection of butt weld on the web and the lower flange of the stick. The comparisons of these four areas with local enlarged figures are shown in Figures 7–10, respectively.

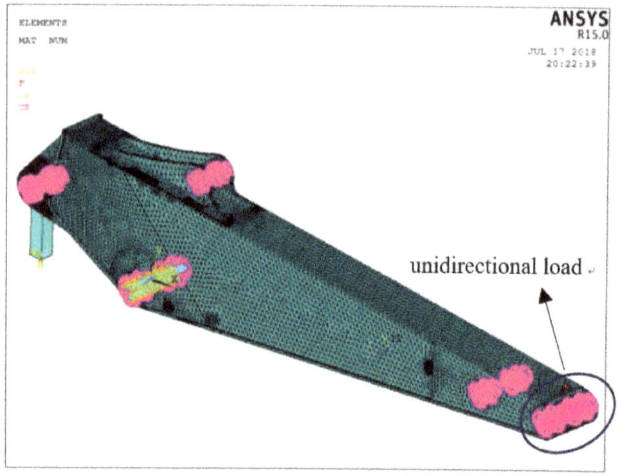

Figure 5. Structural stress analysis model under unidirectional load.

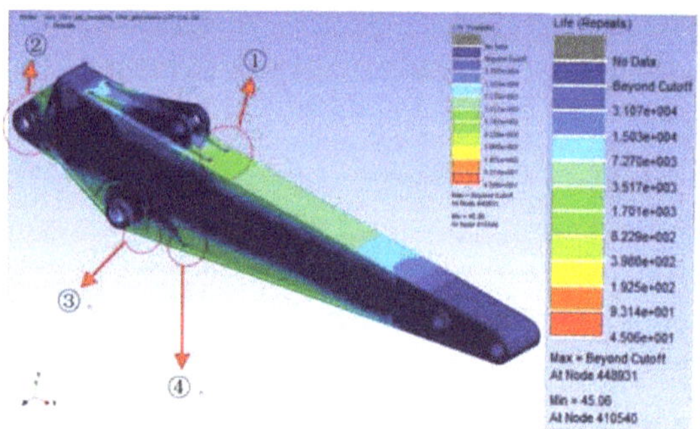

(a)

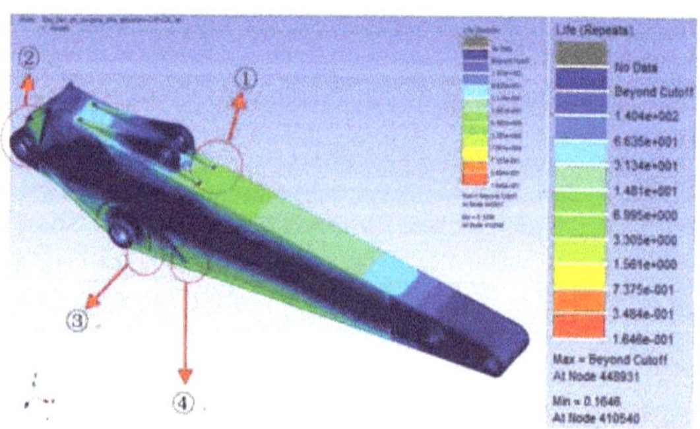

(b)

Figure 6. Comparison of the overall life distribution of the stick structure under the action of the two load spectra. (**a**) Random spectrum. (**b**) Program load spectrum.

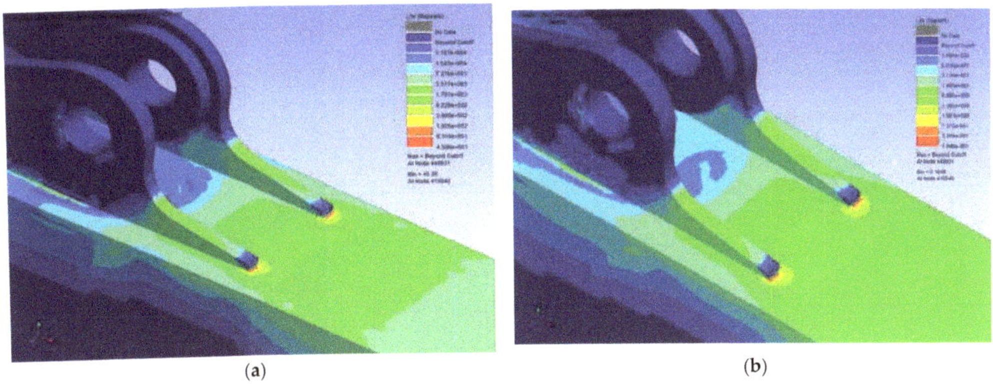

Figure 7. Fatigue life distribution cloud diagram of the front end of the bucket cylinder support ear plate located on the upper wing plate of the stick (Local area ①). (**a**) Random spectrum. (**b**) Program load spectrum.

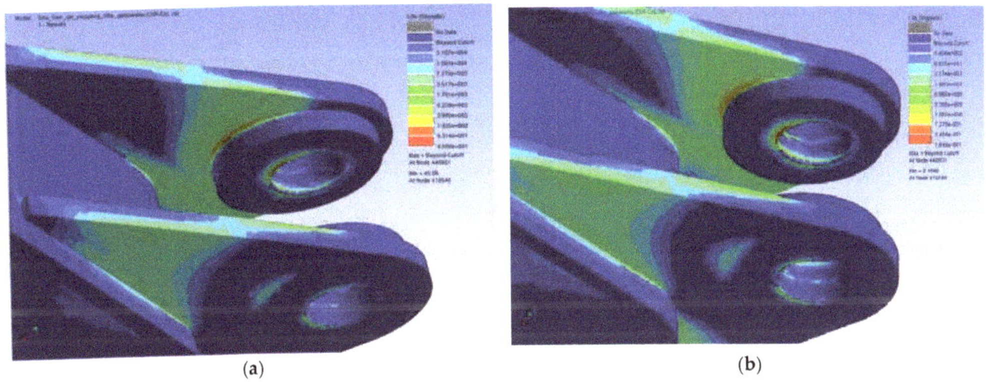

Figure 8. Fatigue life distribution cloud diagram of the stick cylinder support ear plate (Local area ②). (**a**) Random spectrum. (**b**) Program load spectrum.

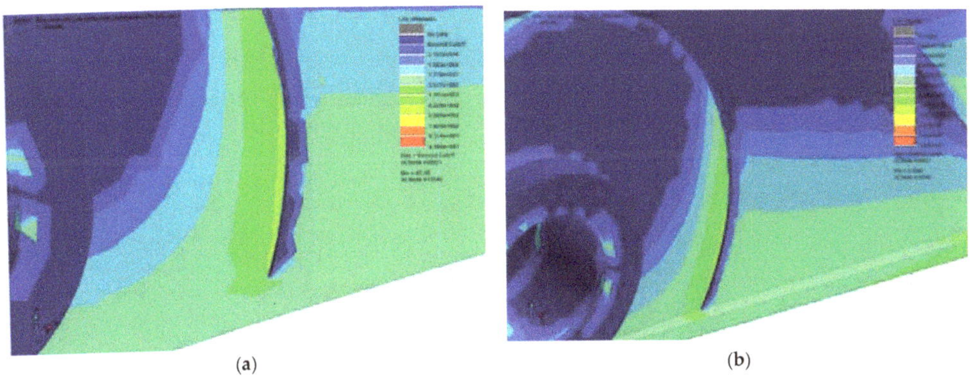

Figure 9. Fatigue life distribution cloud diagram of the weld near the reaming hole of the stick and the boom (Local area ③). (**a**) Random spectrum. (**b**) Program load spectrum.

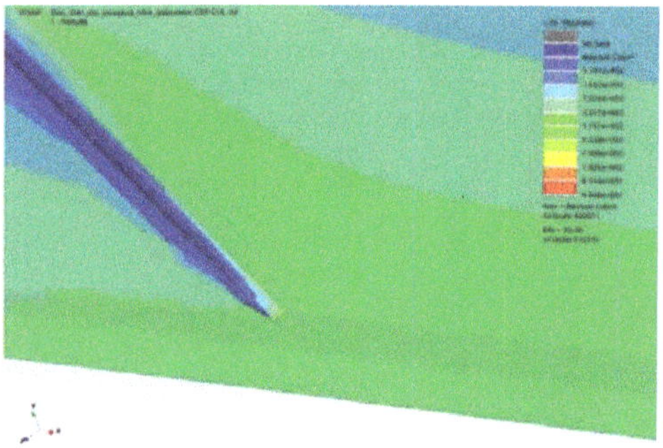

(a)

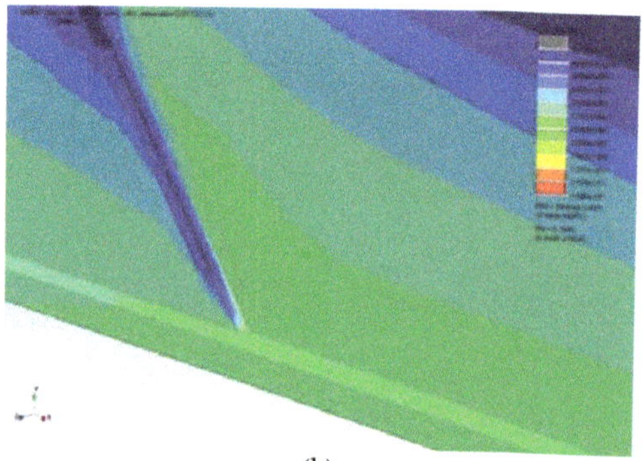

(b)

Figure 10. Fatigue life distribution cloud diagram of the intersection of butt weld on the web and the lower flange of the stick (Local area ④). (**a**) Random spectrum. (**b**) Program load spectrum.

The fatigue life conversion values of the above four local areas of the stick structure under the two load spectra are presented in Table 7. It can be seen that the life distribution of the stick structure and the life values were relatively consistent, and the maximum relative deviation was merely about 8.8%, indicating that the programming method of the load spectrum was reasonable. Assuming that the maximum loadable frequency of the test bench is 1Hz, the fatigue lifetime test using the random spectrum takes 1000 h to represent 1000 h of working time. In contrast, the fatigue lifetime test using the proposed program load spectrum only consumes 60.31 h, which largely shortens the time expense. Hence, the program load spectrum outperforms the random spectrum in terms of time efficiency.

Table 7. Comparison of the fatigue life of several local details of the stick under two load spectra.

Local Details		Random Spectrum (kN)	Program Load Spectrum (kN)	Relative Deviation
① Front end of the bucket cylinder support ear plate located on the upper wing plate of the stick	left	4603.69	4364.22	−5.2%
	right	4745.47	4492.22	−5.3%
② Near the reaming hole on the ear plate of the stick cylinder support		3472.9	3298.83	-5.0%
③ Part of the weld near the reaming hole of the stick and the boom		6603.18	6304.32	-4.5%
④ Intersection of butt weld on the web and the lower flange of the stick		9924.8	10,800.54	8.8%

4. Conclusions

This paper focuses on solving several key issues in the compilation of the bench fatigue test program load spectrum of the hydraulic excavator stick:

(1) A simple and convenient calculation method to determine the omitting threshold value of small amplitude cyclic load was proposed. The threshold value was determined as the critical value to keep the fatigue damage at the dangerous point unchanged before and after the small load was omitted from the full load-time history.

(2) The load extremum extrapolation criterion, the synthetic method of various digging objects, and the correction and acceleration method based on the damage consistency criterion and the damage equivalent principle were proposed. Taking a medium-sized excavator's stick as an example, its programmed load spectrum for fatigue bench test was obtained and further normalized to adapt to other excavators with different tonnages.

(3) The validity of the program load spectrum is verified by using the finite element method and fatigue analysis theory. The comparison showed that the stick structure had the same damage position and life distribution under the action of the bench fatigue test program load spectrum and the random spectrum. The program load spectrum compilation method also had a certain reference value for compiling the load spectrum of other multi-condition hydraulic construction machinery.

Author Contributions: Conceptualization, Q.X. and Z.L.; methodology, P.W., G.K., and P.L.; software, B.W.; validation, Q.X., P.W., and Z.L.; formal analysis, P.W., G.K., and P.L.; investigation, Q.X.; resources, Z.L.; data curation, B.W.; writing—original draft preparation, Q.X. and Z.L.; writing—review and editing, G.K.; visualization, B.W.; supervision, Q.X.; project administration, G.K.; funding acquisition, Q.X. and Z.L. All authors have read and agreed to the published version of the manuscript.

Funding: This work was supported by the Natural Science Basis Research Plan in Shaanxi Province of China (No. 2021JQ-282), the Natural Science Basis Research Plan in Shaanxi Province of China (No. 2021JQ264), and the Fundamental Research Funds for the Central Universities (No. 300102251104).

Data Availability Statement: Not applicable.

Acknowledgments: The research leading to these results has received funding from the Norwegian Financial Mechanism 2014–2021 under Project Contract No. 2020/37/K/ST8/02748. The authors would like to thank Lei Deng for his comments and suggestions on this manuscript.

Conflicts of Interest: The authors declare no conflict of interest.

References

1. Bošnjak, S.M.; Arsić, M.A.; Gnjatović, N.B.; Milenović, I.L.; Arsić, D.M. Failure of the Bucket Wheel Excavator Buckets. *Eng. Fail. Anal.* **2018**, *84*, 247–261.
2. Bošnjak, S.M.; Arsić, M.A.; Zrnić, N.Đ.; Rakin, M.P.; Pantelić, M.P. Bucket Wheel Excavator: Integrity Assessment of the Bucket Wheel Boom Tie-Rod Welded Joint. *Eng. Fail. Anal.* **2011**, *18*, 212–222.
3. Bošnjak, S.; Petković, Z.; Zrnić, N.; Simić, G.; Simonović, A. Cracks, Repair and Reconstruction of Bucket Wheel Excavator Slewing Platform. *Eng. Fail. Anal.* **2009**, *16*, 1631–1642.
4. Bošnjak, S.M.; Petković, Z.D.; Atanasovska, I.D.; Milojević, G.Z.; Mihajlović, V.M. Bucket Chain Excavator: Failure Analysis and Redesign of the Counterweight Boom Supporting Truss Columns. *Eng. Fail. Anal.* **2013**, *32*, 322–333.
5. Arsić, D.; Gnjatović, N.; Sedmak, S.; Arsić, A.; Uhričik, M. Integrity assessment and determination of residual fatigue life of vital parts of bucket-wheel excavator operating under dynamic loads. *Eng. Fail. Anal.* **2019**, *105*, 182–195.
6. Yin, Y.; Grondin, G.Y.; Obaia, K.H.; Elwi, A.E. Fatigue Life Prediction of Heavy Mining Equipment.Part 1: Fatigue Load Assessment and Crack Growth Rate Tests. *J. Constr. Steel Res.* **2007**, *63*, 1494–1505.
7. Bae, H.R.; Ando, H.; Nam, S.; Kim, S.; Ha, C. Fatigue Design Load Identification Using Engineering Data Analytics. *J. Mech. Des.* **2015**, *137*, 011001.
8. Gao, Y.; Gao, Z.; Qin, S. Load Spectrum of Bucket Rod of R961 Hydraulic Excavator. *Constr. Mach. Equip.* **1980**, *9*, 34–40.
9. Shi, L.; Cao, S.; Yu, L. Research on Load Spectrum of Simulation Experiment for Single Bucket Hydraulic Excavator Model. *J. Tongji Univ. (Nat. Sci.)* **1992**, *4*, 395–402.
10. Xiang, Q.; Lu, P.; Wang, B. Identification of cross-section loads for steel beam with simple cross-sectional shape based on measured strain and its application. *J. Mech. Sci. Technol.* **2020**, *34*, 5017–5028.
11. Xiang, Q.; Lu, P.; Wang, B.; Zhang, H.; Xue, L. Load Spectrum Test Method for Working Device of Hydraulic Excavator. *China J. Highw. Transp.* **2017**, *30*, 151–158.
12. Xiang, Q.Y.; Lu, P.M.; Wang, B.H.; Xue, L. Research Characteristics of Actual Hinge Joint Load of Hydraulic Excavator Working Device. *J. Mech. Strength* **2018**, *40*, 1063–1070.
13. Xiang, Q.; Lu, P.; Wang, B.; Ren, X. Research of Load Equivalent Method for Fatigue Bench Test of Hydraulic Excavator Stick. *China J. Highw. Transp.* **2018**, *31*, 318–326.
14. Wang, C.J.; Yao, W.X.; Xia, T.X. A Small-Load-Omitting Criterion Based on Probability Fatigue. *Int. J. Fatigue* **2014**, *68*, 224–230.
15. Xie, F.; Yao, W.; Jin, J.; Xu, L. Small Load Omitting Approach in Load Spectra for Aluminum-Alloy Notched Specimens and Experiment Verification. *J. Nanjing Univ. Aeronaut. Astronaut.* **2017**, *49*, 60–66.
16. Li, F.; Wu, P.; Zeng, J. Compilation Method of Fatigue Test Load Spectrum for Underfloor Equipment Bearing Structure. *J. Mech. Eng.* **2016**, *52*, 99–106.
17. Bao, R.; Zhang, X. Fatigue Crack Growth Behaviour and Life Prediction for 2324-T39 and 7050-T7451 Aluminium Alloys under Truncated Load Spectra. *Int. J. Fatigue* **2010**, *32*, 1180–1189.
18. Yu, J.W.; Zheng, S.L.; Zhao, L.H.; Zhao, Z.S. Research on Spectrum Development Methodology for Vehicle Indoor Road Simulation Test. *J. Mech. Eng.* **2015**, *51*, 93–99.
19. Heuler, P.; Seeger, T. A Criterion for Omission of Variable Amplitude Loading Histories. *Int. J. Fatigue* **1986**, *8*, 225–230.
20. Shi, L. Finite Life Design and Testing of Machinery Lesson 15 Fatigue Life Testing of Machinery Parts. *Constr. Mach.* **1989**, *9*, 37–43.
21. Chen, Y.; Liu, Y.; Lu, P. Operation Medium and Type of Load Spectrum Test of Excavator. *Road Mach. Constr. Mech.* **2018**, *35*, 95–100.

machines

MDPI

Article

Modified Drive Structure of a Knitting Machine

Josef Skrivanek *, Martin Bilek and Ondrej Batka

Department of Textile Machine Design, Faculty of Mechanical Engineering, Technical University of Liberec, 461 17 Liberec, Czech Republic; martin.bilek@tul.cz (M.B.); ondrej.batka@tul.cz (O.B.)
* Correspondence: josef.skrivanek@tul.cz; Tel.: +420-48535-3764

Abstract: This paper deals with the problematic of the drive system of small-diameter knitting machines (hereinafter "knitting machines"). An alternative design of a new drive system is presented that addresses the effective working position of selected operating elements, i.e., a cylinder, a dial, and a roller cutter. The design is protected by patent No. 303578, and the authors of this paper also authored the patent. For the description and simulation of the drive system, mathematical models have been elaborated of the existing and new drive system, which include an additional design of a more suitable stroke for the movement of the cylinder. The output of the simulations are the courses of kinematic quantities, torque, and electric current, elucidating the benefits of the new drive system. Based on the structure of the new drive system, a test device has been designed and manufactured to verify functionality and reliability. The new drive system consumes less energy, simplifies the machine frame and reduces noise emissions. The obtained results can help in researching the issue of knitting machine frames, and in addition, the new system brings new technological possibilities for knitting.

Keywords: knitting machine; stroke; drive; simulation; cylinder

Citation: Skrivanek, J.; Bilek, M.; Batka, O. Modified Drive Structure of a Knitting Machine. *Machines* **2022**, *10*, 615. https://doi.org/10.3390/machines10080615

Academic Editors: Marek Kočiško and Martin Pollák

Received: 21 June 2022
Accepted: 23 July 2022
Published: 27 July 2022

1. Introduction

Knitting machines are an essential part of traditional textile engineering with a rich history, though knitting technology is relatively young [1,2] compared to other areas of this industry. In addition to classic knitting technologies and their products, extensions of classic knitted fabrics with higher added value are increasingly coming to the fore [3–7].

Previously used in the drive systems of small-diameter knitting machines were mainly an asynchronous motor with short circuit "anchor" and graduated pulleys for changing speed [2]. The term "drive system" in this paper includes the motor, its control, and the transmission that propels the required elements, while the term "drive" includes only the motor and its control. After the development of semiconductor rectifiers, DC brush control motors were used relatively briefly. Today, frequency converters or control units for servo drives are used to change speeds in asynchronous motors [8,9]. In the past, the drive systems of small-diameter knitting machines contained relatively complex mechanisms with numerous mechanical transmissions installed in intricate, voluminous frames [10]. Mechanical transmissions set the limits of use of the operating elements forming the knitted fabric in terms of effective use throughout the entire work process [11]. Working elements mean the cylinder, dial plate (abbreviated as "dial") and the roller cutter [12]. The dial in knitting practice is called a rotary element, used mainly to create an overhanging hem especially on socks, as well as to produce a solid beginning and in some types of machine also double-faced knit fabric [1]. An integral part of the dial is a roller cutter used for cutting yarn according to technological requirements. In the overhanging mode, i.e., the formation of an overhanging hem on the sock, it is essential to uphold the technological condition of mutual rotation between the cylinder and the dial [13]. This is due to the sliding out of the sinkers from the dial between the needles during rotation. The limit of the maximum rotation is given by the sum of the maximum plays in the gearing of the drive system given by its stiffness.

Theoretically, from the technological standpoint, the dial would be synchronized with the cylinder only for two revolutions during manufacture of the sock. The first revolution would be to catch the beginning of the knitted fabric and after knitting the hem, hanging the knitted fabric back on the needles [2]. The roller cutter is not dependent in principle on the movement of the knitting roller or the dial itself, so there is no need for synchronization [1]. In the past, there were knitting machines, especially for manufacturing delicate women's stockings, which did not include the dial. On these machines, it was possible to create an overhanging hem even on the front side [8].

The percentage of utilization of the aforementioned operating elements (cylinder, dial and roller cutter) is significantly different during the process of creating sock goods, ranging from 5% to 100% [8]. The mentioned workload and length of the process during knitting was based on the creation of one classic sock with the top hem designed for the foot size EU 44, Figure 1. The knitting time of one sock is different and especially depends on its size, shape, material used and the variety of uses of colors and patterns required by the customer [14]. The current concepts of the drive systems do not allow flexible control of individual operating elements according to their current needs for inclusion in the technological process. This problem occurs due to the direct mechanical connection between the dial and the roller cutter; their movement is standardly derived by constant transmission by gears from the movement of the cylinder [15].

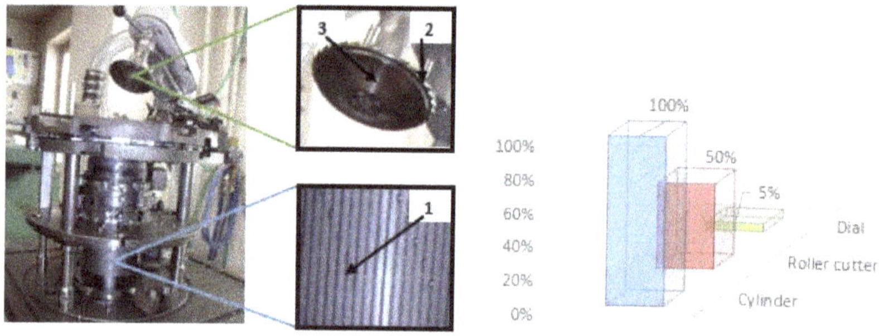

Figure 1. Cylinder 1, roller cutter 2, dial 3 and their percentage load during the knitting process.

The application of controlled drives for the individual work elements mentioned above offers the possibility of increasing the efficiency of the knitting process and the associated significant savings in electricity consumption [16]. An indisputable advantage is the reduction of demands on the installation space and the overall simplification of the machine design due to the reduction of moving elements of the drive system [17]. It is worth mentioning that the gears making up the current drive system are also a key vibration exciter and noise source, so one expected benefit of using controlled drives is the reduction of noise pollution and ensuring improvement of health conditions for machine attendants [18]. Another important criterion for assessing the suitability of employing controlled drives is the demands on the dynamic properties of the machine, as their applications can bring an improvement in their properties, thanks to the possibility of a suitable choice of lift dependence [19,20].

This paper focuses on assessment of the new drive system compared to the existing one. The new drive system will respect the targeted inclusion of the cylinder, dial, and roller cutter in the technological process. The use of controlled drives in this area appears to be a possible way of increasing the degree of technical sophistication of small-diameter knitting machines. There are a number of varying structures of the drive system of small-diameter knitting machines. One of the most commonly used methods of designing the drive system of small-diameter knitting machines is shown in Figure 2 [2]. The principle of the system is based on the transmission of power to the cylinder (4) and to the dial (10) from the servo drive (M1) by a system of gears (1,2,3,5,6,7,8,9) and shafts (11,12,13).

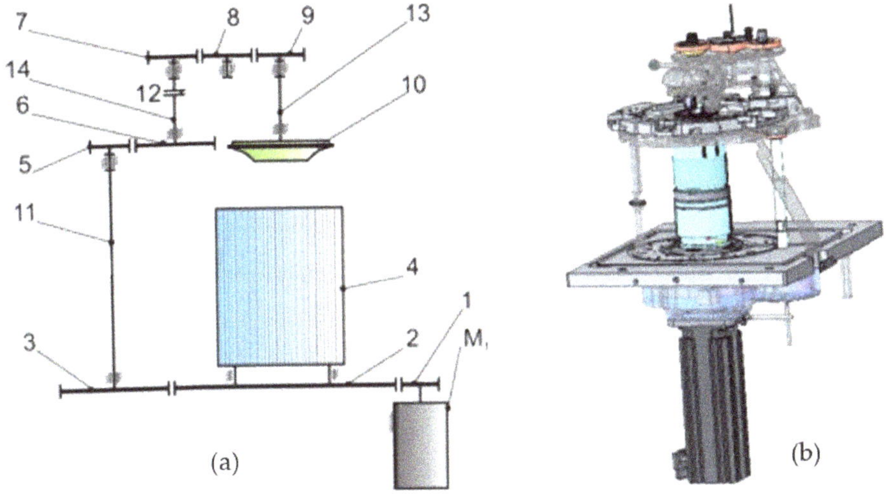

Figure 2. Structure of the present concept of the small diameter knitting machine: (**a**) schematic view (**b**) 3D model.

The main movement of the knitting machine is performed by a brushless alternating servo drive (M_1), a pinion (1), and a toothed rim (2), which is part of the cylinder (4). The dial drive system begins by transmission from the toothed rim (2) to the gear (3). The system continues through the shaft (11) to the gearing (5 and 6). The shaft (12) is fitted with a clutch. This clutch must ensure that its original position is precisely reset after the dial arm has been tilted (10), i.e., after disconnection of the drive system. Furthermore, the system connects the gears (7,8,9 (V)) to the shaft (13) of the dial (10), which is mounted and firmly connected to the roller cutter. The gears ensure the same angular velocity of the dial (10) and the cylinder (4). The delimitation of plays between gear teeth is realized by changing the axial distance between the gears in combination with split gears [21].

2. Mathematical Models

2.1. Existing Drive System of a Small-Diameter Knitting Machine

An analysis was performed of the existing design to assess the impact of the changes to the new drive system. The results of the analysis are compared with those obtained by analyzing the modified drive system in the following chapter.

To investigate the dynamic behavior of the drive system of a small-diameter knitting machine, a discrete model was elaborated, including the torsional compliance of all main shafts and the cylinder. The actual mathematical model of the knitting machine drive system in Figure 3 was formulated while considering the following assumptions:

- Gears (0–7) and the dial (8) are considered as perfectly rigid bodies
- Each of the shafts (R0, 23, 45, 78) is replaced in the model by two geometrically identical shapes whose moments of inertia are included in the moments of inertia of the corresponding gears. In the case of the shaft (78), the moment of inertia of its lower part is included in the moment of inertia of the dial (8). The relevant shapes are connected by non-material elastic elements with torsional stiffnesses and viscous damping elements.
- The cylinder (9) is replaced in the model by two geometrically identical shapes, whereas the moment of inertia of the lower shape is included in the moment of

inertia of the gear (1). Both shapes are connected by non-material elastic elements with torsional stiffnesses and viscous damping elements.

- The moments of inertia of the needles and sinkers are included in the moment of inertia of the cylinder (9).
- The moments of inertia of the sinkers are included at the moment of inertia of the dial (8).
- In the calculation, plays in gears are considered, which are simply entered into the mathematical model using angular deflections of shafts R0, 19, 23, 45, 78; the play in the shaft bearing is not considered by the model.
- The model does not consider the bending compliance of shafts and the thermal expansion of materials.
- The model does not consider gravitational and friction forces and forces caused by the tension of the yarn.
- The model does not consider the force required for cutting the ends of the yarn.

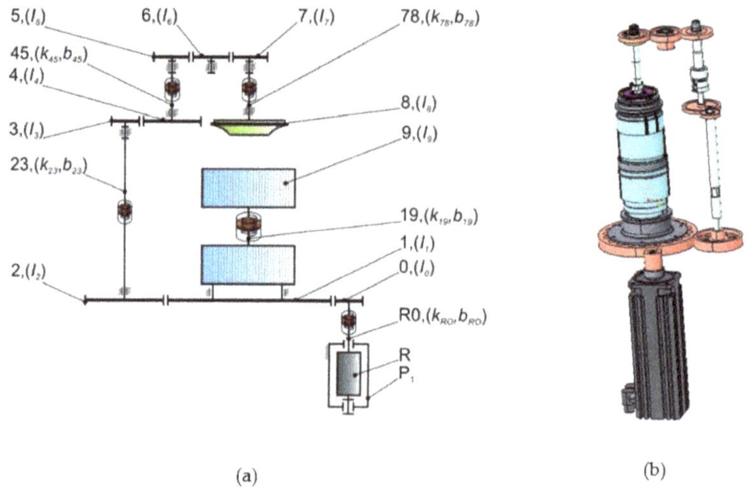

Figure 3. Existing small-diameter knitting machine drive system-(**a**) Mathematical model, (**b**) 3D model.

The required moments of inertia of the individual elements of the drive system were determined using the cad model of the drive system, see Table 1. Conversions appearing in the equations of motion are given in Table 2. The designation and position of the elements correspond to the diagram in Figure 4.

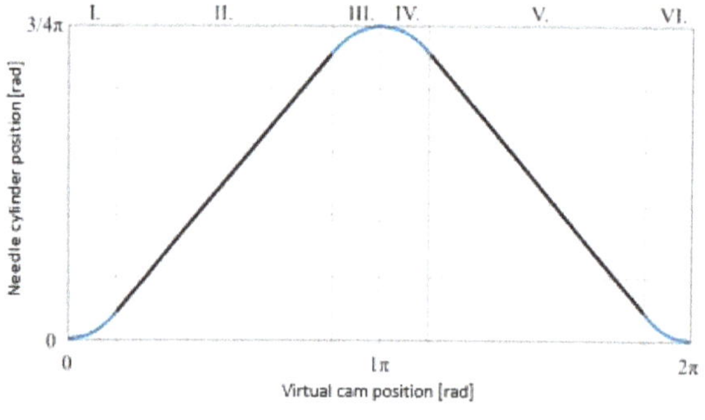

Figure 4. Input stroke.

Table 1. Moments of inertia of the considered parts of the drive system.

Component	Position	Moment of Inertia I [kg·m²]
gear 0	0	$9,83 \times 10^{-6}$
gear 1	1	$2,13 \times 10^{-2}$
gear 2	2	$1,34 \times 10^{-3}$
gear 3	3	$3,07 \times 10^{-4}$
gear 4	4	$2,87 \times 10^{-4}$
gear 5	5	$1,86 \times 10^{-4}$
gear 6	6	$2,52 \times 10^{-4}$
gear 7	7	$1,49 \times 10^{-4}$
dial	8	$2,40 \times 10^{-4}$
rotor	R	$8,10 \times 10^{-4}$
needle roller	9 (V)	$3,38 \times 10^{-2}$
shaft R0	R0	$9,02 \times 10^{-6}$
shaft 23	23	$7,26 \times 10^{-5}$
shaft 45	45	$8,42 \times 10^{-5}$
shaft 78	78	$1,06 \times 10^{-5}$

Table 2. Gears appearing in equations of motion.

Gearing	0–1	0–2	3–4	5–6	5–7
Designation	μ_{01}	μ_{02}	μ_{34}	μ_{56}	μ_{57}
Value	1/6	1/3	1/2	1	1

Lagrange equations of the second kind were used to elaborate the actual equations of motion of the discrete model of the drive system. This is among the most widely used methods of analytical mechanics, especially in more complex mechanical systems [22,23].

2.1.1. Form of the Lagrange Equation of the Second Kind

E_K is the kinetic energy of the system, E_P is the potential energy of the system, R_d is the Rayleigh dissipation function, and q_i is the generalized coordinate.

$$\frac{d}{dt}\left(\frac{\partial E_K}{\partial \dot{q}_i}\right) - \frac{\partial E_K}{\partial q_i} + \frac{\partial E_P}{\partial q_i} = -\frac{\partial R_d}{\partial \dot{q}_i} \tag{1}$$

for: $i = 1, 2, \ldots, n$.

Matlab R2022a software and its Simulink module were used to solve the system of equations of motion. The individual equations were written in matrix form (1) and then solved by gradual integration in the Simulink environment, see Figure 4.

$$[I]\cdot\{\ddot{q}\} + [B]\cdot\{\dot{q}\} + [K]\cdot\{q\} = 0 \tag{2}$$

2.1.2. Listing of Matrices Used in the Equation (2)

Moment of Inertia Matrix I

$$\begin{bmatrix} I_0 + I_1\cdot\mu_{01}^2 + I_2\cdot\mu_{02}^2 & 0 & 0 & 0 & 0 \\ 0 & I_9 & 0 & 0 & 0 \\ 0 & 0 & I_3 + I_4\cdot\mu_{34}^2 & 0 & 0 \\ 0 & 0 & 0 & I_5 + I_6\cdot\mu_{56}^2 + I_7\cdot\mu_{57}^2 & 0 \\ 0 & 0 & 0 & 0 & I_8 \end{bmatrix}$$

Damping Matrix B

$$\begin{bmatrix} b_{R0} + b_{19}\cdot\mu_{01}^2 + b_{23}\mu_{02}^2 & -b_{19}\cdot\mu_{01} & -b_{23}\cdot\mu_{02} & 0 & 0 \\ -b_{19}\cdot\mu_{01} & b_{19} & 0 & 0 & 0 \\ -b_{23}\cdot\mu_{02} & 0 & b_{23} + b_{45}\cdot\mu_{34}^2 & -b_{45}\cdot\mu_{34} & 0 \\ 0 & 0 & -b_{45}\cdot\mu_{34} & b_{45} + b_{78}\cdot\mu_{57}^2 & -b_{78}\cdot\mu_{57} \\ 0 & 0 & 0 & -b_{78}\cdot\mu_{57} & b_{78} \end{bmatrix}$$

Stiffness Matrix K

$$\begin{bmatrix} k_{R0} + k_{19}\cdot\mu_{01}^2 + k_{23}\mu_{02}^2 & -k_{19}\cdot\mu_{01} & -k_{23}\cdot\mu_{02} & 0 & 0 \\ -k_{19}\cdot\mu_{01} & k_{19} & 0 & 0 & 0 \\ -k_{23}\cdot\mu_{02} & 0 & k_{23} + k_{45}\cdot\mu_{34}^2 & -k_{45}\cdot\mu_{34} & 0 \\ 0 & 0 & -k_{45}\cdot\mu_{34} & k_{45} + k_{78}\cdot\mu_{57}^2 & -k_{78}\cdot\mu_{57} \\ 0 & 0 & 0 & -k_{78}\cdot\mu_{57} & k_{78} \end{bmatrix}$$

Listing of vectors of generalized coordinates q and their derivatives

$$q = \begin{Bmatrix} \varphi_0 \\ \varphi_9 \\ \varphi_3 \\ \varphi_5 \\ \varphi_8 \end{Bmatrix} \quad \dot{q} = \begin{Bmatrix} \dot{\varphi}_0 \\ \dot{\varphi}_9 \\ \dot{\varphi}_3 \\ \dot{\varphi}_5 \\ \dot{\varphi}_8 \end{Bmatrix} \quad \ddot{q} = \begin{Bmatrix} \ddot{\varphi}_0 \\ \ddot{\varphi}_9 \\ \ddot{\varphi}_3 \\ \ddot{\varphi}_5 \\ \ddot{\varphi}_8 \end{Bmatrix}$$

where I is the moments of inertia matrix, B is the damping matrix, K is the stiffness matrix, and q is the rotation vector of the selected elements of the drive system. The coefficients of torsional stiffness k, viscous damping b of the selected elements of the drive system were taken from a study [24]. Furthermore, in the equations of motion, conversions μ appear between the individual stages of the drive system. In the mathematical model of the drive system, the toothing plays are respected. These plays were included in the drive system using the conditions specified in the study, which replace the actual plays with angular deflections [25].

For Ange series knitting machines, a lift dependence is often used containing a polynomial of the 3rd degree describing the desired position of the cylinder in the start-up and run-out areas of the drive system. Linear sections appear in the lift dependence in which a constant velocity of the cylinder is required. The position of the cylinder is a function of a virtual cam corresponding to one of its revolutions, during which the cylinder performs a symmetrical reciprocating rotational movement between two dead centers from the start position to the end position in the range of 0–3/4 π, see Figure 4 [26].

Through the use of the elaborated solver [27] (see Figure 5) of differential equations, the gained motion Equation (2) was solved [28], and responses were observed in the relationship between the drive system and the kinematic excitation of the system, which was the desired course of acceleration of the drive system drive element during reverse motion. The reverse motion mode was selected for analysis due to the highest degree of dynamic loading of the entire knitting machine [24].

The output of the analysis are the courses of kinematic quantities of individual elements of the drive system (see Figures 6–8), the designation of the elements corresponds to the diagram in Figure 3. The achieved angular velocity of the 16.8 rad·s^{-1} cylinder technologically corresponds to the highest recommended operating velocity of the Ange series small-diameter knitting machine in the reverse motion of the drive system using standard materials for production of sock goods [2].

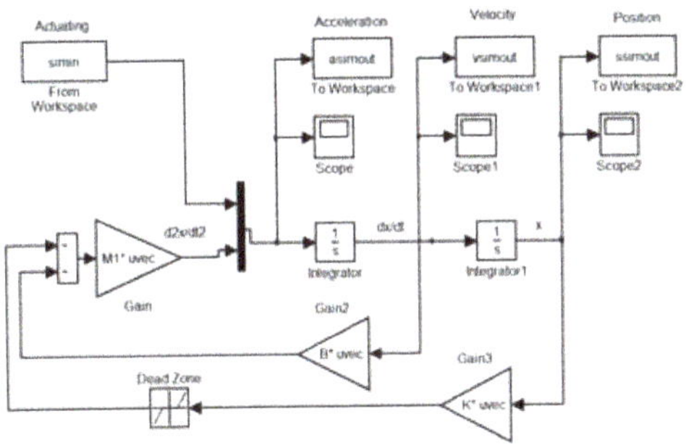

Figure 5. Schematic of the differential equation solver in Matlab R2022a-Simulink software, MathWorks, Natick, MA, USA, Humusoft s.r.o. (distributor for the Czech Republic).

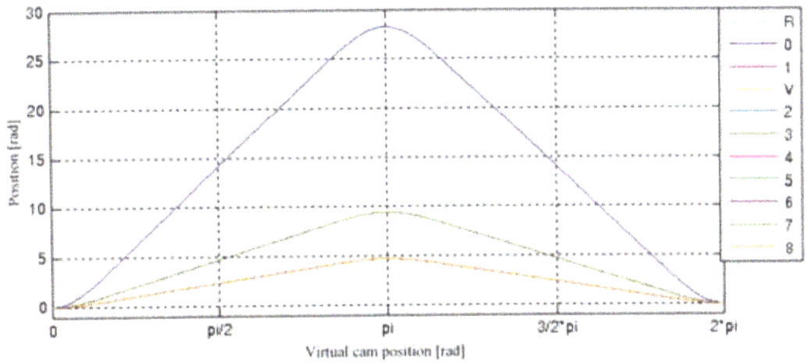

Figure 6. Courses of positions of individual drive system elements.

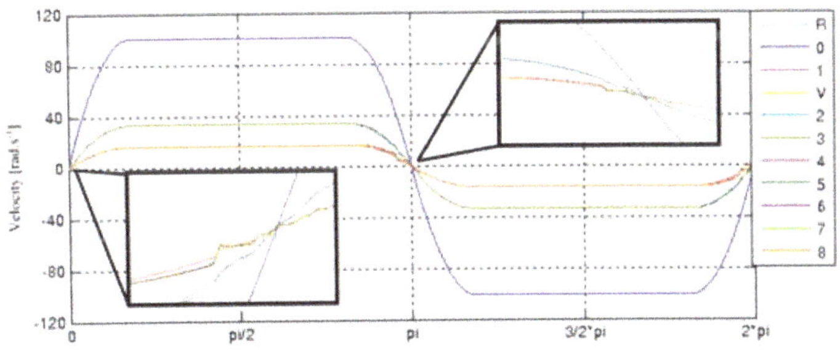

Figure 7. Courses of angular velocities of individual drive system elements.

The results of the analysis point to the theoretical influence of plays in the system of the knitting machine drive. These cause shock forces during their delimitation; this phenomenon can be clearly observed during the acceleration of the system in the form of maximum oscillation values (see Figure 8). Similarly, although not so significantly, delimitation of plays appears in the course of angular velocities of Figure 7, and there are evident oscillations, especially in the areas of motion dead points. In addition to play, these oscillations are also influenced by dynamic system parameters. Some of the curves overlap in the courses, and this phenomenon is particularly evident in the course of the position.

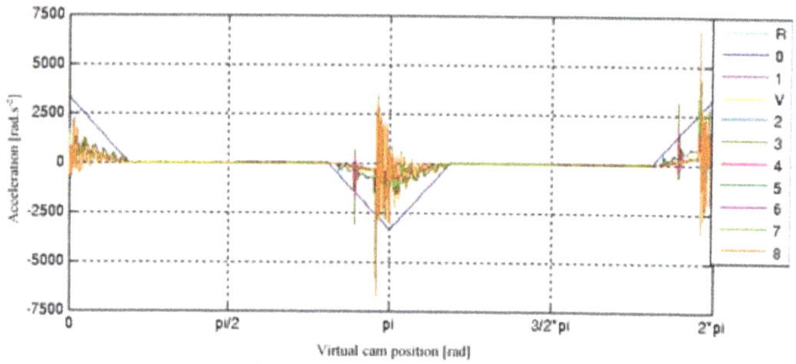

Figure 8. Courses of angular accelerations of individual drive system elements.

2.2. Modified Drive System of a Small-Diameter Knitting Machine

The newly designed structure (see Figure 9a) is protected by a patent [17]. The structure removes the part of the existing drive system that sets the dial with roller cutter in motion and further extends the basic driving unit (P_1) by two additional controlled drives (P_2) and (P_3) for individual control of the dial (10) and the roller cutter (11). The functional model of the roller cutter drive system comprises a friction transmission mechanism in which the power transfer between the drive (P_3) and the roller cutter is performed by a flexible shaft (13) fitted with a drive wheel (12).

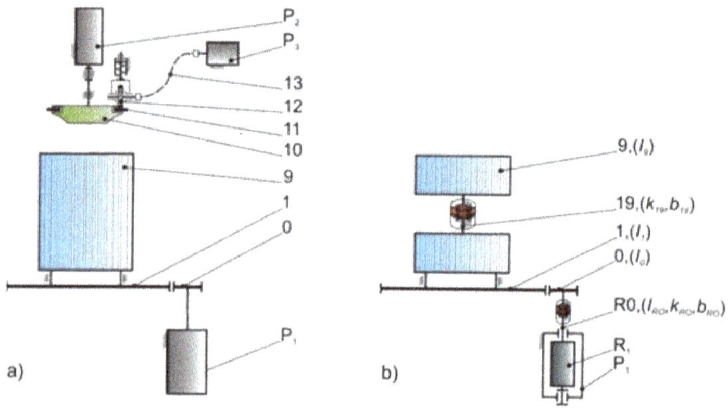

Figure 9. (**a**) Structure of the modified drive system of the cylinder, dial and roller cutter with controlled drives (**b**) considered mathematical model of the modified cylinder drive system.

The gearbox containing the gearing (0–1) and the drive (P_1) were taken from the existing machine structure to control the cylinder (9 (V)). In reverse motion of the modified drive system, only the cylinder is driven. This fact corresponds to the elaborated mathematical model of the part of the modified drive system in which only the cylinder is driven, see Figure 9b. The dynamic analysis of this part of the modified drive system was carried out in a similar way, considering the same assumptions and using the same solver of differential equations as the dynamic analysis of the existing structure of the drive system (Section 3), see the following list of matrices.

Matrices moment of inertia I, damping B and stiffness K

$$I = \begin{bmatrix} I_0 + I_1 \cdot \mu_{01}^2 & 0 \\ 0 & I_9 \end{bmatrix} \; B = \begin{bmatrix} b_{R0} + b_{19} \cdot \mu_{01}^2 & -b_{19} \cdot \mu_{01} \\ -b_{19} \cdot \mu_{01} & b_{19} \end{bmatrix} \; K = \begin{bmatrix} k_{R0} + k_{19} \cdot \mu_{01}^2 & -k_{19} \cdot \mu_{01} \\ -k_{19} \cdot \mu_{01} & k_{19} \end{bmatrix}$$

Listing of vectors of generalized coordinates q and their derivatives

$$q = \left\{ \begin{array}{c} \varphi_0 \\ \varphi_9 \end{array} \right\} \dot{q} = \left\{ \begin{array}{c} \dot{\varphi}_0 \\ \dot{\varphi}_9 \end{array} \right\} \ddot{q} = \left\{ \begin{array}{c} \ddot{\varphi}_0 \\ \ddot{\varphi}_9 \end{array} \right\}$$

The modified drive system is conceptually divided into three independent subsystems, each of which has its own drive (P_1, P_2 and P_3). Individual subsystems are placed in service with different requirements for the prescription of movement depending on the technology being implemented. For this reason, individual theoretical movement cycles have been introduced for the cylinder, the machine and the roller cutter.

Based on the previous analysis of the existing drive system, a lift dependence was proposed containing a polynomial of the 7th degree for the system start-up and run-out areas, see Equations (3)–(5). The chosen polynomial degree appears to be a suitable means to suppress shocks in the drive system, to achieve more favorable courses of kinematic quantities, and it is also assumed that the drive is more capable of maintaining the required course.

$$\begin{aligned} \varphi(\psi) = {} & C_0 + C_1\cdot\psi + C_2\cdot\psi^2 + C_3\cdot\psi^3 + C_4\cdot\psi^4 + C_5\cdot\psi^5 + C_6\cdot\psi^6 + \\ & + C_7\cdot\psi^7 \end{aligned} \tag{3}$$

$$\begin{aligned} \dot{\varphi}(\psi) = {} & C_1\cdot\dot{\psi} + 2C_2\cdot\dot{\psi}\cdot\psi + 3C_3\cdot\dot{\psi}\cdot\psi^2 + 4C_4\cdot\dot{\psi}\cdot\psi^3 + 5C_5\cdot\dot{\psi}\cdot\psi^4 + \\ & 6C_6\cdot\dot{\psi}\cdot\psi^5 + 7C_7\cdot\dot{\psi}\cdot\psi^6 \end{aligned} \tag{4}$$

$$\begin{aligned} \ddot{\varphi}(\psi) = {} & 2C_2\cdot\dot{\psi}^2 + 6C_3\cdot\dot{\psi}^2\cdot\psi + 12C_4\cdot\dot{\psi}^2\cdot\psi^2 + 20C_5\cdot\dot{\psi}^2\cdot\psi^3 + \\ & 30C_6\cdot\dot{\psi}^2\cdot\psi^4 + 42C_7\cdot\dot{\psi}^2\cdot\psi^5 \end{aligned} \tag{5}$$

Figures 10–12 show the courses of the position, angular velocity and angular acceleration of the elements of the modified drive system. The dial with roller cutter and their corresponding part of the drive system are no longer in the dynamic analysis of the drive reverse motion mode, so the results do not indicate their corresponding courses. By using a polynomial of the 7th degree in the stroke, the course of kinematic quantities was smoothed. This benefit contributed to a significant reduction of the oscillation amplitudes of the modified drive structure compared to the existing structure in which a polynomial of the 3th degree was used. This decrease is best apparent on the courses of angular velocity and angular acceleration, see details in Figures 11 and 12, in these courses there was a significant reduction of oscillations, especially in transition areas.

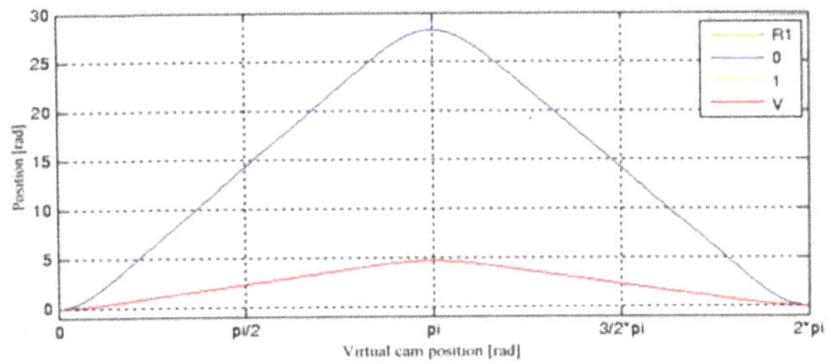

Figure 10. Courses of positions of individual drive system elements.

The working speed is intended for classic materials of sock goods (Cotton, Lycra, Polypropylene, Moira and others). The knitting machine would be able to work at higher speeds and therefore it would be possible to achieve higher productivity, but the classic sock materials set a limit in the form of yarn strength [1].

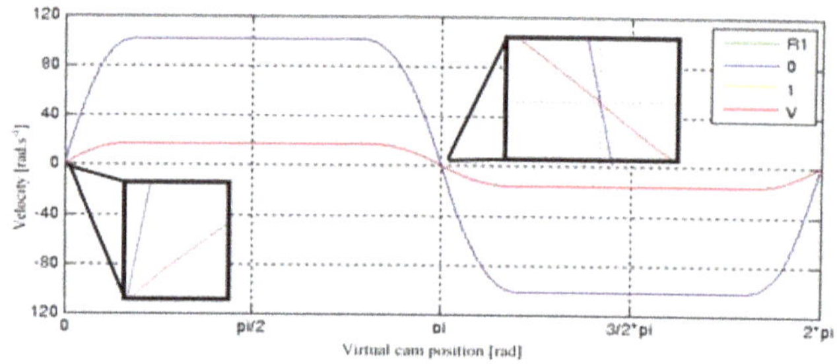

Figure 11. Courses of angular velocities of individual drive system elements.

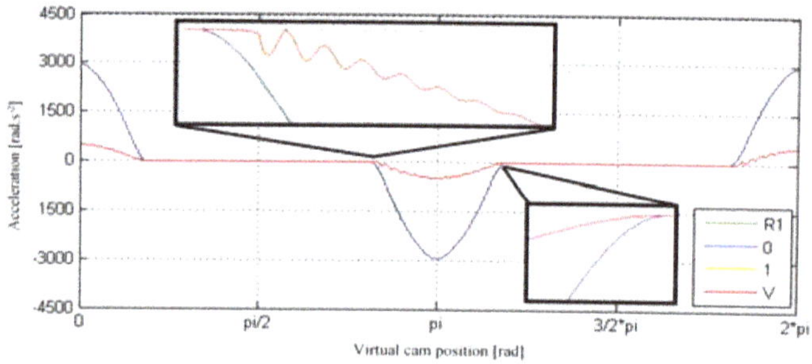

Figure 12. Courses of angular accelerations of individual drive system elements.

2.3. Modified Drive System of a Small-Diameter Knitting Machine

A reduction in the number of elements in the modified drive system caused a decrease in the moment of inertia and plays, which then led to a noticeable suppression of transient oscillating phenomena resulting from delimitation of plays Differences between drive systems are particularly pronounced in angular acceleration courses. The maximum values during acceleration decreased significantly by up to 51%, and proportionally thereby the dynamic forces generated in the drive system as well. By increasing the degree of the polynomial in the lift dependence, we achieved more favorable courses of kinematic quantities of the drive system elements, as well as enhanced the ability of the drive to maintain the desired value.

The aforementioned adjustments appeared in more favorable courses of theoretical torques and theoretical electricity consumption of the modified drive system in comparison with the existing system, see Figures 13 and 14, using different lift dependences. The decrease in total electricity consumption reached 10%.

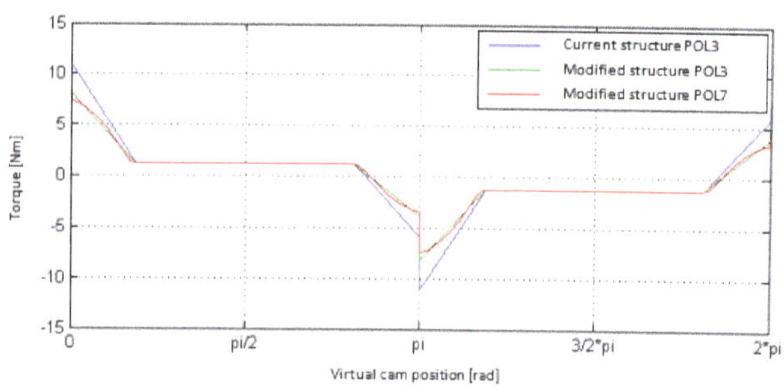

Figure 13. Course of torques during one period of reverse motion of the P_1 drive of the investigated systems, application of polynomials of the 3rd and 7th degree in the lift dependences describing the position of the cylinder.

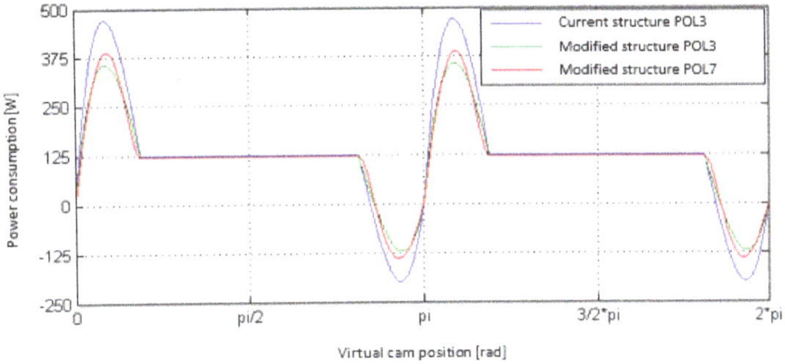

Figure 14. Course of power inputs during one period of reverse motion of the P_1 drive of the investigated systems, application of polynomials of the 3rd and 7th degree in the lift dependences describing the position of the cylinder.

3. Experimental Verification of the Drive System

For testing purposes and verification of the modified drive system of the small diameter knitting machine, a mechatronic test dial was assembled, see Figure 15, on which the previously acquired knowledge was applied. The test equipment is primarily adapted to verify the applicability of the modified system, i.e., to simulate the overhanging mode with emphasis on meeting the requirement for the maximum position deviation between the rotation of the cylinder and dial, and for application of lift dependences to the cylinder drive. The issue of cylinder and dial deviation with the replacement of mechanical bond by electronics was addressed in the study [26]. The test equipment contains no transmissions compared to the real machine. To capture the transmission between the cylinder and its motor, the flywheel 1 representing the cylinder rotates at a six-fold angular speed. Flywheel 2 represents the machine. Using the law of conservation of (kinetic) energy, we determined their reduced moments of inertia [29]. Flywheel 1 has two geometric designs. In the case of the existing drive system, the moment of inertia is 5.09×10^{-4} kg·m^2, and in the case of the modified drive system, it is 4.15×10^{-4} kg·m^2. The flywheels are mounted on the output shafts of servo drives 3 and 4, which are mounted to the frame 5. Type of servo motors used CTM4-07 (Control Techniques). Another part of the measuring dial were frequency converters 6, 7 UNI1405 (Control Techniques) and a computer 8 with the corresponding control system of the MotionPerfect V4 Program, Control Techniques software, Stafford Park 4, United Kingdom, distributor for the Czech Republic Control Techniques Brno s.r.o.

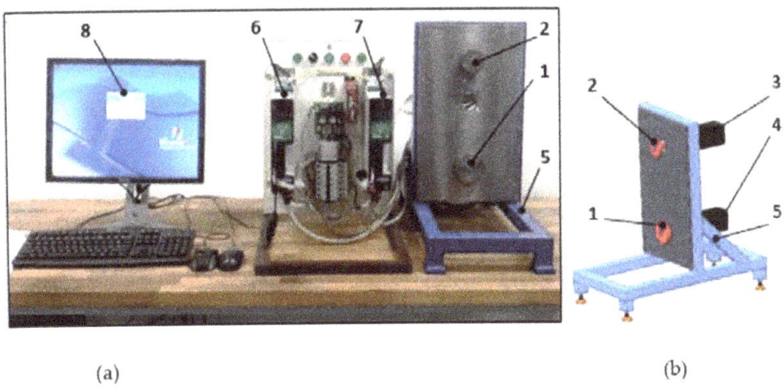

(a) (b)

Figure 15. (**a**) Apparatus used during measurement-measuring workplace (**b**) CAD model of the test equipment of the modified knitting machine drive system.

The structure of the control and simultaneously measuring system is composed of two complementary parts. The lower control level contains the basic settings of motion feedback control parameters using CTSoft software. These low-level settings allow the user to control strokes, properties and speed feedback loops and current controller parameters, but are not able to ensure the generation of more complex trajectories. For that reason, the higher program platform was created for the Motion Processor Plus module using the SyPTPro (System programming toolkit for drive automation systems) development environment. The APC (Advanced Position Controller) programming tool was used to create user programs. A complex control program was created implementing the synchronization of servo drives, automated parameter setting, measuring and recording system, and the desired position generator. The basis for obtaining information for both used servo drives (CTM4-07, Control Techniques, Stafford Park 4, Telford, UK, distributor for the Czech Republic Control Techniques Brno s.r.o.) is their own integrated sin-cos encoder, the output of which is information about the position of the shaft depending on the position of the virtual cam. The sin-cos encoder has a 32-bit resolution.

A sampling step of one tenth of an angular degree was chosen as a compromise solution. One revolution of the virtual cam was thus sampled into 3600 discrete values. Programs created in the APC requiring the sampling period to be set to 500 μs so that the control unit is able to properly complete all operations related to the main control loop period with generator, synchronization, measurement and data storage. The sampling frequency during the measurement was therefore corresponding to 2 kHz or 0.5 kHz. For individual measurements, the length of the measured section was chosen for 5000 samples, which corresponds to 2.5 s or 10 s in the time domain at the given sampling frequency. The measured data was stored in the integrated memory of the module and after stopping the run, it was downloaded separately via a serial communication channel to the PC, where it was interpreted using the mathematical programming tool Matlab. In the program created in the APC interface, the only controlled quantity is the position, which is also given by the feedback.

The movement mode of reverse motion of the machine was investigated on the test equipment. During reverse motion measurement, two flywheels differing in mass were applied, which simulated the existing and modified system, as well as two different lift dependences.

Measurements were used to verify the findings of previous analyses, which confirm that a reduction in the moment of inertia of the modified roller drive system in combination with an appropriate lift dependence can pave the way for a reduction in the maximum acceleration at the limits of movement intervals during reverse motion of the drive system. These changes resulted in a 38% decrease in the maximum value of position deviation (see Figure 16) and in a 31% decrease in the value of active current (see Figure 17). The biggest differences were observed in the area of transitions at the limits of movement intervals, especially when changing the sense of rotation. Kinematic values were obtained from the data of the connected sin-cos encoder.

Figure 18 shows the courses of the position deviation in the standard velocity range during steady cylinder running conditions. The velocity range (160–200 min^{-1}) recommended by the manufacturer [21] is used during the knitting of the tubular part of the sock, depending on the type of material used. When operating the drive in the specified velocity range, it can be concluded that the value of deviations depending on RPMs is negligible.

In a real knitting machine, a transmission is installed due to the technological and dynamic parameters of the cylinder drive system, therefore the resulting value is usually extended by the value of the play in the transmission. This play may reflect in the position control accuracy of the cylinder. The possibility of eliminating this play to achieve higher accuracy can be achieved, for example, by direct control of the position of the cylinder using appropriate position sensors, or by using a gearbox without play. It can be assumed that the obtained measurement results, in particular the deviations of the position between the cylinder and the dial, will also correspond to the real modified drive system using optimal drives, their control and adjustment.

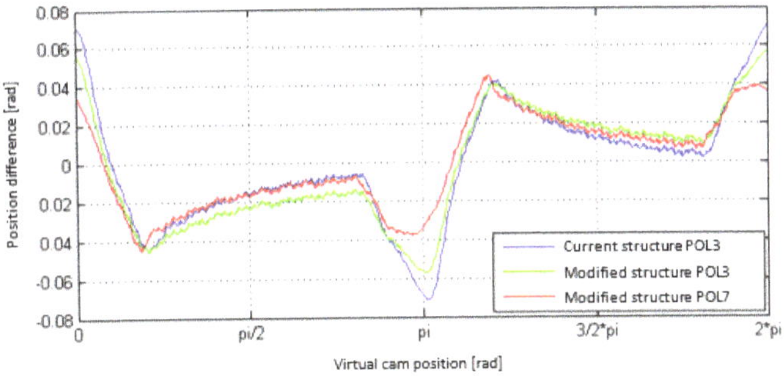

Figure 16. Courses of needle cylinder position deviations, during reverse motion, in all designs at a velocity of 160 min^{-1}.

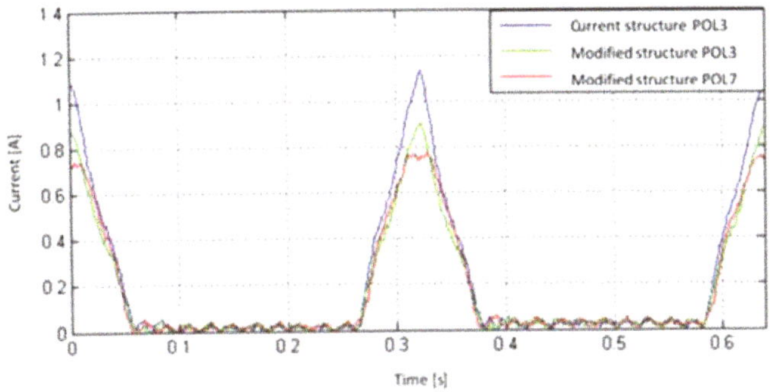

Figure 17. Course of current in drive M1, during reverse motion, in all designs at a velocity of 160 min^{-1}.

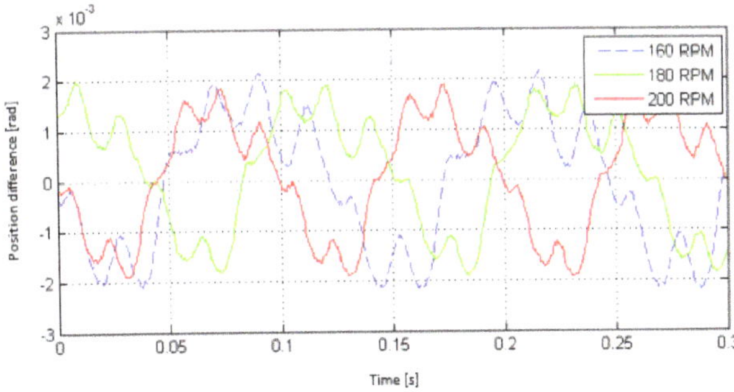

Figure 18. Courses of needle cylinder and dial position deviations after stabilization of their movement, for velocities of 160–200 min^{-1}.

4. Verification of the Implementation of the Individual Roller Cutter Drive System into the Existing Structure

Figure 19 shows the implementation of a functional model of an individual roller cutter drive system into the existing installation of a small-diameter knitting machine. The CAD model is comprised of several subgroups and was created in the software Creo parametric 7.0.1.0. The first structural unit is a modified dial 1 serving as a bearing for the roller cutter 2. Another subgroup is formed of the pinion 3 fitted with a rubber ring in the circumferential groove and its installation in the bearing 4 built into the defined machine space. The roller cutter and the pinion form the whole of the friction transmission. The last part is the pinion drive, which is implemented by means of a stepping motor 5 and flexible shaft 6. A flexible type of shaft was used due to the position of the motor outside the lift arm 7 on the joint. This solution led to preventing an increase in arm weight, which would mean a reduction in ergonomic comfort for the machine attendant.

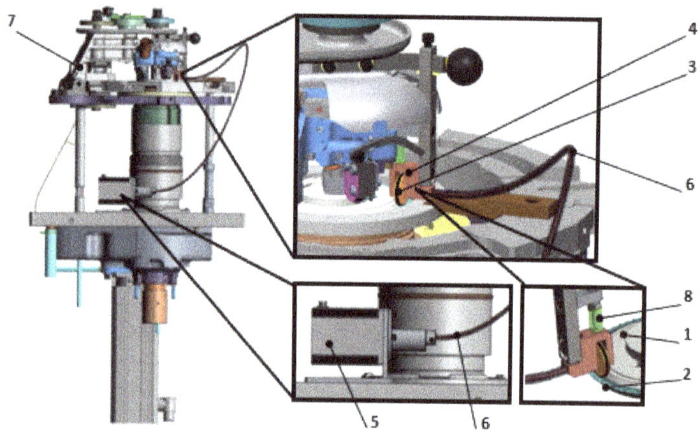

Figure 19. CAD-model of the installation of a functional model of the individual roller cutter drive system into the Ange 18.1 small diameter knitting machine.

The pinion placement allows vertical movement of the fork, which is influenced by the pressure of the spring 8. This design variant has been designed for two reasons. The first ensures the regulation of the normal force by pre-stressing the spring and its optimum setting, so as to avoid pinion slipping or unnecessary overload. The second reason is the function of creating a flexible element compensating for manufacturing inaccuracies, in particular the flatness of the contact surfaces of the roller cutter and the pinion.

Figure 20a shows the installation of an individual roller cutter drive system on a real machine. A small-diameter knitting machine Ange 18.1 (Uniplet Trebic a.s., Czech Republic) was selected for positioning the drive system; its drive structure corresponds to the diagram in Figure 2, Section 1. The Ange knitting machine is designed for knitting classic and sports socks. It is a single cylinder fully electronic knitting machine with several knitting systems. The option of one or several additional auxiliary knitting systems for the production of sock classic or plush goods. Pattern production is possible in all parts of the knit and the machine includes a device for knitting wrapped elastomer. The machine includes electronically controlled retraction locks with electronically selected heel and toe. The lever selection of needles for patterning is controlled electromagnetically.

The installation of the main part of the system (see Figure 20b) consisted of fitting a part of the dial 1 with a modified supporting flange 2 and a lower pan to form a bearing for the roller cutter 3. The test cutter had the same internal and external diameter and thickness defining its functional surfaces for proper placement in the bearing in the dial as a real roller cutter. Figure 20c shows a detailed view of the drive 4 and its used accessories needed to connect the flexible shaft 5.

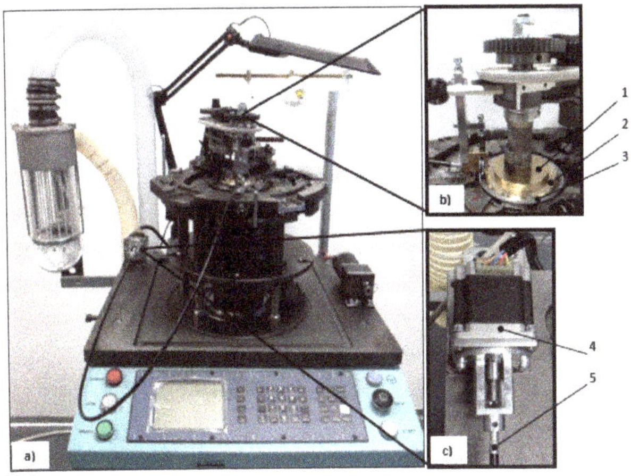

Figure 20. Installation of the individual roller cutter drive system into the Ange 18.1 small diameter knitting machine and detailed display of the basic nodes of the individual roller cutter drive system.

The aim of the measurement was to verify the function of the designed individual drive system of the roller cutter at maximum velocity and the suitability of the drive used. The 103H7123-1710 stepping motor (Sanyo Denki, Tokyo, Japan) was selected to drive the cutter due to the assumption of power sufficiency and good speed control. The motor was controlled using the stepping motor drive box NDC69A (R.T.A., Düsseldorf, Germany), which is equipped with a built-in oscillator. It therefore does not require an external clock signal to determine the velocity for the constant velocity running function. The drive can then only be controlled with a start/stop signal. The driver furthermore provides start-up functions, and the drive current can also be adjusted. Another function is the automatic reduction of the drive current when stopping.

The time course of the voltage from the sidewall was recorded by the measuring control panel (HBM) and converted to the current value, see Figure 21. The outputs from the measurement of electrical values of the individual roller cutter drive system are presented in Table 3. Practical verification of the aforementioned design solution revealed that the roller cutter can move independently within the range of normal working velocities (0–320 min^{-1}) of the considered knitting machine independently of the cylinder and the dial.

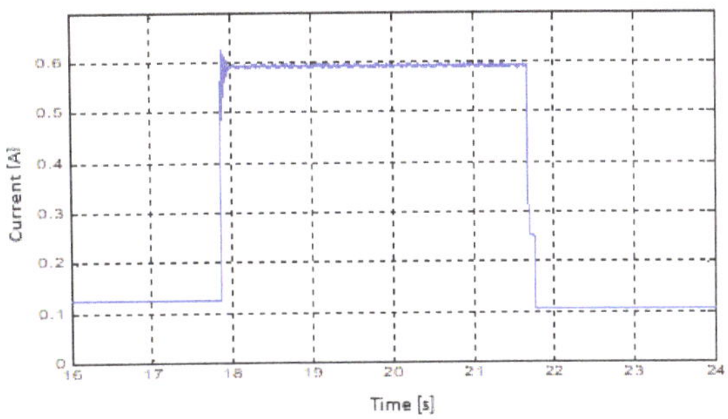

Figure 21. Current course measurement.

Table 3. Outputs from the measurement of the individual cutting disc drive system.

Parameter	Designation	0–2	3–4
Driver supply voltage at rest Value	U_{no}	42.1	V
Driver supply voltage-during runnig	U_n	40.7	V
Driver current consumption-off runnig	I_{no}	125	mA
Driver current consumption-during runnig	I_n	594	mA
Power input-off runnig	P_{no}	5.2	W
Power input-during runnig	P_n	24.1	W

5. Conclusions

In the context of the production of classic stocking and sock goods, knitting is divided in practice into two modes. The first involves knitting the heel or toe of the sock, and the second generally involves making tubular knitted fabric. The first mode was subjected to a dynamic analysis, as the system performs rotational reverse motion within the range of the cylinder movement 0–3/4 π. This movement method of the drive system is classified as the dynamically least favorable load case of the entire machine. For a better understanding of the dynamic behavior of the drive system in reverse motion, two mathematical models have been elaborated for the existing and for the (patented) modified structure that now includes use of controlled drives. Mathematical models were created in the Matlab-Simulink programming environment. The findings of the performed dynamic analysis of the existing drive structure indicate (a) a significant influence of plays, whereas shocks occur upon their delimitation, and (b) the accompanying influence of the moment of inertia.

As a rule, a linear acceleration course cannot be achieved in the true configuration of the drive system, especially in transition areas of movement intervals due to sudden changes occurring in these areas. In the modified structure of the drive system, a reduction was achieved in the moment of inertia and the targeted use of the cylinder, dial and roller cutter according to the technological process. These adjustments appeared in more favorable courses of torques and power consumption of the modified drive system in comparison with the existing system while using more suitable lift dependences.

The use of fewer elements in the modified drive system achieved a significant suppression of the impact of plays and a decrease in dynamic effects adversely affecting not only this system, but also the entire machine and the technology performed. Another monitored parameter was the load on the drives in the individual phases of system movement and determination of the courses of their power inputs, especially in the area of rotational reverse motion. Here, the reduction of moment of inertia in the modified drive system was most evident, which significantly affected the decrease in electricity consumption during the start-up and run-out of the drive system. There was a 10% decrease in total electricity consumption in the creation of one sock in the modified drive system compared to the current one. The value of real electricity consumption will be higher due to the exclusion of certain effects; see Section 2.

In the past, the design of an individual dial drive system has been addressed by various studies, but its movement was always tied to the movement of the roller cutter, which still did not provide efficient use of the drive system of small-diameter knitting machines. For this reason, a design has been elaborated of an individual drive system of the roller cutter. The design was conceived in such a way as to influence the other construction nodes of the knitting machine as little as possible with respect to the existing installation possibilities. The result of the design is a drive system structure capable of independently driving the roller cutter and can be used modularly with possible structural modifications on the existing knitting machines of the analyzed type. Changing the structure of the drive brings the possibility of simplifying the overall concept of the knitting machine. With a reduction of rotating elements, especially gears—which generate noise, vibrations, and reduce the efficiency of power transmission from the drive to the main individual operating elements (dial, roller cutter and needle cylinder) and increase the demands on installation space and maintenance of the machine—the possibility of a conceptually completely different machine is offered. In light of the obtained results, the new concept of the knitting machine would bring, above all, a more energy-saving solution with a lower noise level on operators. Another benefit can be gained in the individual and programmable control of the mentioned operating elements and the expansion of technological possibilities in terms of production knitted products.

Author Contributions: Conceptualization, J.S. and M.B.; methodology, J.S.; software, J.S. and M.B.; validation, M.B.; formal analysis, J.S. and M.B.; investigation, J.S. and M.B.; resources, O.B. data curation, M.B. and O.B.; writing—original draft preparation, J.S.; writing—review and editing, J.S.; visualization, M.B.; supervision, O.B.; project administration, J.S.; funding acquisition, J.S. All authors have read and agreed to the published version of the manuscript.

Funding: This research was funded by the Student Grant Competition of the Technical University of Liberec under the project No. SGS-2022-5046 and the Institutional Endowment for the Long Term Conceptual Development of Research Institutes, as provided by the Ministry of Education, Youth and Sports of the Czech Republic.

Institutional Review Board Statement: Not applicable.

Informed Consent Statement: Not applicable.

Data Availability Statement: The data presented in this study is available on request from the cor-responding author.

Conflicts of Interest: The authors declare no conflict of interest concerning the research, authorship and publication of this article.

References

1. Kopal, J. *Knitting, Stitch-Knitting and Stranding 1*, 1st ed.; Technical University of Liberec: Liberec, Czech Republic, 2006; ISBN 80-7372-119-8.
2. Kopal, J. *Knitting, Stitch-Knitting and Stranding 2*, 1st ed.; Technical University of Liberec: Liberec, Czech Republic, 2007; ISBN 978-80-7372-246-3.
3. Abbas, A.; Asim, A.; Zahid, A.; Anas, S.; Basra, S.A.; Azam, Z. Development of Knitted Reinforced Structures with Self-Diagnostic Function for Composite Applications. *J. Ind. Text.* **2022**, *51*, 15280837221098438. [CrossRef]
4. Ertekin, G. Analysis of Thermo-Physiological Comfort and Moisture Management Properties of Flat Knitted Spacer Fabrics. *Tekst. Konfeksiyon* **2017**, *27*, 241–250.
5. Oglakcioglu, N. Design of Functional Knitted Fabrics for Medical Corsets with High Clothing Comfort Characteristics. *J. Ind. Text.* **2016**, *45*, 1009–1025. [CrossRef]
6. Supuren, G.; Oglakcioglu, N.; Ozdil, N.; Marmarali, A. Moisture Management and Thermal Absorptivity Properties of Double-Face Knitted Fabrics. *Text. Res. J.* **2011**, *81*, 1320–1330. [CrossRef]
7. Zhou, L.; Feng, X.; Du, Y.; Li, Y. Characterization of Liquid Moisture Transport Performance of Wool Knitted Fabrics. *Text. Res. J.* **2007**, *77*, 951–956. [CrossRef]
8. Spencer, D.J. 12-Electronics in Knitting. In *Knitting Technology*, 3rd ed.; Spencer, D.J., Ed.; Woodhead Publishing Series in Textiles; Woodhead Publishing: Sawston, UK, 2001; pp. 134–144, ISBN 978-1-85573-333-6.
9. Jindal, H.; Kaur, S. Robotics and Automation in Textile Industry. *Int. J. Sci. Res. Sci. Eng. Technol.* **2021**, *8*, 40–45. [CrossRef]
10. Skřivánek, J.; Bílek, M. Dynamics of the Frame of Small-Diameter Knitting Machine. Dynamika ramy małośrednicowej maszyny dziewiarskiej. *ACC J.* **2010**, *14*, 1–7.
11. Goch, G. Gear Metrology. *CIRP Ann.-Manuf. Technol.* **2003**, *52*, 659–695. [CrossRef]
12. Ray, S.C. 7-Circular Weft Knitting Machine and Mechanism. In *Fundamentals and Advances in Knitting Technology*; Ray, S.C., Ed.; Woodhead Publishing: Delhi, India, 2012; pp. 69–85, ISBN 978-0-85709-108-6.
13. Power, E.J. Chapter 12-Yarn to Fabric: Knitting. In *Textiles and Fashion*; Sinclair, R., Ed.; Woodhead Publishing Series in Textiles; Woodhead Publishing: Sawston, UK, 2015; pp. 289–305, ISBN 978-1-84569-931-4.
14. Lucia, T. *Consiglia: A Complete Guide to Machine Knitting*; Hoakibooks S.L.: Barcelona, Spain, 2021; ISBN 84-17412-86-7.
15. Subhankar, M.; Sohel, R.; Pintu, P.; Kunal, S. *Advanced Knitting Technology | Libristo-Czech Republic*; Elsevier Books: Amsterdam, The Netherlands, 2021; ISBN 978-0-323-85534-1.
16. Paprocki, M.; Erwiński, K. Synchronization of Electrical Drives via EtherCAT Fieldbus Communication Modules. *Energies* **2022**, *15*, 604. [CrossRef]
17. BILEK, M.; Skřivánek, J.; Kopal, J.; Váša, P. *Okrouhlý Pletací Stroj, No. 303578*; Czech Pattent Office: Prague, Czech Republic, 2012.
18. Zaw, A.K.; Myat, A.M.; Thandar, M.; Htun, Y.M.; Aung, T.H.; Tun, K.M.; Han, Z.M. Assessment of Noise Exposure and Hearing Loss Among Workers in Textile Mill (Thamine), Myanmar: A Cross-Sectional Study. *Saf. Health Work.* **2020**, *11*, 199–206. [CrossRef]

19. Dini, P.; Saponara, S. Design of an Observer-Based Architecture and Non-Linear Control Algorithm for Cogging Torque Reduction in Synchronous Motors. *Energies* **2020**, *13*, 2077. [CrossRef]
20. Zheng, B.; Jiang, G.; Dong, Z.; Liu, H. Design of Warp Knitting Electronic Shogging System Based on Mixed-Velocity Planning Curve. *Text. Res. J.* **2021**, *91*, 1594–1608. [CrossRef]
21. Intellectual Property of Uniplet Třebíč a.s. *Technical Documentation of the Small-Diameter Knitting Machine Series Ange 18.1*; Intellectual Property of Uniplet Třebíč a.s.: Třebíčv, Czech Republic, 2014.
22. Negrean, I.; Crișan, A.-V.; Vlase, S. A New Approach in Analytical Dynamics of Mechanical Systems. *Symmetry* **2020**, *12*, 95. [CrossRef]
23. Negrean, I.; Crișan, A.; Șerdean, F.; Vlase, S. New Formulations on Kinetic Energy and Acceleration Energies in Applied Mechanics of Systems. *Symmetry* **2022**, *14*, 896. [CrossRef]
24. Skřivánek, J.; Bílek, M. Analysis of Dynamic Model of the Drive of Small Diameter Knitting Machines Ange 18.1. Analiza modelu dynamicznego napędu małośrednicowej maszyny dziewiarskiej Ange 18.1. *ACC J.* **2011**, *15*, 1–8.
25. Mrázek, J. Theoretical Analysis of Dynamics of Four-Bar Beat-up Mechanism of a Loom. *Mech. Mach. Theory* **1992**, *27*, 331–341. [CrossRef]
26. Skřivánek, J.; Bílek, M.; Kašpárek, M. Control Optimisation of the New Drive of Small Diameter Knitting Machines. *Autex Res. J.* **2016**, *17*, 1–5. [CrossRef]
27. Alam, S.N.; Islam, S.; Patel, S.K. *Advanced Guide to MATLAB: Practical Examples in Science and Engineering*; IK International Publishing House Pvt. Ltd.: Delhi, India, 2015; ISBN 978-93-84588-35-9.
28. Hairer, E.; Nørsett, S.; Wanner, G. *Solving Ordinary Differential Equations I, Nonstiff Problems*; Springer: Berlin, Germany, 1993; ISBN 978-3-540-56670-0.
29. Rotational Kinetic Energy—An Overview | ScienceDirect Topics. Available online: https://www.sciencedirect.com/topics/engineering/rotational-kinetic-energy (accessed on 3 June 2022).

machines

MDPI

Article

Machining of Wood Plastic Composite Using AWJ Technology with Controlled Output Quality

Zuzana Mitaľová [1,*], František Botko [1], Radoslav Vandžura [1], Juliána Litecká [2], Dušan Mitaľ [1] and Vladimír Simkulet [1]

1 Faculty of Manufacturing Technologies of Technical University of Košice with Seat in Prešov, 080 01 Prešov, Slovakia; frantisek.botko@tuke.sk (F.B.); radoslav.vandzura@tuke.sk (R.V.); dusan.mital@tuke.sk (D.M.); vladimir.simkulet@tuke.sk (V.S.)

2 Faculty of Humanities and Natural Science, University of Prešov, 080 01 Prešov, Slovakia; juliana.litecka@unipo.sk

* Correspondence: zuzana.mitalova@tuke.sk; Tel.: +421-55-602-6465

Abstract: The paper deals with the application of abrasive water-jet cutting to composite material containing natural reinforcement—wood plastic composite. The specimens were cut through the application of four flows of different abrasive mass: 150, 200, 250, and 300 g·min^{-1}, respectively, and under different traverse speeds required to achieve the (expected) quality level Q1–Q5 (according to the SN 214001: 2010 standard). The output quality of Q1–Q5 was set in the CNC cutting programs and the real traverse speed values were calculated by machine control system according to change in the flow of the abrasive mass. The quality of surface topography was assessed using a tester (contact roughness) and an Inspectis digital zoom microscope. The results of topography–surface roughness parameters *Ra* presented here are compared with the values normalized for individual samples sets. The applied technology, i.e., the AWJ, eliminated the problem of tool wear and adhesion of the thermoplastic matrix to tool surfaces (compared to standard machining).

Keywords: abrasive water jet; wood plastic composite; natural reinforcement

Citation: Mitaľová, Z.; Botko, F.; Vandžura, R.; Litecká, J.; Mitaľ, D.; Simkulet, V. Machining of Wood Plastic Composite Using AWJ Technology with Controlled Output Quality. *Machines* **2022**, *10*, 566. https://doi.org/10.3390/machines10070566

Academic Editor: Gianni Campatelli

Received: 15 June 2022
Accepted: 13 July 2022
Published: 14 July 2022

1. Introduction

Since the 1990s, the professional literature and technologists have been paying increased attention to the application of natural fiber/particle reinforcement materials (lignocellulosic, mineral, and animal) in the automotive, construction, and flooring industries [1–4]. Natural reinforcements can replace synthetic/glass fibers in plastic matrices (PE, PVC, PP, and epoxy) based on three attributes:

- Reduced input costs, shorter production cycles of composite materials, and lower weight of materials;
- Identical/similar mechanical properties compared to components reinforced with glass fiber, good dimensional stability and soundproofing;
- Recyclability of primary raw materials, non-toxicity, and CO_2 neutrality (carbon neutrality means achieving a balance between carbon emissions and their absorption from the atmosphere into carbon traps) [5–8].

In the commercial processing of WPC materials, wood flours with various types of wood, softwood and hardwood, are applied and, alternately, also plant fibers (lignocellulosic fibers) [9]. Many scientific studies point to the possibilities of using waste from the agricultural industry as an alternative to wood flour and achieving similar properties of the WPC product. More than 95% of WPC products originating in China are made from agricultural waste [10–15]. The choice of matrix material determines the processing temperature. Reinforcements consist of organic materials (cellulose, hemicellulose, and lignin) susceptible to thermal degradation upon reaching the boundary temperature of about

200 °C [16]. Based on their melting point, the following materials are suitable: polyethylene, polypropylene, polyvinylchloride, or polystyrene. Studies are currently underway to verify the application of PMMA or Nylon 6 as matrices of WPC products [17–21]. To improve the mechanical properties of the final product, to provide chemical bonding, and to achieve a simpler process of mixing components (rheological properties), additives are applied in different percentages (Figure 1: components of wood-filled plastics) [16,22–30].

Figure 1. Wood-filled plastic components [31–33].

The main conventional processes used in WPC manufacturing are: extrusion process, injection molding, compression molding, calendaring, laser sintering, and fused laser modeling [17,34–36]. Technologies applied in the products manufacturing in individual industries are presented in Table 1 (++ very important and gaining rapid traction, or + important and on the rise, or 0 less important and small, respectively). The WPC raw materials should undergo preprocessing treatments to enhance the interfacial bonding between the wood flour and the polymer matrix [37].

Table 1. The most important industries and their preferred technologies [22].

Applied Technologies	Construction Industry	Automotive Industry	Furniture Industry	Consumer Goods
Extrusion	++	0	++	0
Injection Molding	0	++	+	++
Compression Molding and Press Flow	0	++	+	0

The earliest developments of WPCs date back to the 1970s, with Gruppo Ovattifici Riuniti producing a product named "*Woodstock*" for Fiat cars in 1972 (composite material made of plastic matrix and wood filling in the ratio of 50:50) and Sonesson AB producing a PVC/wood-flour composite for use as flooring tiles in 1973 [38–41]. In 1991, the first conference on the mentioned composite materials was held, with participation of 50 researchers and WPC manufacturers themselves. At present, demand for WPC products is expected to increase by 14% year on year in European countries. Research focusing on improving the compatibility of components, reduction in their density and price, and application of biopolymers for matrices is becoming ever more important [39].

From the machinability point of view, WPC materials can be characterized as easily machinable (the authors Wilkowski et al. made the comparison with standard MDF boards with their determined machinability index) [42]. Many current authors deal with classic machining technologies, which are oriented on selection of tool geometry/material, study of suitable conditions of machining technologies in relation to the emerging surface quality, the emergence of cutting forces during machining, etc. [43–51]. However, they encounter a fundamental problem—melting the thermoplastic matrix and adhesion to the tool surfaces during the engagement (functional surfaces). An option for WPC machining is the application of water jet/abrasive water jet technology. There are a number of studies dealing with the machining of a wide range of materials (including different types of woods) [52–54]. However, only a small percentage of publications are available in the given area of interest (machining of WPC by AWJ/WJ); in the year 2012, the study of wood slabs was conducted by Hutyrová et al., examining the influence of a selected technological parameter: traverse speed and the MESH abrasive applied on micro-geometric and macro-geometric surface characteristics (basic surface roughness parameters *Ra*—arithmetic mean deviation of the profile and *Rz*—maximum height of the profile were determined in accordance with ISO 4287 and the parameters of the emerging thread of the turned surface). The best surfaces were achieved at lower traverse speed: 40 mm·min^{-1} [55]. In 2022, Boopathi et al. dealt with the influence of process parameters (traverse speed and water jet pressure) and the makeup of the composite in terms of the percentage representation of its individual components on surface roughness and kerf angle using the Taguchi method. It was observed that kerf angle and surface roughness were greatly impacted by the percentage of neem wood saw powder, table traveling speed, and water-jet pressure [56].

As the number of publications in the field is scanty, the article focuses on the influence of variable factors of selected technological parameters on the surface topography of composite materials containing natural reinforcement after cutting by a hydro abrasive water jet. The controlled parameter is the final surface roughness (Q1–Q5). Currently, the Swiss standard SN 214001: 2010 (the standard applies to materials up to a thickness of 300 mm) defining individual quality levels is in force for assessing surface topography by hydro abrasive water jet. The mentioned standard is applicable to materials suitable for AWJ cutting. Based on the determined measured parameters, the machined surface is divided into five levels (Q5 up to Q1), while the cut quality Q1 is defined as the lowest and the cut quality Q5 as the highest [57].

2. Materials and Methods

The experiment used wood–plastic composite material (beam with a rectangular cross-section, dimensions 40 × 60 × 150 mm), consisting of a polyethylene matrix reinforced with wood fibers (ratio 25/75 vol.%). The material contains cracks at the interface: wood versus plastic; wood fibers/particles copy the flow of the extruded matrix (Figure 2). Mechanical properties were determined in the laboratories of VÚHŽ-Dobrá, in accordance with relevant ISO standards (tensile test/triaxial bending test). Average values were Rm = 9.5 MPa, A = ca 3%, Z = 0.9%, flexural strength 16.8 MPa, and deformation work 0.7 mJ (statistically verified values). Density of extruded profile was between 500 and 700 kg·m^{-3} (at the temperature of 20 °C), and swelling after water storage was ≤7.0% (volume weight). Applications of WPC materials: construction (exterior cladding), flooring and furniture industry (floor coverings), and automotive/shipbuilding industry (dashboards, trunk bottoms, etc.).

WPC materials represent an ecological alternative (in relation to wood products) with a positive contribution to sustainable natural resources (forest areas). The ever-increasing environmental pressure, new ways of using natural-based composite materials, as well as technical innovations are leading manufacturers to apply them. Goal: to reduce the consumption of costly, nonrecyclable types of reinforcement, e.g., glass fibers.

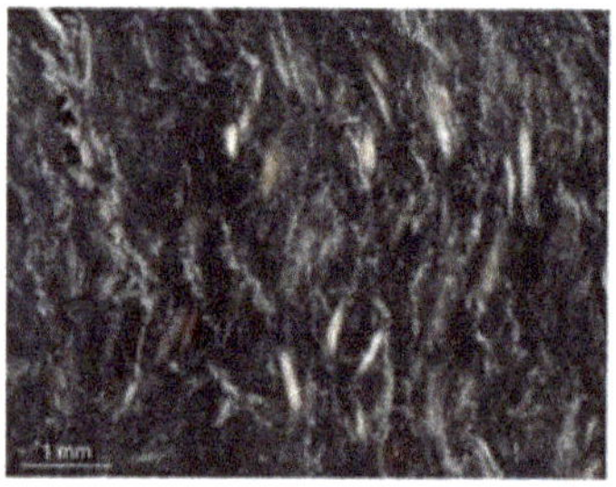

Figure 2. Microscopic structure of wood–plastic composite material—cracks at the interfaces (Nikon Eclipse 80i, Tokyo, Japan) [55].

2.1. Sample Preparation and Analysis of Experimental Surface Topography Measurement

The measurement was performed on 20 samples (5 sets of samples—Figure 3) divided by different sliding speeds of the cutting head (sliding speeds of the cutting head were set based on the required cut quality according to SN 214001: 2010, see Table 2) with different abrasive flow (variable factors: head feed rate v_f/abrasive mass flow m_a). To study the influence of process parameters on the quality of the machined surface during AWJ cutting, a conventional device for cutting flat materials using Water Jet 3015 RT-3D was used. The required water pressure of 400 MPa was generated by the PTV JETS—3.8/60 Classic pump. Australian Garnet with MESH 80 was used as the abrasive material. The angle of inclination of the abrasive head = 90°, the diameter of the water nozzle d_0 = 0.3 mm, the diameter of the focusing tube d_f = 0.9 mm, and the standoff distance of the nozzle was 4 mm.

Table 2. Marking of samples (sets based on process conditions).

Q-Level	Sample No.	Abrasive Mass Flow m_a (kg·min^{-1})	Traverse Speed v_p (mm·min^{-1})
Q1 (set 1)	1	150	346
	2	200	387
	3	250	423
	4	300	455
Q2 (set 2)	5	150	229
	6	200	257
	7	250	281
	8	300	302
Q3 (set 3)	9	150	144
	10	200	161
	11	250	176
	12	300	190
Q4 (set 4)	13	150	103
	14	200	116
	15	250	126
	16	300	136
Q5 (set 5)	17	150	80
	18	200	90
	19	250	100
	20	300	105

Figure 3. Clamping the workpiece on the table of the AWJ machine, divided profile—a semi-finished product.

2.2. *Topography of Horizontal Areas of the Material Surface Created by Abrasive Cutting (Experimental Setup)*

Currently in force, the Swiss standard SN 214001: 2010 applies to materials up to a thickness of 300 mm, defining individual quality levels for the evaluation of surface topography by hydro abrasive current. Based on the set measured parameters, the machined surface is divided into 5 levels (Q1–Q5), while the cut quality Q1 is defined as the lowest and the cut quality Q5 as the highest. The corresponding values of the parameter *Ra* defined by the standard are given in Table 3.

Table 3. Examined samples, real values of *Ra* in line L8, and the standard value for the presented QX level.

Q-Level	Real Value of *Ra* (μm) Measured in Line L8
Q1 (set 1) 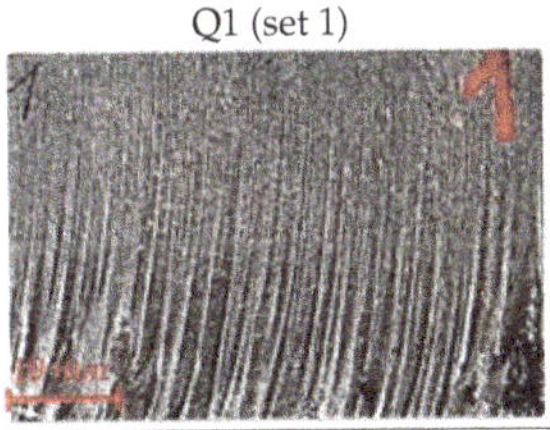	1 (N/A) 2 (N/A) 3 (N/A) 4 (N/A) (standard value for level Q1) *Ra* = 50 μm
Q2 (set 2) 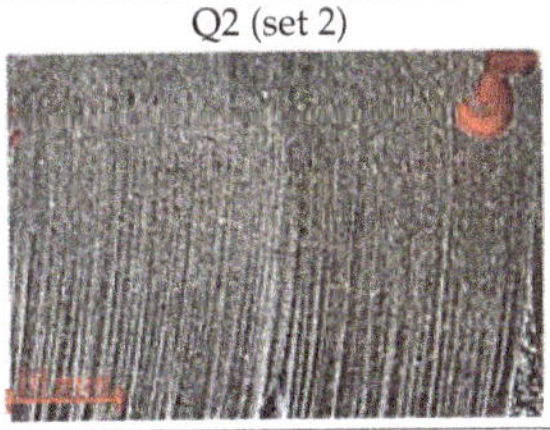	5 (N/A) 6 (N/A) 7 (*Ra* = 13.09 μm) 8 (*Ra* = 9.842 μm) (standard value for level Q2) *Ra* = 25 μm
Q3 (set 3) 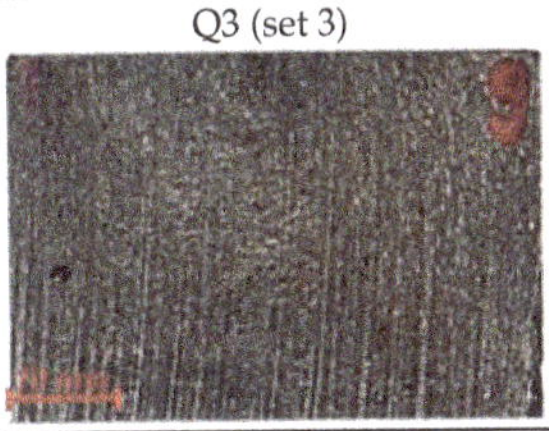	9 (*Ra* = 12.54 μm) 10 (*Ra* = 11.28 μm) 11(*Ra* = 11.32 μm) 12 (*Ra* = 7.83 μm) (standard value for level Q3) *Ra* = 12.5 μm

Table 3. *Cont.*

Q-Level	Real Value of *Ra* (μm) Measured in Line L8
Q4 (set 4) 	13 (*Ra* = 7.04 μm) 14 (*Ra* = 6.71 μm) 15 (*Ra* = 5.6 μm) 16 (*Ra* = 6.9 μm) (standard value for level Q4) *Ra* = 6.3 μm
Q5 (set 5) 	17 (*Ra* = 7.48 μm) 18 (*Ra* = 7.04 μm) 19 (*Ra* = 5.57 μm) 20 (*Ra* = 7.18 μm) (standard value for level Q5) *Ra* = 3.2 μm

Ra value specified by standard SN 214001: 2010, samples from 1 to 6, measured value of *Ra* is not specified, impossible to measure surface roughness parameter *Ra* in L8 (N/A—measurements not available). Pictures—Inspectis digital zoom microscope.

Parameters of surface roughness (*Ra* and *Rz*) were determined by the contact roughness tester: MITUTOYO at the Faculty of Manufacturing Technologies seated in Prešov. The standard EN ISO 4287: 1997 is adopted for the quantitative evaluation of surface roughness [58]. Surface roughness is a geometric property and there are no direct methods for its assessment. Suitable characteristics and parameters are taken into account, which are considered to constitute surface roughness criteria. The most common evaluation criterion is the mean arithmetic deviation of the surface irregularities *Ra* (or the greatest height of the profile *Rz*) evaluated with respect to the baseline in the base length l_r. In order to identify and analyze factors affecting the surface topography, measurements were performed stepwise through the horizontal area of the sample according to Figure 4 (in relation to the shape of the profile, produced by extrusion, the first measured section was placed outside the rounded area of the sample). Line L8 is located 2 mm from the bottom edge of the profile (Figure 4). Values shown in the graphical dependences are averages from 9 repeated measurements (evaluated characteristic: *Ra*). The Grubbs test with a probability of p = 95% was applied to exclude values subjected to gross error.

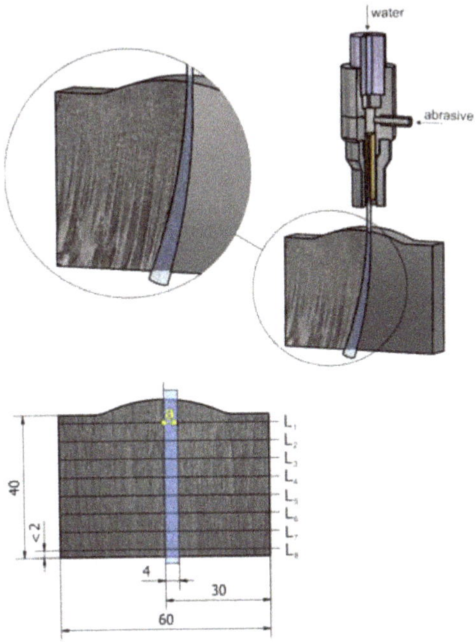

Figure 4. Measurement principle, displayed lines from L1 to L8, direction of travel of the roughness measuring probe (measured length l_r = 4 mm, direction of tip movement—in the direction of the arrow "a"), and the actual surface after the AWJ machining (sample 1, set Q1).

3. Results and Discussion

After the performed experimental measurements and processing of the measured parameters, information was obtained about the height inequalities of the surface topography created under the changing experimental conditions. In relation to the controlled result parameter QX (categorization of the resulting surface Q1 to Q5 with varying abrasive mass flow and feed rate), each set must be assessed separately. Measurement was conducted nine times for each measuring line; subsequently, graphical dependences and tables show arithmetical values for individual lines.

Set Q1 (Figure 5): the surfaces shown consist of two distinctive zones: smooth and striation zones. In the case of four evaluated samples, it is not possible to perform measurements in all defined lines L1 to L8, where, in samples 1 and 2, missed measurements are in lines L7 and L8; in sample 3, missed measurements are in lines L6 to L8; and, in sample 4, a missed measurement is in line L8. The surface of the mentioned lines is defined as too rough. The values of the parameter *Ra* have a significantly increasing tendency from the line L4 (about half of the cut made). In the case of samples of the Q1 set, there was no complete division and the samples had to be separated manually (see right corner of sample 1—phenomena is caused by jet lag in the direction of traverse speed, Table 3). As the depth of the cut increases, the values of the *Ra* parameters increase. After passing line L4, increased deformation in the lower part of the workpiece is visible in all samples—formation of defects belonging to a macroscopic (scratches and grooves) group, Table 3. In relation to the inhomogeneous nature of the material, samples 1 and 2 (in addition to significant traces of lagging water flow) also contain traces/grooves/deformities. The mechanism of material removal using the AWJ is characterized as a micro-erosion process. With the increasing thickness of machined material and increased traverse speed, significant traces in the material can be observed, caused by kinetic energy loss.

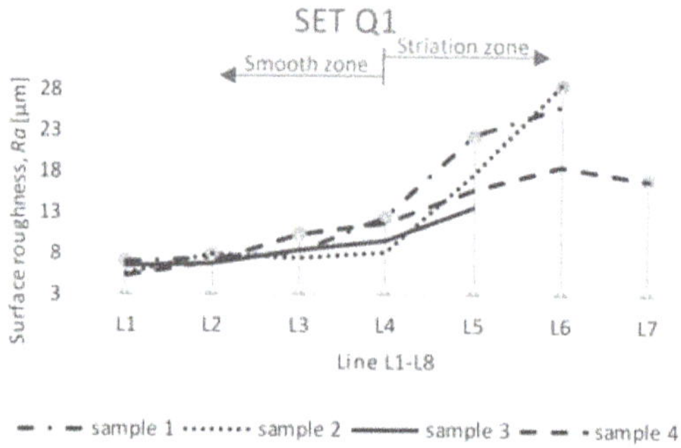

Figure 5. Evaluation of surface topography created by the AWJ (set Q1).

Set Q2 (Figure 6): the surfaces shown also consist of two distinctive zones: smooth and striation zones. In the case of the two evaluated samples, it is not possible to perform measurements in all eight lines under assessment (sample 5: missed measurements in lines L7, L8; sample 6: missed measurement in line L8). The highest value of the *Ra* parameter achieved was recorded in the case of sample 5, namely, of 14.64 μm in line L7.

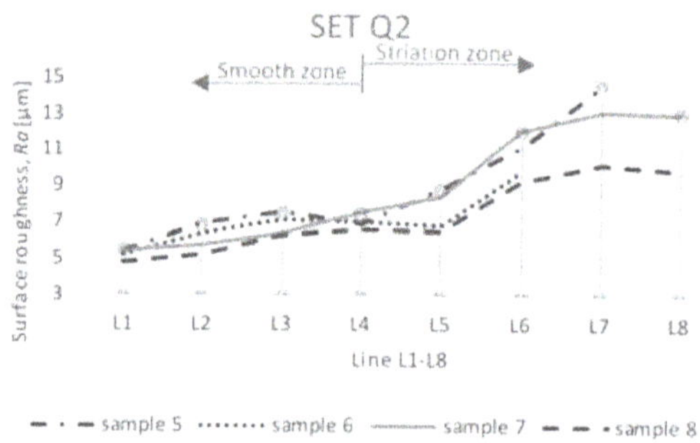

Figure 6. Evaluation of surface topography created by the AWJ (set Q2).

In the case of sets Q3 and Q4 (Figures 7 and 8), it was possible to measure the assessed parameter *Ra* in all selected lines. Smooth and striation zones are defined by line L5. For samples 9 to 12, the highest *Ra* values were recorded in the last line L8 (located 2 mm from the bottom edge of the sample).

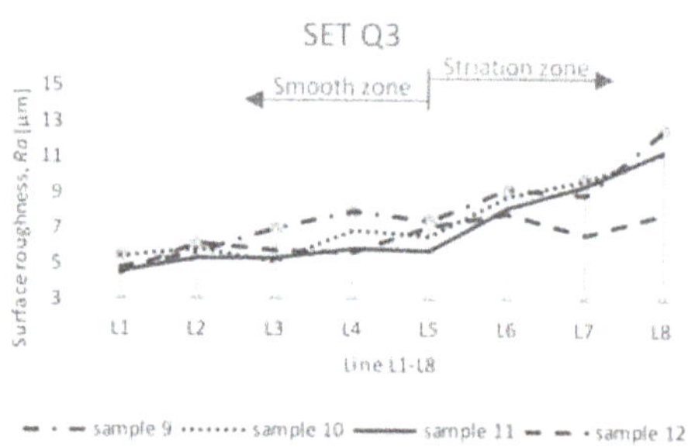

Figure 7. Evaluation of surface topography created by the AWJ (set Q3).

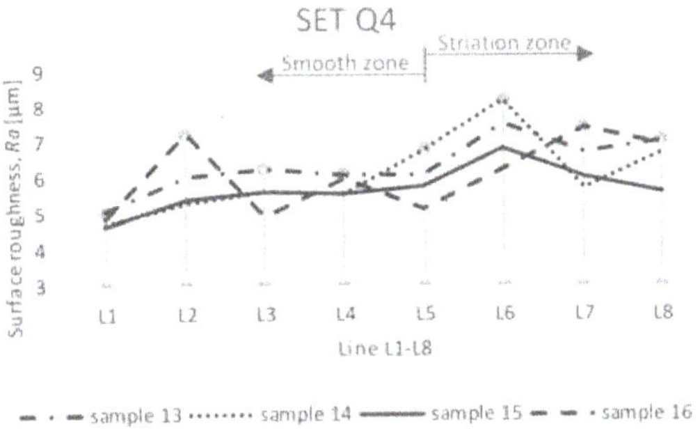

Figure 8. Evaluation of surface topography created by the AWJ (set Q4).

In the case of set Q5 (Figure 9), the values of the parameter *Ra* for all samples were in the range from 3.93 to 7.59 μm (from sample 17 to sample 20). The surface is relatively smooth (compared to other sets). For comparison with the value specified by the standard, we present the table below. Only in the case of six divided areas (samples 7, 8, 10, 11, 12, and 15), the *Ra* values measured in line L8 comply with the relevant Q-level. The set Q3 contains the most samples corresponding to the prescribed value.

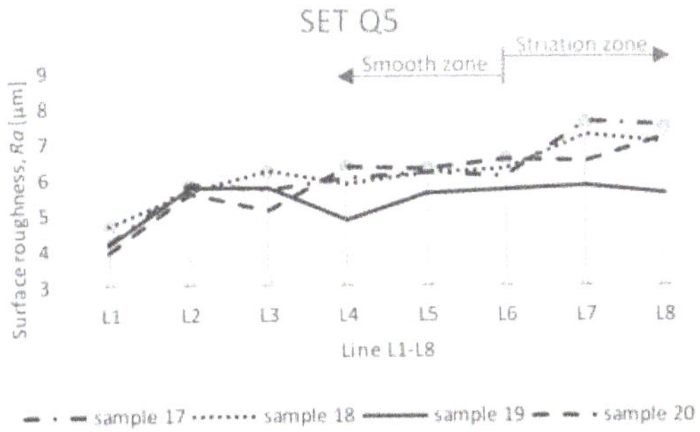

Figure 9. Evaluation of surface topography created by the AWJ (set Q5).

Pictures in Table 3 were observed on the Inspectis digital zoom microscope (samples marked 1, 5, 9, 13, and 17). The individual samples show a visible difference in the quality of the cut trace after the AWJ machining under different technological conditions. Samples marked 9, 13, and 17 showed lowest "ripple" as opposed to samples marked 1 and 5. Samples 1 and 5 show the resistance that the material exerts on the nozzle transverse trace. The greatest resistance can be observed in the sample marked no. 1. In this sample 1, small cavitation points of wood fillings are also observed, which can be observed in smaller evaluated samples.

4. Conclusions

The applied cutting technology by means of the AWJ eliminated the problem of tool wear and adhesion of the thermoplastic component to the tool surfaces in the engagement (so-called functional surfaces). The resulting split areas of sets Q1–Q5 can be divided into two zones: smooth and striation zones, depending on the specific sample of each of the five sets. Based on the ISO 4287 standard, the surface roughness parameter *Ra*, mean arithmetic deviation of the profile (assessed in eight lines of machined surfaces), was assessed. In the case of six samples, sample 1 to sample 4, sample 5, and sample 8, measurements in the specified lines could not be performed (in the lower lines, the surface was defined as too rough). The SN 214001: 2010 standard was applied for the qualitative evaluation of the machined surface. When compared with the prescribed value of *Ra* (determined by the standard according to the quality level Q1 to Q5), it is possible to say that the most compliant with the standard are values of *Ra* of the Q3 set (samples: 10, 11, and 12). The *Ra* values achieved for samples 10 to 12 are lower than the value set by the standard (standard *Ra* value = 12.5 μm). However, it remains questionable whether the standard and the assignment of quality levels Q1–Q5 are satisfactory for all types of assessed materials. The aim of the present contribution was to verify whether it is possible to achieve the surface quality prescribed by the standard on heterogeneous WPC materials through a controlled AWJ cutting process.

Author Contributions: Conceptualization, Z.M.; methodology, Z.M.; software, F.B. and R.V.; investigation, Z.M. and D.M.; resources, F.B.; data curation, Z.M.; writing—original draft preparation, F.B. and Z.M.; writing—review and editing, F.B.; visualization, J.L.; supervision, V.S. All authors have read and agreed to the published version of the manuscript.

Funding: This research was funded by APVV-20-0514 and KEGA no. 032TUKE-04/2021.

Data Availability Statement: Not applicable.

Acknowledgments: This work was supported by the Slovak Research and Development Agency under the contract No. APVV-20-0514 (Research of the influence of technological parameters of abrasive water jet machining on the surface integrity of tool steels) and project KEGA no. 032TUKE-4/2021—Transfer and implementation of knowledge in the field of energy beam technologies into study programs of technical secondary schools supporting dual education.

Conflicts of Interest: The authors declare no conflict of interest. The founding sponsors had no role in the design of the study; in the collection, analyses, or interpretation of data; in the writing of the manuscript; or in the decision to publish the results.

References

1. Sanjay, M.R.; Suchart, S.; Parameswaranpillai, J.; Jawaid, M.; Ozbakkaloglu, T. Lignocellulosic fiber reinforced composites: Progress, performance, properties, applications, and future perspectives. *Polym. Compos.* **2021**, *43*, 645–691.
2. Puglia, D.; Biagiotti, J.; Kenny, J.M. Applications of Natural Reinforcements in Composite Materials for Automotive Industry. *J. Nat. Fibers* **2005**, *1*, 23–65. [CrossRef]
3. Ferreira, F.V.; Pinheiro, I.F.; de Souza, S.F.; Mei, L.H.I.; Lona, L.M.F. Polymer Composites Reinforced with Natural Fibers and Nanocellulose in the Automotive Industry: A short Review. *J. Compos. Sci.* **2019**, *3*, 51. [CrossRef]
4. Khalid, M.Y.; Rashid, A.A.; Arif, Z.U.; Ahmed, W.; Arshad, H.; Zaidi A., A. Natural fiber reinforced composites: Sustainable materials for emerging applications. *RINENG* **2021**, *13*, 100263. [CrossRef]
5. Dilpreet, S.B.; Bhattacharjee, S. Current Progress, trends and Challenges in the Application of Biofiber Composites by Automotive Industry. *J. Nat. Fibers* **2016**, *13*, 660–669.
6. Azman, A.; Asyraf, M.; Khalina, A.; Petru, M.; Ruzaidi, C.M.; Sapuan, S.M.; Nik, W.B.W.; Ishak, M.R.; Ilyas, R.A.; Suriani, M.J. Natural Fiber Reinforced Composite Material for Product Design: A short Review. *Polymers* **2021**, *13*, 1917. [CrossRef]
7. Lotfi, A.; Huaizhong, L.; Dao, D.V.; Prusty, G. Natural fiber-reinforced composites: A review on material, manufacturing and machinability. *J. Thermoplast. Compos. Mater.* **2019**, *34*, 238–284. [CrossRef]
8. Campilho, R.D.S.G. *Natural Fiber Composites*; Taylor & Francis Group: Boca Raton, FL, USA, 2016; pp. 16–22.
9. Fabiyi, J.S.; McDonald, A.G. Effect of wood species on property and weathering performance of wood plastic composites. *Compos. Part A Appl. Sci. Manuf.* **2010**, *41*, 1434–1440. [CrossRef]
10. Nourbakhsh, A.; Ashori, A. Wood plastic composite from agro-waste materials: Analysis of mechanical properties. *Bioresour. Technol.* **2010**, *101*, 2525–2528. [CrossRef]
11. Abba, H.A.; Nur, I.Z.; Salit, S.M. Review of agro waste plastic composites production. *J. Met. Mater. Miner.* **2013**, *1*, 271–279. [CrossRef]
12. Chen, J.; Teng, Z.; Wu, J. Recycling of waste FRP and corn straw in wood plastic composite. *Polym. Compos.* **2017**, *38*, 2140–2145. [CrossRef]
13. Chun, K.S.; Fahamy, N.M.Y.; Yeng, H.Y.; Choo, H.L.; Ming, P.M.; Tshai, K.Y. Wood plastic composites made from corn husk fiber and recycled polystyrene foam. *J. Eng. Sci. Technol. Rev.* **2018**, *13*, 3445–3456.
14. Rahman, M.R.; Bakri, M.K.B. *Recycled Plastic Composites*, 1st ed.; Elsevier: Amsterdam, The Netherlands, 2022; pp. 21–24.
15. Binhussain, M.A.; El-Tomsy, M.M. Palm leave and plastic waste wood composite for out-door structures. *Constr. Build. Mater.* **2013**, *47*, 1431–1435. [CrossRef]
16. Klyosov, A.A. *Wood-Plastic Composites*, 1st ed.; John Wiley & Sons, Inc.: Hoboken, NJ, USA, 2007; pp. 50–51.
17. Gardner, D.J.; Han, Y.; Wang, L. Wood-Plastic composite technology. *Curr. For. Rep.* **2015**, *1*, 139–150. [CrossRef]

18. Vovk, M.; Sernek, M. Aluminium Trihydrate-filled Poly(methyl methacrylate) (PMMA/ATH) Waste Powder Utilization in Wood-plastic Composite Boards Bonded by MUF Resin. *Bioresources* **2020**, *15*, 3252–3269. [CrossRef]
19. Laming, P.B. Some anatomical and physical aspects of wood-plastic (pMMA) combination of spruce. *Leiden Bot. Ser.* **1976**, *3*, 266–280.
20. Rowell, R.M. *Wood Chemistry and Wood Composites*, 1st ed.; Taylor & Francis: Boca Raton, FL, USA, 2005; pp. 366–377.
21. Liu, H.Z.; Yao, F.; Xu, Y.J.; Wu, Q.L. A novel wood flour-filled composite based on microfibrillar high-density polyethylene (HDPE)/Nylo-6 blends. *Bioresource Technol.* **2010**, *101*, 3295–3297. [CrossRef]
22. Niska, K.O.; Sain, M. *Wood-Polymer Composites*, 1st ed.; Woodhead Publishing Limited: Cambridge, UK, 2008; pp. 23–40.
23. Najafi, S.K. Use of recycled plastics in wood plastic composites—A review. *J. Waste Manag.* **2013**, *33*, 1898–1905. [CrossRef]
24. Harper, D.; Wolcot, M. Interaction between coupling agents and lubricants in wood-polypropylene composites. *Compos. Part A Appl. Sci. Manuf.* **2010**, *101*, 3295–3297. [CrossRef]
25. Bhaskar, K.; Jayabalakrishnan, D.; Kumar, M.V.; Sendilvelan, S.; Prabhahar, M. Analysis on mechanical properties of wood plastic composite. *Mater. Today Proc.* **2022**, *61*, 5886–5891. [CrossRef]
26. Markarin, J. Additive developments aid growth in wood-plastic composites. *Plast. Addit. Compd.* **2017**, *25*, 361–367. [CrossRef]
27. Schirp, A.; Ibach, R.E.; Pendleton, D.E.; Wolcott, M.P. Biological degradation of wood-plastic composites (WPC) and strategies for improving the resistance of WPC against biological decay. In Proceedings of the 229th National Meeting of the American-Chemical Society, San Diego, CA, USA, 13–17 March 2005.
28. Khamedi, R.; Hajikhani, M.; Ahmaditabar, K. Investigation of maleic anhydride effect on wood plastic composites behavior. *J. Compos. Mater.* **2018**, *53*, 1955–1962. [CrossRef]
29. Huuhilo, T.; Martikka, O.; Butylina, S.; Kärki, T. Impact of Mineral Fillers to the Moisture Resistance of Wood-Plastic Composites. *Balt. For.* **2010**, *16*, 126–131.
30. Dai, L.; Wang, X.; Zhang, J.; Wang, F.; Ou, R.; Song, Y. Effect of lubricants on the rheological and mechanical properties of wood flour/polypropylene composites. *J. Appl. Polym. Sci.* **2019**, *136*, 47667. [CrossRef]
31. Available online: https://www.pawanplastics.com/orange-dana-3850200.html (accessed on 15 June 2022).
32. Available online: https://www.dreamstime.com/royalty-free-stock-photos-wood-shavings-image38097888 (accessed on 15 June 2022).
33. Available online: https://balticday.com (accessed on 15 June 2022).
34. Benoit, N.; Rodrigue, D. Extrusion-calendering of foamed and unfoamed wood/plastic composites. In Proceedings of the Annual Technical Conference—ANTEC, Orlando, FL, USA, 2–4 April 2012.
35. Gacitua, W.; Bahr, D.; Wolcott, M. Damage of the cell wall during extrusion and injection molding of wood plastic composites. *Compos. Part A Appl. Sci. Manuf.* **2010**, *41*, 1454–1460. [CrossRef]
36. Wlsheik, A.H.; Palchal, H.; Shanmugan, S.; Muthuramalingam, T.; El-Kassas, A.M.; Ramesh, B. Recent progresses in wood-plastic composites: Pre-processing treatments, manufacturing, techniques, recyclability and eco-friendly assessment. *Clean. Eng. Technol.* **2022**, *8*, 100450.
37. Imken, A.A.P.; Brischke, C.; Kögel, S.; Krause, K.C.; Mai, C. Resistance of different wood-based materials against mould fungi: Comparison of methods. *Eur. J. Wood Wood Prod.* **2020**, *78*, 661–671. [CrossRef]
38. Jones, J.; Brischke, K. *Performance of Bio-Based Building Materials*; Woodhead Publishing Limited: Cambridge, UK, 2017; pp. 21–96.
39. Ansel, M.P. *Wood Composites*; Woodhead Publishing Limited: Cambridge, UK, 2015.
40. Clemons, C.M. Woodfiber-plastic composites in the United States—History and current and future markets. In Proceedings of the 3rd International Wood and Natural Fibre Composites Symposium, Kassel, Germany, 19–20 September 2000.
41. Zefang, X.; Linbo, Z.; Yanjun, X.; Qingwen, W. Review for development of wood Plastic Composites. *J. Northeast For. Univ.* **2003**, *31*, 39–41.
42. Wilkowski, J.; Borysiuk, P.; Górski, J.; Czarniak, P. Analysis of relative machinability indexes of wood particle boards bonded with waste thermoplastic. *Drewno* **2013**, *56*, 139–144.
43. Hutyrová, Z.; Harničárová, M.; Zajac, J.; Valíček, J.; Mihok, J. Experimental study of surface roughness of wood plastic composite after turning. *Adv. Mater. Res.* **2013**, *856*, 108–112. [CrossRef]
44. Wei, W.; Yingli, L.; Li, Y.; Xu, Y. Tool wear during high-speed milling of Wood-plastic composites. *Bioresources* **2019**, *14*, 8678–8688.
45. Zhu, Z.; Buck, D.; Wang, J.; Wu, Z.; Xu, W.; Guo, X. Machinability of different Wood-plastic composites during peripheral milling. *Materials* **2022**, *15*, 1303. [CrossRef]
46. Guo, X.; Ekevad, M.; Marklund, B.; Li, R.; Cao, P.; Grönlund, A. Cutting forces and chip morphology during wood plastic composite orthogonal cutting. *Bioresources* **2014**, *9*, 2090–2106. [CrossRef]
47. Buehlmann, U.; Saloni, D.; Lemaster, R.L. Wood fiber-plastic composites: Machining and surface quality. In Proceedings of the 15th International Wood Machining Seminar, Aaheim, CA, USA, 30 July–1 August 2001.
48. Pritchard, G. Two technologies merge: Wood plastic composites. *Plast. Addit. Compd.* **2004**, *6*, 18–21. [CrossRef]
49. Rajmohan, T.; Vinayagamoorthy, R.; Mohan, K. Review on effect machining parameters on performance of natural-fibre reinforced composites (NFRCs). *J. Thermoplast. Compos. Mater.* **2018**, *32*, 1282–1302. [CrossRef]
50. Vinayagamoorthy, R.; Rajmohan, T. Machining and its challenges on bio-fibre reinforced plastics: A review. *J. Reinf. Plast. Compos.* **2018**, *37*, 1037–1050. [CrossRef]

51. Svetlik, J.; Baron, P.; Dobránsky, J.; Kočiško, M. Implementation of computer system for support of technological preparation of production for technologies of surface processing. In Proceedings of the 13th International Conference on Industrial, Service and Humanoid Robotics, Strbské Pleso, Slovakia, 15–17 May 2014.
52. Pelit, H.; Yaman, O. Influence of processing parameters on the surface roughness of solid wood cut by abrasive water jet. *Bioresources* **2020**, *15*, 6135–6148. [CrossRef]
53. Haskish, M. Characteristics of surfaces machined with abrasive-waterjets. *J. Eng. Mater. Technol.* **1991**, *113*, 354–362. [CrossRef]
54. Hoogstrate, A.; Karpuchewski, B.; van Luttervelt, C.A.; Kals, H.J.J. Modellig of high velocity, loose abrasive machining processes. *CIRP Ann.* **2002**, *1*, 263–266. [CrossRef]
55. Hutyrová, Z.; Sčučka, J.; Hloch, S.; Hlaváček, P.; Zeleňák, M. Turning of wood plastic composites by water jet and abrasive water jet. *J. Adv. Manuf. Technol.* **2016**, *84*, 1615–1623. [CrossRef]
56. Boopahi, S.; Thillaivanan, A.; Azeem, M.A.; Shanmugam, P.; Pramod, V.R. Experimental investigation on abrasive water jet machining of neem wood plastic composite. *Funct. Compos. Struct.* **2022**, *4*, 25001. [CrossRef]
57. *SN 214001:2010*; Contact-Free Cutting—Water Jet Cutting-Geometrical Specification and Quality. Swiss Standards-Vereiningung: Zurich, Switzerland, 2013.
58. *ISO 4287:1997*; Geometrical Product Specification (GPS)—Surface Texture: Profile Method-Terms, Definitions and Surface Texture Parameters. International Organization for Standardization: Geneva, Switzerland, 1998.

machines

MDPI

Article

Online Monitoring of Surface Quality for Diagnostic Features in 3D Printing

Natalia Lishchenko [1], Ján Piteľ [1,*] and Vasily Larshin [2]

1 Department of Industrial Engineering and Informatics, Faculty of Manufacturing Technologies, Technical University of Kosice, Bayerova 1, 080 01 Prešov, Slovakia; natalia.lishchenko@tuke.sk
2 Department of Digital Technologies in Engineering, Odessa Polytechnic National University, 1 Shevchenko Ave., 65044 Odessa, Ukraine; vasilylarshin@gmail.com
* Correspondence: jan.pitel@tuke.sk; Tel.: +421-55602-6455

Abstract: Investigation into non-destructive testing and evaluation of 3D printing quality is relevant due to the lack of reliable methods for non-destructive testing of 3D printing defects, including testing of the surface quality of 3D printed parts. The article shows how it is possible to increase the efficiency of online monitoring of the quality of the 3D printing technological process through the use of an optical contactless high-performance measuring instrument. A comparative study of contact (R130 roughness tester) and non-contact (LJ-8020 laser profiler) methods for determining the height of irregularities on the surface of a steel reference specimen was performed. It was found that, in the range of operation of the contact method (Ra 0.03–6.3 μm and Rz 0.2–18.5 μm), the errors of the contactless method in determining the standard surface roughness indicators Ra and Rz were 23.7% and 1.6%, respectively. Similar comparative studies of contact and non-contact methods were performed with three defect-free samples made of plastic polylactic acid (PLA), with surface irregularities within the specified range of operation of the contact method. The corresponding errors increased and amounted to 65.96% and 76.32%. Finally, investigations were carried out using only the non-contact method for samples with different types of 3D printing defects. It was found that the following power spectral density (PSD) estimates can be used as diagnostic features for determining 3D printing defects: Variance and Median. These generalized estimates are the most sensitive to 3D printing defects and can be used as diagnostic features in online monitoring of object surface quality in 3D printing.

Keywords: signal processing; monitoring system; laser profiler; surface roughness; quality assessment; non-contact method; vision-based method; frequency analysis

Citation: Lishchenko, N.; Piteľ, J.; Larshin, V. Online Monitoring of Surface Quality for Diagnostic Features in 3D Printing. *Machines* **2022**, *10*, 541. https://doi.org/10.3390/machines10070541

Academic Editor: César M. A. Vasques

Received: 28 May 2022
Accepted: 30 June 2022
Published: 4 July 2022

1. Introduction

The statistics show the growth in size of the global 3D printing market from 2013 to the present. For example, in 2020, the most popular use cases for 3D printing were prototyping (68%), proofs of concept (59%), production (49%) and R&D (42%) [1]. Nowadays, 3D printing is used in the aerospace, automotive, military, food, healthcare and medicine, architecture, fashion and electrical and electronics industries. The following materials can be used in 3D printing: metals, polymers and plastics, ceramics and composites [2].

Fused deposition modeling (FDM), associated with the Stratasys trademark, is currently the most widely used 3D printing technology. FDM printers use a thermoplastic filament, which is heated to its melting point and then extruded layer by layer to create a three-dimensional object. One of the key strengths of the FDM technique is its compatibility with various types of thermoplastic polymers. The most popular and stable materials are acrylonitrile butadiene styrene (ABS) and polylactic acid (PLA). FDM printers have demonstrated the ability to print other thermoplastics, which currently include polycarbonates (PCs), polystyrene (PS), polyamide, polyetherimide (PEI) and polyetheretherketone (PEEK).

There is also a demand for construction of composite filaments by adding certain materials into polymer matrices, as they can offer improved mechanical properties, biocompatibility, or conductivity [3–5].

To address the differences between additive manufacturing (AM) and conventional or subtractive manufacturing (SM), the science and engineering community is gravitating toward an AM solution centered on three pillars [6]:

1. Quality assurance (QA) derived from build planning (the use of advanced modeling and simulation to develop a plan for a machine to produce a specific part);
2. Build monitoring and inspection (monitoring the build process with sensors while the part is being constructed);
3. Feedback control to link the previous pillars together (using data from the build monitoring sensors to iteratively update the build plan).

In the right applications, additive manufacturing delivers a perfect trifecta of improved performance, complex geometries, and simplified fabrication [3].

Fundamentally, quality is about a part's ability to perform the task for which it has been designed while maintaining structural integrity. Contributing factors are usually included in a part's specifications and typically include geometry (the shape of the finished part and how it fits with other parts), surface finish (the desired smoothness, roughness, or other functional surface treatment of the finished part), and material properties (a variety of attributes, including mechanical strength, stiffness, and fatigue life) [6].

The advantages of the FDM method include easy handling, high printing speed, cost efficiency, the variety of types of thermoplastic polymers, the capacity for freeform fabrication without the use of expensive moulding sand tools, and the low cost of machines and consumables. The disadvantages of the FDM method include the different types of defects:

- Surface defects (over-extrusion [7], overheating [7], high roughness [8], stringy first layer [8]);
- Structure defects (cracking [9–12], under-extrusion [7,13,14], porosity [15,16]);
- Form defects (overheating [7], warping [9,10,17], blistering [9], stringing [9], layer shifting [18,19]).

Indirect and direct control methods can be used for the quality assessment of 3D printing. For indirect control, researchers have used acoustic emission (AE) signals [20–23], thermal cameras [24–26], vibration signals [27,28], current sensors [29,30], etc.I Indirect methods of quality control are most often used to detect different printing failures [20–31]; for example, those shown in Table 1, which indicates both the failures and signal (or sensor) types. These methods are indirectly related to the surface quality of printed 3D objects.

Table 1. Signals and sensors used for indirect monitoring [20–31].

Sensor (Signal) Type	Quality Characteristics
Accelerometer	Nozzle clogging Filament jamming Material leakage
Thermal camera	Extrusion stopping Nozzle clogging Irregular material flow
Acoustic emission	Filament breakage Extruder state Fan activity Nozzle clogging
Current, force, and pressure	Sabotage attacks in G-code Extrusion pressure Material flow rate

To implement direct methods quality control that can be embedded into the online surface quality monitoring system, a computer vision approach has become widespread. This is due to:

- The diverse methods available for assessing the quality of images;
- The possibility of using multiple cameras simultaneously to capture images from several sides of the printed object;
- The possibility of using advanced machine learning algorithms for image processing and quality assessment, such as support vector machines (SVMs), convolutional neural networks (CNNs), decision trees, etc. [32,33];
- The ease with which cameras can be integrated into equipment;
- The wide variety of image quality characteristics that can be evaluated with cameras (geometric deviations, infill structures, layer shifting, and surface defects, such as voids, overfill, underfill, blobs, cracks, misalignment, warping, detachment (delamination), etc.).

In computer vision systems, three main methods for assessing the quality of a manufactured object (that is, 3D printing system output parameters, which constitute a subjective classification as opposed to a scientific, statistical one and machine-learning algorithms) can be distinguished and used, namely:

1. Surface quality assessment; i.e., the presence of defects, such as voids, cracks, blobs, and misalignment;
2. Determination of geometric deviations in the dimensions of a printed object through comparison with the object's CAD model;
3. Determination of the print output parameters, such as layer, height, layer contour, material color, etc.

Among the disadvantages of this direction, researchers have noted the need to interrupt the printing process to capture an image, the large amount of data needed for training models, and the need for separate algorithms to identify different printing defects.

A review of the literature on the direct method for printed product quality assessment showed that most studies are devoted to the detection of surface defects that determine surface integrity. Therefore, this is an actual direction of quality control in additive manufacturing technologies. Since the FDM method is characterized by the periodicity of the pattern, the application of Fourier analysis is one possible direction that could be interesting, both here as well as in other applications, including for drilling [34] and milling [35]. For example, Fourier analysis has been used for no-reference estimation of 3D printed surface quality [36]. For this, two different fragments with "good" and "bad" quality are selected from the same sample. The image is converted into shades of gray (grayscale) and the spectrum is built from the original signal perpendicular to the layers of the filament. For high-quality printing, the regular pattern results in explicit peaks in the Fourier domain but, in the presence of structural distortion, there are no peaks. As the numerical value allowed the determination of the quality of the analyzed fragments of the samples, the authors used the integral of the spectra values calculated using the trapezoidal rule. High values for the specified parameter indicated low quality for the fragment and low values indicated high quality. The authors noted that the results for the integral were highly dependent on the color of the sample. It also became necessary to use different threshold values to distinguish between high and low quality samples. They analyzed images only in the frequency domain, ignoring the time domain, analysis of which can contribute, for example, to identifying various types of defects and distortions in the sample geometry, determining the layer thickness, etc.

Frequency analysis is considered to improve quality in Big Area Additive Manufacturing [37]. The authors of this study also noted the rationale for using Fourier analysis for horizontal stacked lines with a given periodicity in the y-direction. Fourier analysis provides insights into the layer thickness and its variation. The presence of a tall, thin spike at the dominant frequency means that the print is regular. An increase in peak width with a decrease in its amplitude indicates slumping of material and extrusion at an excessively high temperature. Further slumping leads to a more than twofold increase in the width of the peak in comparison to the reference peak. Thus, the image analysis in [36] was performed only in the frequency domain, and no quantitative indicator was proposed that could be used as a diagnostic indicator for the monitoring system.

Laser profilometry has recently been increasingly used for non-contact measurement of profile irregularities; e.g., surface roughness parameters (Figure 1a). The influence of both the material type and cutting speed on surface roughness using an abrasive water jet has been studied with a special measurement scheme [38], but more modern equipment was used for plastics in this study (Figure 1b).

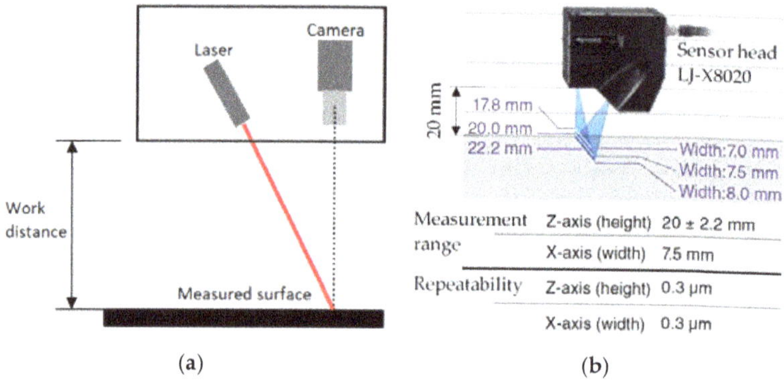

Figure 1. Measurement schemes using laser profilometry: (**a**) composite [38]; (**b**) integrated.

The surface roughness parameters of the metal samples were measured using both a laser profilometer (Contactless LPM, 2015, KVANT spol. s.r.o., Bratislava, Slovakia) and a conventional contact roughness meter (Mitutoyo SJ 400, Mitutoyo America Corporation, Aurora, FL, USA). It was found that the Ra values measured by the laser profilometer lay within the range of the values measured by the Mitutoyo SJ 400, while the obtained Rz values were two to three times lower than the values measured with the laser profilometer. The reasons were that noise occurred when measuring some areas of the surface with the laser profilometer due to the reflection of the complementary metal–oxide–semiconductor (CMOS) camera and because the measuring tip of the contact meter could not sense all the surface irregularities.

The LPM 4 laser profilometer was used for the unevenness measurement of the beech sample surface. When doing so, the reference material in the experiment was an aluminum standard with a surface roughness value Ra = 10 μm. The optical non-contact method is a much faster, more accurate, and more reliable solution compared to manual measurement systems. Furthermore, the optimal wavelength was identified experimentally and higher accuracy and sensitivity with the blue light were confirmed.

A constructed laser profiler has been presented and investigations were carried out to verify the developed device's accuracy through the comparison of the measured data with the standard [37]. It was noted that the value obtained for the Ra parameter with the laser profiler turned out to be higher than the standard value. The pros of the laser profiler were its measurement speed and acceptable price. Its cons were that the shape of the object to be measured had to not be too curved and the surface of the object had to not be glossy. The latter was because, when measuring glossy surfaces, a large amount of the laser light incident on the surface components was reflected back to the charge-coupled device (CCD) camera.

Thus, the review demonstrates that non-contact methods for surface quality control have advantages for 3D printing. Laser profilometry (2D measurement and 3D inspection) and vision-based methods are the most promising in this context. Laser profilers have been used to determine the absolute values of many surface profile parameters [39,40]. However, the influence of the reflectivity of the material to be used may introduce an error in the measurement results. However, the differentiated initial quality of the surface under study requires that certain settings be implemented on the laser profilometer to obtain the correct result. For these reasons, the numerical values of the profile parameters obtained using a laser profiler and a contact device may differ. This is why it is necessary to investigate this difference and establish appropriate links between the results from contact and non-contact measurements.

To develop mathematical software for an online quality monitoring system for 3D parts, it is necessary to create information signals containing the diagnostic signs that characterize the change in quality of the 3D part during 3D printing, and which can be used to monitor the state of the 3D printing system [41]. In addition, these diagnostic signs should provide information about the presence or absence of defects that may occur during the printing of plastic objects, as the roughness parameters do not provide such information. A possible reason for this is that the roughness parameters represent an integral assessment of the surface quality. This also needs to be investigated. In addition, given the periodic texture of the printed surface, it would be useful to investigate evaluation functions for online monitoring in the time and frequency domains using Fourier analysis.

Existing in situ monitoring techniques differ depending on the features of the laser additive manufacturing. For example, laser-induced breakdown spectroscopy (LIBS) has been suggested for in situ and real-time elemental analysis of cladding, as well as for cladding process failure detection [42]. In situ monitoring and ex situ elasticity mapping were introduced in [43], where an ultrasonic time-of-flight measurement monitoring technique was numerically and experimentally used to study the behavior of laser-induced melting pools. Furthermore, in situ X-ray imaging of defects has been deployed in laser additive manufacturing [44], as well as in situ thermal imaging for single-layer build-time alteration [45].

The aim of this research was therefore to develop information signals in the time and frequency domains that contain the surface quality diagnostic features (signs) for in situ monitoring of the 3D printing system state.

2. Materials and Methods

The test samples were made using the FDM method. A Creality Ender-3 3D printer, CREALITY, Shenzhen, China (Figure 2a) with a PLA filament was used for the research. The samples were designed with Autodesk Inventor Professional 2021 software. To generate the G-code for object printing, the conversion of the CAD model into the stereolithography (STL) format was undertaken with the Ultimaker Cura slicer software (a popular 3D printing software). The printing parameters were selected as shown in Table 2 [46].

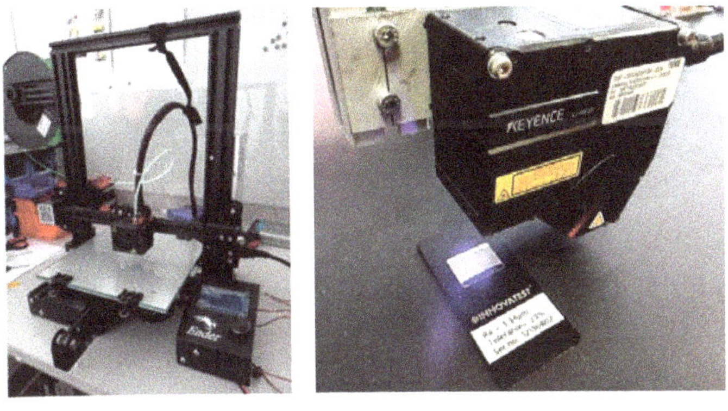

(a) (b)

Figure 2. (**a**) Creality Ender-3 3D printer; (**b**) LJ-8020 laser profiler with a metal sample.

Table 2. Printing process parameters.

Parameter	Value	Units
Nozzle diameter	0.4	mm
Filament size	1.75	mm
Layer thickness	0.2	mm
Raster angle	45.90	degree
Raster width	0.4	mm
Bed temperature	60	°C
Printing temperature	210	°C
Printing speed	45	mm/s
Infill density	20	%
Infill flow	100	%

A KEYENCE LJ-8020 laser profiler, KEYENCE INTERNATIONAL, Mechelen, Belgium (Figure 2b) was used to measure and evaluate the special samples (Figure 3) made of polylactic acid (PLA).

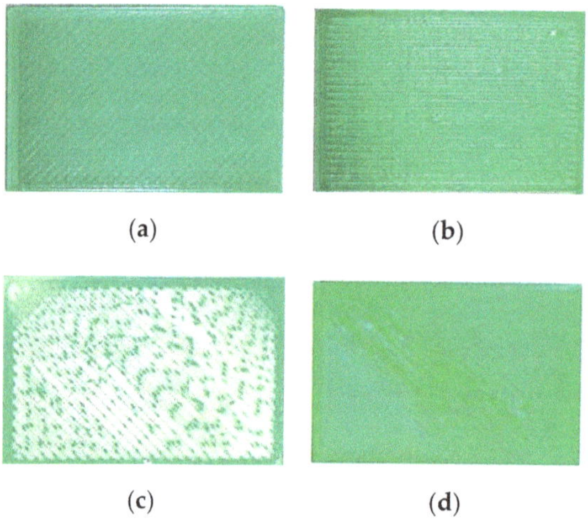

Figure 3. Samples under investigation: (**a**) reference; (**b**) under-extrusion; (**c**) stringy first layer; (**d**) high roughness.

First, for verification, the surface roughness parameters of the metal specimen (Figure 4a) were determined using an R130 roughness tester (Figure 4b) and the LJ-8020 laser profiler (Figure 2b).

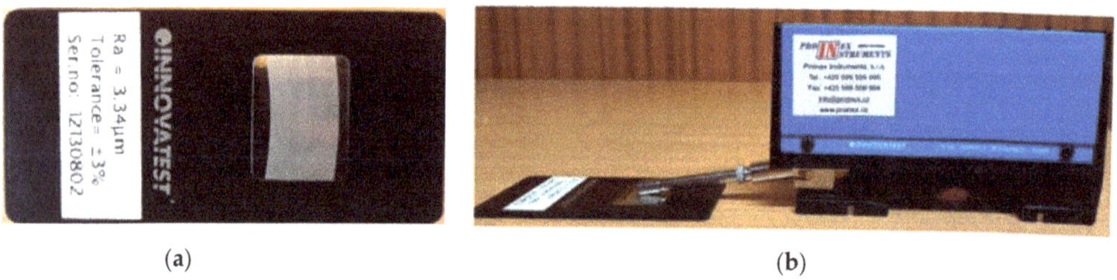

Figure 4. (**a**) Specimen measured to determine roughness parameters; (**b**) R130 roughness tester at work.

Several defect-free 3D printed samples (Figure 5) were simultaneously measured and evaluated with the Innovatest R130 roughness tester (INNOVATEST Europe BV, Maastricht, The Netherlands) and the KEYENCE LJ-8020 laser profiler (Figure 6a) to compare the results obtained using contact and contactless instruments, respectively.

(**a**) (**b**) (**c**)

Figure 5. Defect-free samples for study: (**a**) green sample with +45° raster angle; (**b**) green sample with 90° raster angle; (**c**) white sample with 90° raster angle.

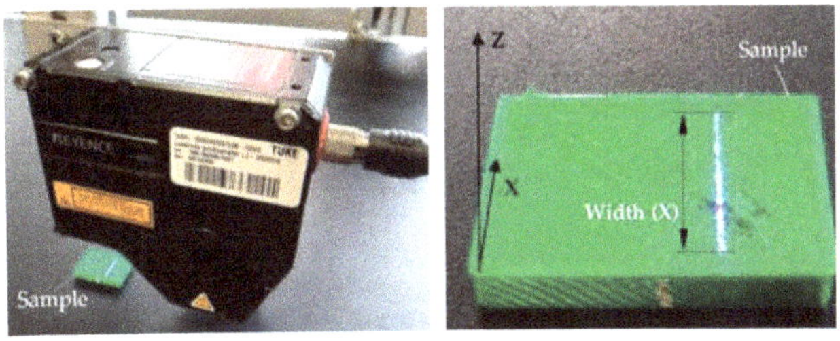

(**a**) (**b**)

Figure 6. Setup for measuring a plastic object: (**a**) laser profiler with a sample; (**b**) defective green 3D sample (30 × 20 × 5 mm) under study.

The following KEYENCE equipment was used for the profile analysis: an LJ-8020 laser profiler (see Table 3 for specifications) and, as a separate unit, an LJ -X8000A controller (KEYENCE INTERNATIONAL, Mechelen, Belgium). The equipment was connected to LJ-X Navigator and LJ-X Observer software (KEYENCE CORPORATION OF AMERICA, Itasca, IL, USA) to show the profile parameters obtained by the laser profiler. Further, MS Excel, MatLAB, and NI-LabVIEW software were used for processing and presentation of measurement results.

Table 3. LJ-X8020 sensor head specifications.

Specification	Name	Value
Reference distance	*z*-axis (height)	20 mm
Measurement range	*x*-axis (width), near side *x*-axis (width), reference distance *x*-axis (width), far side	7 mm 7.5 mm 8 mm
Light source	Blue semiconductor Laser wavelength Class 2M laser product (IEC60825-1, FDA (CDRH) Part 1040.10) Output	405 nm (visible light) 10 mW
Spot size		Approx. 16 mm × 32 μm
Repeatability	*z*-axis (height) *x*-axis (width)	0.3 μm 0.3 μm
Profile data interval	*x*-axis (width)	2.5 μm
Profile data count		3200 points

3. Results and Discussion

3.1. Calibration of R130 Roughness Tester

The Innovatest R130 roughness tester had the following specifications:

- Ra and Rz measurement ranges: 0.03 μm–6.3 μm and 0.2 μm–18.5 μm, respectively;
- Display resolution: 0.01 μm;
- Cut-off: 0.25 mm; 0.8 mm; 2.5 mm;
- ANSI 2RC filter, Sino Age Development Technology, Ltd., Beijing, China.

Prior to operation, the R130 must be calibrated and checked using the reference specimen (Figure 4a). Calibration settings: cut-off—0.8 mm, traverse length—4.5 mm, evaluation length—4.0 mm, number of cut-offs—5. The results were as follows: Ra (1) = 3.24 μm, Ra (2) = 3.23 μm, and Ra (3) = 3.28 μm; i.e., the average mean Ra (ave) = 3.25 μm. According to the manual, if the reading is within ± 0.1 μm (3.24 μm < Ra < 3.34 μm), calibration is within tolerance. In our case, Ra (ave) = 3.25 μm was within the specified range. The Rz values (the maximum roughness depth or the largest of the peak-to-valley roughness depths across the evaluation length) were Rz (1) = 12.1 μm, Rz (2) = 12.3 μm, and Rz (3) = 12.1 μm; i.e., the average mean Rz (ave) = 12.17 μm (Table 4).

Table 4. Roughness parameters obtained for the metal reference specimen with the R130 roughness tester and LJ-8020 laser profiler.

Roughness Parameter	R130 Roughness Tester	LJ-X8020 Laser Profiler
Ra, μm	3.24 3.23 3.28 Average **3.25**	2.684 2.377 2.383 Average **2.481**
Rz, μm	12.1 12.3 12.1 Average **12.17**	12.740 11.554 12.800 Average **12.365**

Note: evaluation length: 4 mm.

Differences in μm and percentages for Ra and Rz between the measurements with LJ-8020 laser profiler and R130 tester are shown in Table 5.

Table 5. Average surface roughness parameters obtained for the metal reference specimen with the R130 roughness tester and LJ-8020 laser profiler.

Instrument and Error	Ra, μm	Rz, μm
LJ-X8020 laser profiler	2.481	12.365
R130 roughness tester	3.250	12.170
Difference, %	23.66	1.60

3.2. Roughness Parameter Measurement on the Metal Reference Specimen with Contact and Non-Contact Methods

The laser profiler (Figure 6a) uses a laser displacement sensor that collects height data across a laser line rather than a single point. This enables 2D/3D measurements, such as height difference, width, or angle, to be obtained using a single sensor; i.e., "three-in-one" measurements. A laser line is emitted from the sensor to illuminate the surface of a target. The reflected light is imaged by the complementary metal–oxide–semiconductor (CMOS) image sensor to create a profile of the surface, which can then be used for measurement and inspection.

The triangulation principle is used for measurement in the following sequence: (1) the laser beam is directed at the sample to be inspected, and then (2) the light reflected by the sample is collected by the receiver lens and (3) reproduced on the light-receiving element. When the distance changes, the collected light is reflected at a different angle, and the position of the beam on the light-receiving element changes accordingly.

To verify and compare the measured values, measurement of the metal specimen (Figure 4a) was also carried out using the R130 contact roughness tester (Table 4). The same part of the surface was selected for each sample with both methods of measurement; i.e., with the R130 contact roughness tester and the LJ-X8020 non-contact laser profiler.

When determining the roughness parameters using the laser profiler, the profile data interval was 2.5 μm and the profile data count was 3200 points; thus, the evaluation length in this case was 8 mm (2.5 μm × 3200 points = 8000 μm; i.e., 8 mm). When determining the roughness parameters using the R130, the evaluation length was 4 mm. Thus, the roughness parameters Ra and Rz, when evaluated using the laser profiler, were determined on two evaluation length segments of 4 mm each (Figure 7).

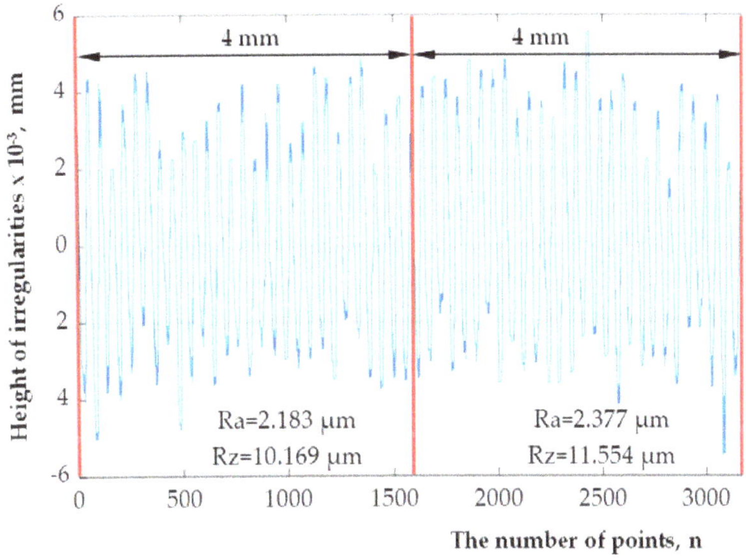

Figure 7. Surface roughness profile for metal specimen as measured by the LJ-8020 laser profiler.

Therefore, the roughness parameters obtained for the metal surface of the reference specimen (with 4 mm evaluation length) using contact (R130 roughness tester) and non-contact (LJ-8020 laser profiler) devices were close: Ra = 3.250 μm and Ra = 2.481 μm, and Rz = 12.170 μm and Rz = 12.365 μm.

3.3. Roughness Parameter Measurement for the Plastic Samples with Contact and Non-Contact Methods

To verify and compare the measured values, measurement of the samples shown in Figure 5 was carried out with both the contact R130 roughness tester (Table 6) and the LJ-8020 laser profiler (Table 7). The same part of the surface was selected for each sample with both methods of measurement. Table 6 shows the surface roughness response obtained with the R130 tester, along with the mean, for three FDM 3D printed parts.

Table 6. Plastic surface roughness parameters obtained with the R130 roughness tester.

Sample	Ra, μm	Rz, μm	Ra (Ave), μm	Rz (Ave), μm
Figure 5a	5.06 4.27 4.20	24.9 22.3 21.2	4.51	22.8
Figure 5b	9.22 8.97 9.32	25.0 25.0 25.0	9.17	25.0
Figure 5c	6.62 6.54 6.18	24.9 24.9 24.9	6.45	24.9

Note: cut-off: 2.5 mm; evaluation length: 2.5 mm; number of cut-offs: 1; Ra (ave) and Rz (ave) means are the averages of three measurements.

Table 7. Plastic surface roughness parameters obtained with the LJ-8020 laser profiler.

Sample	Ra, μm	Rz, μm	Ra (Ave), μm Difference	Rz (Ave), μm Difference
Figure 5a	9.155 8.219 6.984	39.903 41.496 33.211	8.119 **80.02%**	38.203 **67.56%**
Figure 5b	8.855 10.655 9.859	68.500 36.500 27.200	9.790 **6.79%**	44.060 **76.24%**
Figure 5c	14.325 14.135 12.381	57.233 30.464 50.626	13.614 **111.07%**	46.108 **85.17%**
Average difference			**65.96%**	**76.32%**

Note: evaluation length: 2.5 mm.

Table 7 shows the surface roughness response obtained with the LJ-8020 laser profiler, along with the mean, for three FDM 3D printed parts. The table also shows differences in percentages compared to the measurements with the R130 tester (Table 6). For example, when Ra (ave) = 4.51 μm (Table 6) and Ra (ave) = 8.119 μm (Table 7), the difference was 80.02%, because $\frac{|4.51-8.119|}{4.51}100\% = 80.02\%$.

Thus, the transition to another material (from metal to plastic) was accompanied by an increase in the difference between the contact and non-contact measurement results from 23.66% (Table 5) to 65.96% (Table 7)—i.e., 2.8 times—for Ra (ave) and from 1.6% (Table 5) to 76.32% (Table 7)—i.e., 47.7 times—for Rz (ave).

3.4. Diagnostic Feature Detection Using NI-LabVIEW

Modern computer measurement systems allow for the synthesis of new information signals in real time, which results from the mathematical processing of an array of primary digital data. For example, the NI-DAQmx measurement data acquisition system provided with the NI-LabVIEW software package has options for configuring information signals from the primary time signals of sensors. These options can be selected in the **Functions** menu using the **Functions→Express→Signal Analysis** scheme.

Two blocks are used in the **Signal Analysis** category:

1. The **Configure Statistics [Statistics]** block, which includes Range in the time domain and Arithmetic Mean, RMS, Standard Deviation, Variation, Median, Mode, and Summation in the frequency domain.
2. The **Configure Spectral Measurements** block, which includes the signal spectral characteristics Magnitude Peak, Power Spectrum, and Power Spectral Density (PSD) in the frequency domain.

For a comparative assessment of the informativeness of the newly generated signals when solving the problem of detecting defects on the surface of a printed sample in the NI-DAQmax computer data acquisition system provided with the NI-LabVIEW software package, information signals in the time domain (Range) and the frequency domain (Arithmetic Mean, RMS, Standard Deviation, Variation, Median, Mode, Summation) were studied. Using these signal processing tools, a Fourier analysis of the profilograms obtained using the laser profiler for the (**a**) reference, (**b**) under-extrusion, (**c**) stringy first layer, and (**d**) high roughness PLA plastic samples (Figure 3) was performed.

In the NI-LabVIEW software environment, the profilogram output signal (Figure 8) was fed to the **Configure Spectral Measurements** unit of the NI-LabVIEW system, which performs the fast Fourier transform (FFT) procedure. Of the three spectral characteristics of the **Configure Spectral Measurements** block (Magnitude Peak, Power Spectrum, and Power Spectral Density), the most sensitive spectral characteristic, Power Spectral Density (PSD), was used.

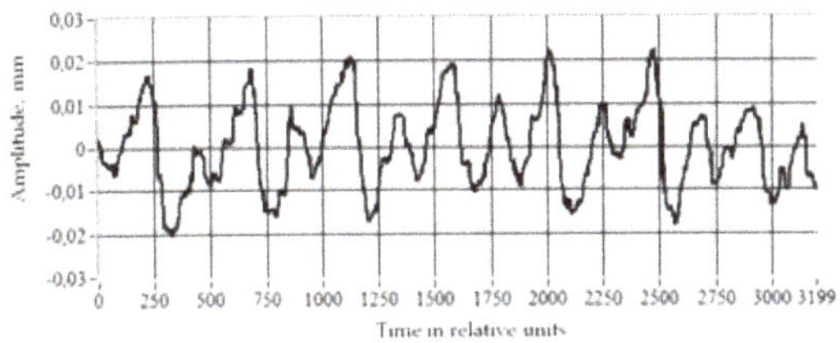

(**a**)

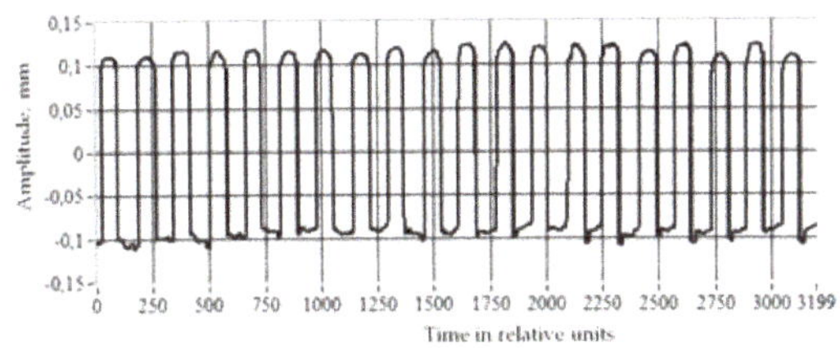

(**b**)

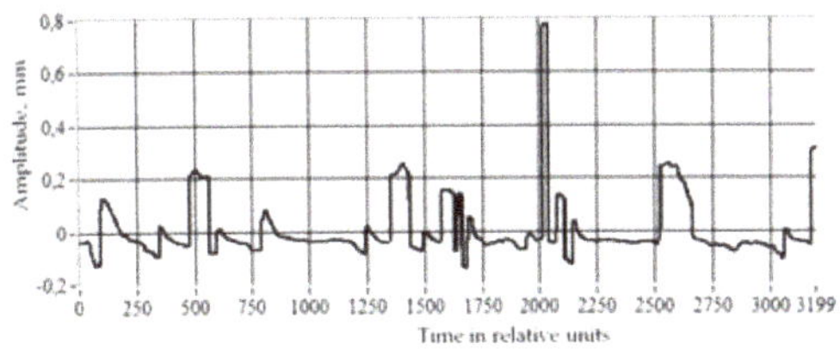

(**c**)

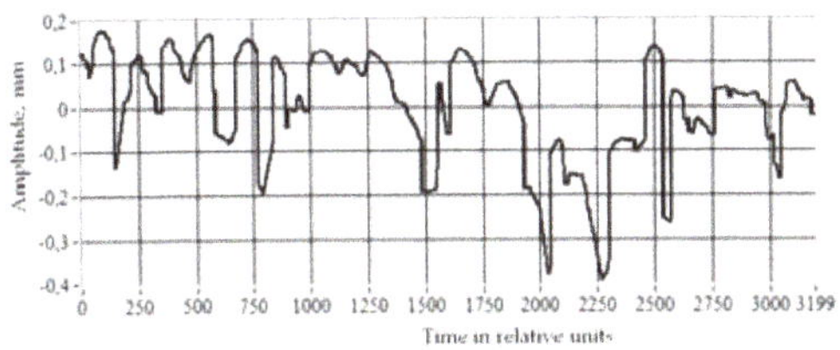

(**d**)

Figure 8. Initial profilograms in the time domain for four samples: (**a**) defect-free reference; (**b**) under-extrusion; (**c**) string first layer; (**d**) high roughness.

The results obtained after the FFT of the PSD configuration as displayed on the front panel of the virtual instrument of the NI-DAQmx system using **Waveform Graph** display units are shown in Figure 9.

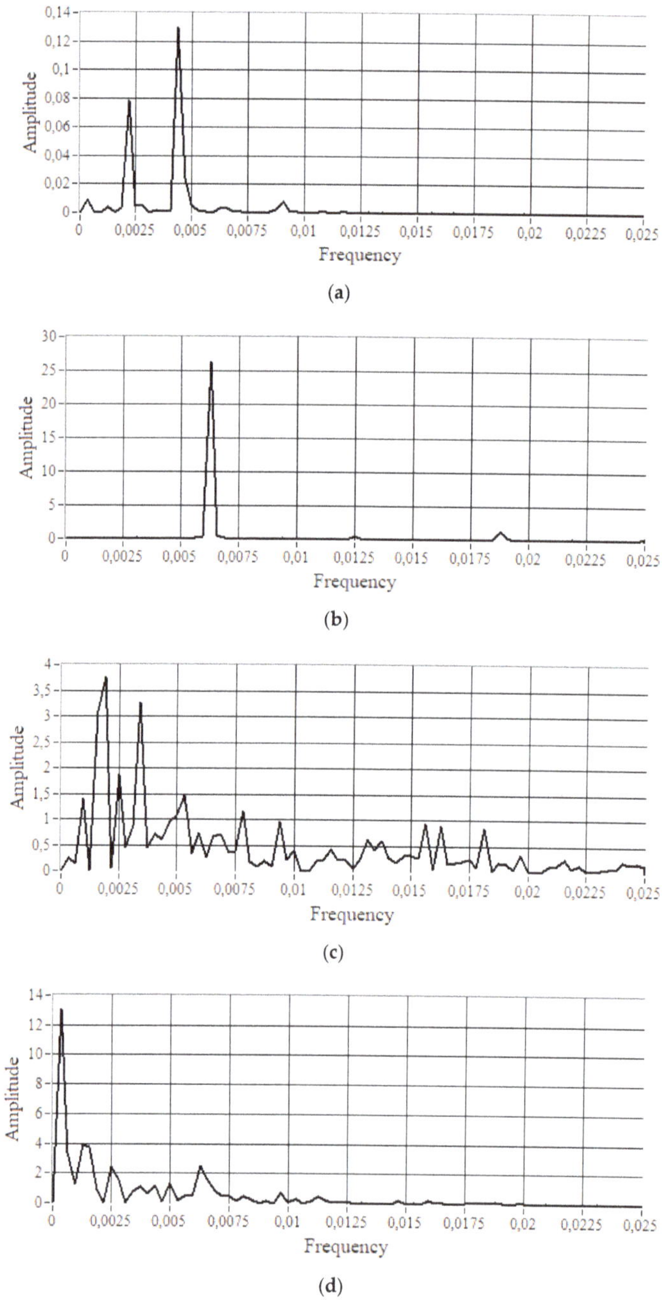

Figure 9. PSD spectra in the frequency domain: (**a**) reference sample; (**b**) under-extrusion; (**c**) stringy first layer; (**d**) high roughness.

Analyzing the PSD configuration data in Figure 9, the following intermediate conclusions can be formulated:

1. The PSD for a reference (defect-free) sample (Figure 9a) was characterized by four pronounced harmonics at the following relative frequencies (signal amplitudes are indicated in parentheses):
   - f1 = 0.000313 (A = 0,00885618);
   - f2 = 7f1 = 0.002188 (A = 0.0781748);
   - f3 = 14f1 = 0.004375 (A = 0.129859);
   - f4 = 28f1 = 0.009062 (A = 0.00748143).
2. The PSD for an under-extrusion-type defect sample (Figure 9b) was characterized by three harmonics at the following frequencies:
   - $f1^{/}$ = 20f1 = 0.00625 (A = 26.376);
   - $f2^{/}$ = 40f1 = $2f1^{/}$ = 0.0125 (A = 0.577246);
   - $f3^{/}$ = 60f1 = $3f1^{/}$ = 0.01875 (A = 1.41452).
3. The PSD for samples with stringy first layer (Figure 9c) and high roughness (Figure 9d) defects were characterized by a significant set of harmonics (with less pronounced amplitudes).

   To quantify the PSD signal, the numerical values for the **Arithmetic Mean**, **RMS**, **Standard Deviation**, **Variance**, **Median**, **Mode**, and **Summation** parameters were determined after repetition of the same types of measurements three times, and then the averaged values of the corresponding signals were calculated (Table 8).

Table 8. Diagnostic feature values.

Diagnostic Sign	The Range- and PSD Signal-Averaged Values for the Plastic Samples in Figure 3			
	Reference	**Under-Extrusion**	**Stringy First Layer**	**High Roughness**
1 Range	0.04342(1)	0.25914(5.97)	0.65727(15.14)	0.92944(21.41)
2 Arithmetic Mean	0.00021(1)	0.02239(106.62)	0.03090(147.14)	0.04158(198.00)
3 RMS	0.00428(1)	0.69723(162.90)	0.36880(86.17)	0.36570(85.44)
4 Standard Deviation	0.00428(1)	0.69709(162.87)	0.36768(85.91)	0.36344(84.92)
5 **Variance**	1.844×10^{-5}(1)	0.48677(**26,397.5**)	0.13720(**7440.34**)	0.14937(**8100.32**)
6 **Median**	3.434×10^{-7}(1)	0.00030(**873**)	0.00020(**582**)	0.00049(**1426**)
7 Mode	0.00066(1)	0.13911(210.77)	0.05692(86.24)	0.04511(68.35)
8 Summation	0.33638(1)	35.8179(106.48)	49.4540(147.02)	66.5230(197.76)
Sum of ratings	8	28.03×10^3	8.60×10^3	10.20×10^3

Note: The numbers in brackets show the ratio of the indicator under consideration (one of three) to the indicator of the same name for the reference sample, taken as a unit.

3.5. Discussion

At the first stage, a comparative analysis of the results of contact (R130 roughness tester) and non-contact (LJ-8020 laser profiler) methods when measuring the same surface irregularities of a metal reference specimen with a known surface roughness was performed. The relative errors of the non-contact method in comparison to the contact method were established. The roughness parameters obtained for the metal surface of the reference specimen (with 4 mm evaluation length) using the contact (R130 roughness tester) and non-contact (LJ-8020 laser profiler) devices were close: Ra = 3.250 μm and Ra = 2.481 μm, and Rz = 12.170 μm and Rz = 12.365 μm. Thus, the contact and non-contact methods for control of standard integral parameters of surface roughness gave similar results for the metal reference specimen. The Errors in the parameters Ra and Rz in the range of operation of the R130 roughness tester were, respectively, 23.7 and 1.6%.

At the second stage, similar investigations were carried out with the plastic polylactic acid (PLA) samples in the range of operation of the contact method. The relative errors increased by 3 and 47 times, respectively, when determining the roughness parameters Ra and Rz. Thus, similar investigations with plastic parts were accompanied by much larger errors for the non-contact optical method (LJ-8020 laser profiler) compared to the contact method (R130 roughness tester): 65.96% and 76.32%, respectively, for the Ra and Rz integral roughness indicators.

At the third stage, diagnostic features were detected for online monitoring of three types of 3D printing defects—under-extrusion, stringy first layer, and high roughness—to produce diagnostic signs of 3D printing defects with the non-contact optical method (LJ-8020 laser profiler). The profilograms of the surfaces of the plastic parts were obtained, with each such profilogram being an analog representation of the change in the corresponding information signal over time. Next, the Fourier transforms of these information signals (i.e., the profilograms) were calculated and secondary information signals obtained. Integral estimates of PSD spectrograms can be produced in terms of the Arithmetic Mean, RMS, Standard Deviation, Variance, Median, Mode, and Summation.

Usually, the quality of 3D printed parts is checked after the completion of printing. As a result of our research, a system for monitoring the surface quality of the nth layer of a printed object in order to detect surface defects during printing can be proposed. The sampling step for 3D printing quality control can take place during the formation of one or more of the layers (Figure 10). In the block diagram below, the term "sign" is used instead of "feature" to highlight the difference between the productive (resulting) and methodological parts of the work.

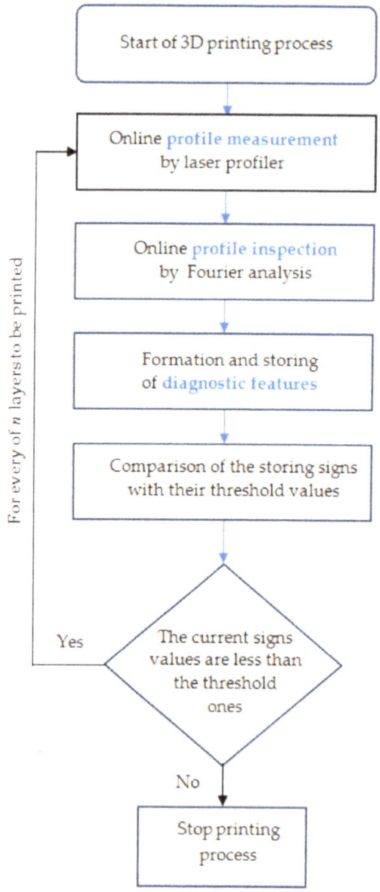

Figure 10. Monitoring algorithm block diagram.

4. Conclusions

The article proposed a new approach for online monitoring of object surface quality in 3D printing, especially for in situ monitoring. This monitoring is important in online printing quality control to prevent poor quality printing. A non-contact high-performance method was used to assess the irregularities on the surface of the part being manufactured. To confirm the feasibility of such an approach, comparative studies of contact (R130 roughness tester) and non-contact (LJ-8020 laser profiler) methods for assessing surface quality were carried out. It was established that the difference in the results of the quality assessment (with contact and non-contact methods) increased when moving from a metal sample to a plastic sample. For example, the transition from metal to plastic was accompanied by an increase in the difference between the measurement results of the contact and non-contact methods from 23.70% to 65.96% (2.8 times) and 1.6% to 76.32% (47.7 times), respectively, for Ra (ave) and Rz (ave).

When monitoring the 3D printing process of plastic parts online, in order to assess the quality of the process and the results of the 3D printing, the following (most defect-sensitive) integral estimates of PSD spectrograms can be used as diagnostic features of printing defects: Variance and Median. For example, the sensitivity of the integral estimate of Variance to defects such as under-extrusion, a stringy first layer, and high roughness was 26,397.50, 7440.34, and 8100.32, respectively. The least sensitive was the Range integral estimate in the initial surface profilogram of the plastic part (an information signal not subjected to the Fourier transform). For this Range estimate, the sensitivity to the previously listed three defects was 5.97, 15.14, and 21.41; i.e., significantly less.

The primary and secondary information signals generated by the non-contact method, as well as the methods for their formation, can be used when processing the topography of 3D printed object surfaces; i.e., when processing images of these surfaces in the online monitoring of 3D printed objects. Using these facts, further research can use the outlined monitoring strategy as a guide.

Author Contributions: Conceptualization, N.L. and V.L.; methodology, N.L. and J.P.; software, N.L.; validation, J.P. and V.L.; formal analysis, V.L.; investigation, N.L. and J.P.; resources, N.L. and J.P.; data curation, N.L.; writing—original draft preparation, N.L.; writing—review and editing, J.P.; visualization, N.L. and V.L.; supervision, V.L.; project administration, J.P.; funding acquisition, J.P. All authors have read and agreed to the published version of the manuscript.

Funding: This work was supported by the Slovak Research and Development Agency under the contract no. APVV-19-0590 and by the Ministry of Education, Science, Research and Sport of the Slovak Republic projects VEGA 1/0700/20 and 055TUKE-4/2020.

Institutional Review Board Statement: Not applicable.

Informed Consent Statement: Not applicable.

Data Availability Statement: The data presented in this study are available on request from the corresponding author.

Conflicts of Interest: The authors declare no conflict of interest.

References

1. Business Data Platform. Available online: https://www.statista.com (accessed on 22 February 2022).
2. Shahrubudin, N.; Lee, T.C.; Ramlan, R. An overview on 3D printing technology: Technological, materials, and applications. *Procedia Manuf.* **2019**, *35*, 1286–1296. [CrossRef]
3. Rahim, T.; Abdullah, A.; Akil, H. Recent developments in fused deposition modeling-based 3D printing of polymer sand their composites. *Polym. Rev.* **2019**, *59*, 589–624. [CrossRef]
4. Kaščak, J.; Gašpár, Š.; Paško, J.; Husár, J.; Knapčíková, L. Polylactic acid and its cellulose based composite as a significant tool for the production of optimized models modified for additive manufacturing. *Sustainability* **2021**, *13*, 1256. [CrossRef]
5. Kaščak, J.; Gašpár, Š.; Paško, J.; Knapčíková, L.; Husár, J.; Baron, P.; Török, J. Design of an atypical construction of equipment for additive manufacturing with a conceptual solution of a printhead intended for the use of recycled plastic materials. *Appl. Sci.* **2021**, *11*, 2928. [CrossRef]

6. Wing, I.; Gorham, R.; Sniderman, B. 3D Opportunity for Quality Assurance and Parts Qualification. Additive Manufacturing Clears the Bar. Available online: https://www2.deloitte.com/content/dam/insights/us/articles/3d-printing-quality-assurance-in-manufacturing/DUP_1410-3D-opportunity-QA_MASTER1.pdf (accessed on 28 February 2022).
7. 44 Issues Explained You're Your Prints Looking Perfect, Not Pathetic. Available online: https://rigid.ink/pages/ultimate-troubleshooting-guide (accessed on 13 March 2022).
8. First Layer Problems. Available online: https://3dprintguides.com/2020/04/3d-printing-first-layer-problems/ (accessed on 22 November 2021).
9. The Definitive Guide to Solving 3D Printing Problems by Bitlab. Available online: https://bitfab.io/blog/3d-printing-problems/ (accessed on 12 March 2022).
10. Fu, Y.; Downey, A.; Yuan, L.; Pratt, A.; Balogun, Y. In situ monitoring for fused filament fabrication process: A review. *Addit. Manuf.* **2021**, *38*, 101749. [CrossRef]
11. Okarma, K.; Fastowicz, J. Improved quality assessment of colour surfaces for additive manufacturing based on image entropy. *Pattern Anal. Appl.* **2020**, *23*, 1035–1047. [CrossRef]
12. Fastowicz, J.; Grudziński, M.; Tecław, M.; Okarma, K. Objective 3D printed surface quality assessment based on entropy of depth maps. *Entropy* **2019**, *21*, 97. [CrossRef] [PubMed]
13. Carlota, V. How to Avoid 3D Printing Problems. Available online: https://www.3dnatives.com/en/how-to-avoid-3d-printing-problems-26102020/ (accessed on 18 March 2022).
14. Jin, Z.; Zhang, Z.; Gu, G.X. Autonomous in-situ correction off used deposition modeling printers using computer vision and deep learning. *Manuf. Lett.* **2019**, *22*, 11–15. [CrossRef]
15. Kadam, V.; Kumar, S.; Bongale, A.; Wazarkar, S.; Kamat, P.; Patil, S. Enhancing surface fault detection using machine learning for 3D printed products. *Appl. Syst. Innov.* **2021**, *4*, 34. [CrossRef]
16. Wang, Y.; Huang, J.; Wang, Y.; Feng, S.; Peng, T.; Yang, H.; Zou, J. A CNN-based adaptive surface monitoring system for fused deposition modeling. *IEEE/ASME Trans. Mechatron.* **2020**, *25*, 2287–2296. [CrossRef]
17. 3D Printing First Layer Problems: Troubleshooting and How to Fix. Available online: https://www.elegoo.com/blogs/learn/3d-printing-first-layer-problems-troubleshooting-and-how-to-fix (accessed on 10 August 2021).
18. 3D Printer Layer Shifting. Available online: https://all3dp.com/2/layer-shifting-3d-printing-tips-tricks-to-solve-it/ (accessed on 23 May 2021).
19. Günaydin, K.; Türkmen, H.S. Common FDM 3D Printing Defects. In Proceedings of the 3rd International Congress on 3D Printing (Additive Manufacturing) Technologies and Digital Industry 2018, Antalya, Turkey, 19–21 April 2018; pp. 368–369.
20. Liu, J.; Hu, Y.; Wu, B.; Wang, Y. An improved fault diagnosis approach for FDM process with acoustic emission. *J. Manuf. Processes* **2018**, *35*, 570–579. [CrossRef]
21. Wu, H.; Wang, Y.; Yu, Z. In situ monitoring of FDM machine condition via acoustic emission. *Int. J. Adv. Manuf. Technol.* **2016**, *84*, 1483–1495. [CrossRef]
22. Wu, H.; Yu, Z.; Wang, Y. Real-time FDM machine condition monitoring and diagnosis based on acoustic emission and hidden semi-Markov model. *Int. J. Adv. Manuf. Technol.* **2017**, *90*, 2027–2036. [CrossRef]
23. Li, F.; Yu, Z.; Yang, Z.; Shen, X. Real-time distortion monitoring during fused deposition modeling via acoustic emission. *Struct. Health Monit.* **2019**, *19*, 412–423. [CrossRef]
24. Kousiatza, C.; Chatzidai, N.; Karalekas, D. Temperature Mapping of 3D Printed Polymer Plates: Experimental and Numerical Study. *Sensors* **2017**, *17*, 456. [CrossRef] [PubMed]
25. Dinwiddie, R.B.; Love, L.; Rowe, J.C.; Stockton, G.R.; Colbert, F.P. Real-time process monitoring and temperature mapping of a 3D polymer printing process. *Proc. SPIE* **2013**, *8705*, 87050L-1–87050L-9.
26. Ketai, H.; Huan, W.; Huaqing, H. Approach to online defect monitoring in fused deposition modeling based on the ariation of the temperature field. *Complexity* **2018**, *2018*, 3426928.
27. Li, Z.; Zhang, D.; Shao, L.; Han, S. Experimental investigation using vibration testing method to optimize feed parameters of color mixing nozzle for fused deposition modeling color 3D printer. *Adv. Mech. Eng.* **2019**, *11*, 1687814019896196. [CrossRef]
28. Li, Y.; Zhao, W.; Li, Q.; Wong, T.; Wang, G. In-situ monitoring and diagnosing for fused filament fabrication process based on vibration sensors. *Sensors* **2019**, *19*, 2589. [CrossRef]
29. Tlegenov, Y.; Feng, L.W.; Soon, H.G. A dynamic model for current-based nozzle condition monitoring in fused deposition modelling. *Prog. Addit. Manuf.* **2019**, *4*, 211–223. [CrossRef]
30. Kim, C.; Espalin, D.; Cuaron, A.; Perez, M.A.; Macdonald, E.; Wicker, R.B. A study to detect a material deposition status infused deposition modeling technology. In Proceedings of the 2015 IEEE International Conference on Advanced Intelligent Mechatronics (AIM), Busan, Korea, 7–11 July 2015; pp. 779–783.
31. Kim, C.; Espalin, D.; Cuaron, A.; Perez, M.A.; Macdonald, E.; Wicker, R.B. Unobtrusive in situ diagnostics of filament-fed material extrusion additive manufacturing. *IEEE Trans. Compon. Packag. Manuf. Technol.* **2018**, *8*, 1469–1476. [CrossRef]
32. Zidek, K.; Maxim, V.; Pitel, J.; Hosovsky, A. Embedded vision equipment of industrial robot for inline detection of product errors by clustering-classification algorithms. *Int. J. Adv. Robot. Syst.* **2016**, *13*, 4901. [CrossRef]
33. Zidek, K.; Pitel, J.; Hosovsky, A. Machine learning algorithms implementation into embedded systems with web application user interface. In Proceedings of the IEEE 21st International Conference on Intelligent Engineering Systems 2017 (INES 2017), Larnaca, Cyprus, 20–23 October 2017; pp. 77–81.

34. Lishchenko, N.; Larshin, V.; Basharov, R. Diagnostics of drilling in numerically controlled machine tools. *Russ. Eng. Res.* **2016**, *36*, 77–80. [CrossRef]
35. Lishchenko, N.; Larshin, V.; Pitel', J. Detecting systematic and random component of surface roughness signal. *Her. Adv. Inf. Technol.* **2020**, *3*, 61–71.
36. Fastowicz, J.; Bak, D.; Okarma, K. Quality Assessment of 3D Printed Surfaces in Fourier Domain. In Proceedings of the International Conference on Image Processing and Communications, Bydgoszcz, Poland, 13–14 September 2017; pp. 75–81.
37. MacDonald, E.; Burden, E.; Walker, J.; Kelly, J.; Conner, B.; Patterson, C.; Schmidt, A.; Bader, A. Spatial Frequency Analysis for Improved Quality in Big Area Additive Manufacturing (BAAM). In Proceedings of the ASME 2017 International Mechanical Engineering Congress and Exposition, Tampa, FL, USA, 3–9 November 2017; p. 2.
38. Mital', G.; Dobránsky, J.; Ružbarský, J.; Olejárová, Š. Application of laser profilometry to evaluation of the surface of the workpiece machined by abrasive water jet technology. *Appl. Sci.* **2019**, *9*, 2134. [CrossRef]
39. Siklienka, J.Š.M. The effect of a laser beam wavelength on accuracy of surface measurement of unevenness of beechwood by laser beam. *Acta Fac. Xylologiae Zvolen* **2018**, *60*, 137–145.
40. Mitaľ, G. Contactless measurement and evaluation machined surface roughness using laser profilometry. *Transf. Inovácií* **2021**, *43*, 19–24.
41. Krehel', R.; Straka, Ľ.; Krenický, T. Diagnostics of production systems operation based on thermal processes evaluation. *Appl. Mech. Mater.* **2013**, *308*, 121–126. [CrossRef]
42. Lednev, V.N.; Sdvizhenskii, P.A.; Asyutin, R.D.; Tretyakov, R.S.; Grishin, M.Y.; Anton, Y.; Stavertiy, A.Y.; Fedorov, A.N.; Pershin, S.M. In situ elemental analysis and failures detection during additive manufacturing process utilizing laser induced breakdown spectroscopy. *Opt. Express* **2019**, *27*, 4612–4628. [CrossRef]
43. Yang, T.; Jin, Y.; Squires, B.; Choi, T.-Y.; Dahotre, N.B.; Neogi, A. In-situ monitoring and ex-situ elasticity mapping of laser induced metal melting pool using ultrasound: Numerical and experimental approaches. *J. Manuf. Processes* **2021**, *71*, 178–186. [CrossRef]
44. Leung, C.L.A.; Marussi, S.; Atwood, R.C.; Towrie, M.; Withers, P.J.; Lee, P.D. In situ X-ray imaging of defect and molten pool dynamics in laser additive manufacturing. *Nat. Commun.* **2018**, *9*, 1355. [CrossRef] [PubMed]
45. Borish, M.; Post, B.K.; Roschli, A.; Chesser, P.C.; Love, L.J.; Gaul, K.T.; Sallas, M.; Tsiamis, N. In-situ thermal imaging for single layer build time alteration in large-scale polymer additive manufacturing. *Procedia Manuf.* **2019**, *34*, 482–488. [CrossRef]
46. Lishchenko, N.; Lazorik, P.; Demčák, J.; Pitel', J.; Židek, K. Quality control monitoring in 3D printing. In *Advances in Design, Simulation and Manufacturing V.*; Springer: Cham, Switzerland, 2022; pp. 31–40.

machines

MDPI

Article

Development of Flexible Fixtures with Incomplete Locating: Connecting Rods Machining Case Study

Vitalii Ivanov [1], František Botko [2,*], Ivan Dehtiarov [1], Marek Kočiško [3], Artem Evtuhov [1], Ivan Pavlenko [4] and Justyna Trojanowska [5]

1 Department of Manufacturing Engineering, Machines and Tools, Faculty of Technical Systems and Energy Efficient Technologies, Sumy State University, 2, Rymskogo-Korsakova St., 40007 Sumy, Ukraine; ivanov@tmvi.sumdu.edu.ua (V.I.); ivan_dehtiarov@tmvi.sumdu.edu.ua (I.D.); evtuhov.a@tmvi.sumdu.edu.ua (A.E.)
2 Department of Automobile and Manufacturing Technologies, Faculty of Manufacturing Technologies with a Seat in Prešov, Technical University of Košice, 1, Bayerova St., 080 01 Prešov, Slovakia
3 Department of Computer Aided Manufacturing Technologies, Faculty of Manufacturing Technologies with a Seat in Prešov, Technical University of Košice, 1, Bayerova St., 080 01 Prešov, Slovakia; marek.kocisko@tuke.sk
4 Department of Computational Mechanics Named after V. Martsynkovskyy, Faculty of Technical Systems and Energy Efficient Technologies, Sumy State University, 2, Rymskogo-Korsakova St., 40007 Sumy, Ukraine; i.pavlenko@omdm.sumdu.edu.ua
5 Department of Production Engineering, Faculty of Mechanical Engineering, Poznan University of Technology, 5, M. Sklodowskej-Curie Sq., 60-965 Poznan, Poland; justyna.trojanowska@put.poznan.pl
* Correspondence: frantisek.botko@tuke.sk

Abstract: The rapid development of manufacturing in recent years has led to a significant expansion of the technological capabilities of modern metal-cutting equipment. Therefore, the modern approach to intensifying production requires an advanced fixture design. Design and manufacture of flexible fixtures capable of machining similar shapes and sizes of complex geometry parts reduce setup time. The article aims to design flexible fixtures for parts such as one-piece connecting rods under incomplete locating conditions. The advantages are the minimum number of parts and tool availability for multi-axis machining connecting rods in one setup. This approach, combined with up-to-date machining centers and industrial robots, can increase the production efficiency of manufacturing non-removable connecting rods. This effectiveness is in a decrease in the number of operations by 5–7 times, fixtures—by 3–4 times, and machine tools—by 3–5 times, depending on the type of a non-removable connecting rod and its design features. The numerical simulation results of the proposed fixture design confirmed the comprehensive technological capabilities and dynamic characteristics. Particularly, a decrease in displacements and oscillation amplitudes up to 7% compared to the full-basing locating chart was provided. It is determined that the system "fixture–workpiece" entirely meets all the strength, accuracy, and rigidity parameters, which allows you to perform machining with intensive cutting modes. The amplitudes of oscillations do not exceed the tolerances on the dimensions of these surfaces, established by requirements for non-removable connecting rods, and all displacements are elastic. During numerical simulation, the workpiece position remained stable at all machining steps.

Keywords: fixture design; machining; sustainable manufacturing; process innovation; complex-shape part

Citation: Ivanov, V.; Botko, F.; Dehtiarov, I.; Kočiško, M.; Evtuhov, A.; Pavlenko, I.; Trojanowska, J. Development of Flexible Fixtures with Incomplete Locating: Connecting Rods Machining Case Study. *Machines* **2022**, *10*, 493. https://doi.org/10.3390/machines10070493

Academic Editor: Gianni Campatelli

Received: 10 May 2022
Accepted: 17 June 2022
Published: 21 June 2022

1. Introduction

At present, in the conditions of active development of the concept of Industry 4.0, manufacturers of spare parts for the automotive industry and other equipment face a rather complicated problem, namely, to ensure rapid production of products of various nomenclature in the shortest possible time in discrete batches [1]. The quality of the finished part/product should meet both the requirements set by the designer in the drawing and international quality standards [2]. Due to the mentioned above, it is necessary to constantly update the equipment and manufacturing technologies and approaches to the production process [3]. Various equipment, including internal combustion engines, is constantly growing in automotive, agricultural, shipbuilding, and military industries. Simultaneously, the time of product entry on the market is reduced, models are updated, and technical characteristics are changed.

In current conditions, the approach to manufacturing spare parts for the above equipment has changed. First, manufacturers are trying to make their products lighter and more efficient. This applies to massive body parts (for example, engine cylinder blocks) [4], and other essential parts of the engine are trying to optimize the weight. In the last 10 years, the technology of manufacturing highly loaded parts from composite materials has been actively developed in the automotive industry [5], but their use has some limitations due to their physical and mechanical properties. For example, parts of the cylinder-piston group cannot be made of composite materials because composites cannot withstand high temperatures without destruction and changes in shape. Secondly, at the enterprises to produce spare parts, much attention is paid to the intensification of production [6,7], which aims to reduce the time of manufacture of the finished part/product and reduce costs. Simultaneously, enterprises are ready to spend money on purchasing advanced metal-cutting machines and other equipment, which allows to reduce the share of manual labor of the worker to a minimum or exclude it altogether. This requires introducing new technical solutions to change the manufacturing process towards the concentration of machining operations of the part.

The work also aims to theoretically prove the hypothesis of achieving the required machining accuracy and approach to fixture design based on numerical modeling of the developed flexible fixtures for one-piece machining rods with an incomplete locating for multiproduct production.

The connecting rod is the main element of a crank mechanism used in all classic configurations of internal combustion engines. Simultaneously, automakers try to make their products as light and durable as possible. This leads to the need to accurately calculate the connecting rods using numerical simulation methods.

In this regard, Shanmugasundar et al. [8] performed a topological optimization of the connecting rod design using the Autodesk software, considering calculations performed using the finite element analysis method using the ANSYS software. Xiao et al. [9] numerically determined the optimal thickness of the smaller connecting rod head to ensure sufficient rigidity when fitting the sleeve and the optimal amount of tension in the joint. Basavaraj et al. [10] determined that the optimal material for rods subjected to high cyclic loads is carbon steel alloyed with chromium and molybdenum.

Optimizing the mass of the connecting rods of large diesel internal combustion engines is an urgent task. Seyedzavvar et al. [11] performed parametric optimization of the connecting rod design using the CATIA software. As a result, the weight was reduced, and the resulting model was tested for strength using the ABAQUS software. Muhammad et al. [12] carried out the topological optimization of the connecting rod design using the ANSYS software. As a result, the total mass was reduced by 60% under the same constant static load.

The analysis confirmed that incorrect calculations lead to accidents during the failure of the connecting rods precisely. Chao [13] and Strozzi et al. [14] gave the main design errors and methods of avoiding them. Jia et al. [15] proved that the marking of connecting rods during stamping should be clearly defined in shape and size and be positioned in specific places not to violate the strength of the whole part.

Balamurugan et al. [16] analyzed the time of manufacture of the connecting rod in each manufacturing process operation and suggested ways to intensify production by combining some operations. Liu et al. [17] also studied the manufacturing process of connecting rod production and proposed a new method for evaluating the manufacturing process that responds to the state of machining in real-time based on digital twins.

In mass production, the connecting rods are manufactured using special multipart fixtures to reduce the machining time [18]. However, connecting rods can be manufactured using universal re-adjustable fixtures in single and small-scale production conditions and sports and special-purpose machines [19].

Many scientists are involved in improving the manufacturing process of machining parts, such as connecting rods. In particular, the manufacturing process of the detachable connecting rod machining was investigated by Kar [20]. Simultaneously, a special expanding mandrel was proposed as a fixture for the operation of machining holes for screws. Raijada and Dudhatra [21] designed a fixture for boring holes in a connecting rod.

Cui et al. [22] proposed a methodology to develop fixtures based on the computer-aided design (CAD) approach. Raja et al. [23] also designed, modeled, and tested fixtures for milling machines based on the CAD software.

Litrop et al. [24] developed an approach to design and optimize fixtures with their consequent testing under cyclic loading. In addition, Lu et al. [25] proposed a multi-objective optimization algorithm for the locating chart in fixture design.

Corrado et al. [26] considered the system "fixture–workpiece" as a separate assembly unit and predicted the accuracy of processing depending on the errors in the connections of the assembly elements.

Parvaz et al. [27] proposed an analytical and algorithmic procedure for designing fixtures for workpieces with free-form geometry of the NURBS type. Wu et al. [28] proposed an algorithm for designing fixtures for automated machining of complex shape parts and software in cases of changes in the shape and number of jet engine blades.

Luo et al. [29] developed geometric theorems and applied DOF workpiece analysis based on a normal constraint line to avoid repositioning when a complex workpiece is installed on many different surfaces.

Kamble and Mathew [30] integrated a segmented fixture design approach into a single methodology for complex design in four stages: installation planning, fixture layout development, detailing, and verification.

Thus, based on the recent advancements in the design, modeling, manufacturing, and intensification of rods production and development of fixtures for their machining, the actual research direction in the development and numerical modeling of fixtures for one-piece machining rods for one step has proposed to be studied.

The scientific novelty is in the fact that the method of numerical modeling theoretically proves the workability of the proposed concept of machining parts such as connecting rods in a fixture with an incomplete location and the possibility of applying this approach to other types of parts of complex shapes, such as forks and brackets.

The practical importance is highlighted by developing a ready-made technical solution that can be used in enterprises where parts such as connecting rods of different designs in small batches with a particular frequency can be processed. Of course, to implement this approach, the company must have a multi-coordinate machine with the function of automatic binding and transfer of the coordinates of the workpiece in the machine's coordinate system.

2. Materials and Methods

2.1. Assignment

The object of study was selected parts such as connecting rods. The connecting rods have a very different design depending on the purpose of the product/machine where the connecting rod works. However, of all the designs of connecting rods, there are four main ones: detachable connecting rods, detachable clip connection rods, one-piece connecting rods, and one-piece clip connecting rods (Figure 1).

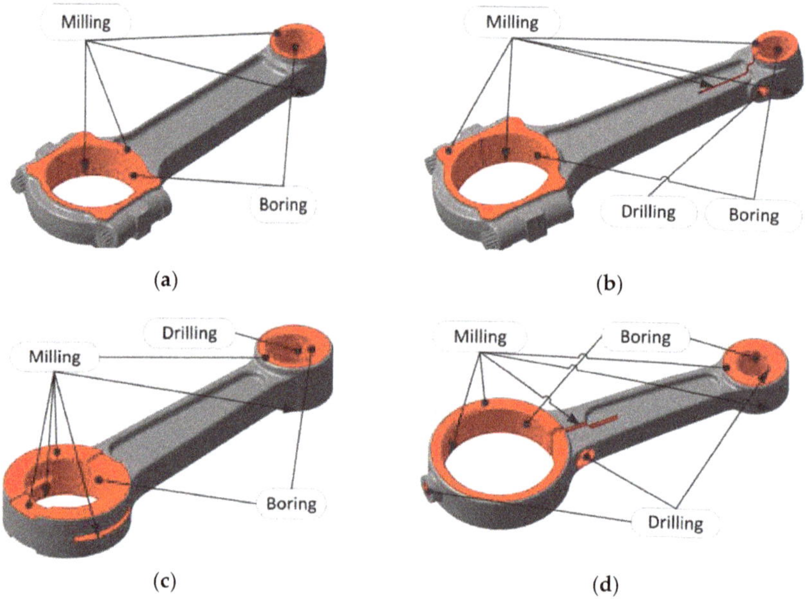

Figure 1. The designs of connecting rods with machined surfaces: (**a**) Detachable connecting rods; (**b**) Detachable clip connecting rods; (**c**) One-piece connecting rods; (**d**) One-piece clip connecting rods.

Detachable connecting rods (Figure 1a) are mainly used in automobile engines and high-power compressors. The connecting rods shown in Figure 1b have an incision in the small head of the connecting rod to allow rigid fixation of the piston pin and prevent its rotation. This design is used for manufacturing large connection rods for low-speed marine engines. One-piece connecting rods (Figure 1c) are usually used in motorcycle engines. One-piece connecting rods (Figure 1d) have a cut in the large head of the connecting rod and are used in reciprocating compressors of low and medium power. The cut is designed to securely lock the rolling bearing in the large connecting rod head by tightening the clip screw.

All the holes and planes of a connecting rod are interconnected by the requirements of center-to-center distances, the axes' parallelism, and the spatial position of the flat surfaces relative to each other.

Due to the widespread application of one-piece connection rods in oversized internal combustion engines (Figure 1c) and compressors, an increase in machining productivity of one-piece clip connection rods (Figure 1d) is an urgent problem. One-piece connecting rods, unlike detachable ones, are one part during the whole manufacturing process of their production, so they were chosen as an object for analysis and further development of the fixture design.

One-piece connecting rods are characterized by many untreated surfaces and complicated spatial geometric shapes. These features cause difficulties in locating and clamping workpieces. They also lead to the increased complexity of machining due to several manufacturing operations.

According to the sequence and principles in designing a progressive manufacturing process, it was established that the number of operations machining for a one-piece connection rod (Figure 1c) is five drilling-milling-boring operations for which you need to perform ten reinstallations of the workpiece. To machine a one-piece clip connection rod (Figure 1d), the manufacturing process consists of seven drilling-milling-boring operations, which require eleven reinstallations of the workpiece.

2.2. Design Requirements

An analysis of typical drawings of non-removable connecting rods and the literature sources, in particular, showed in [31] that the hole and end surfaces of the large connecting rod head are its main design bases in contact with the crankshaft surfaces. Accordingly, the hole and the end surfaces of the small connecting rod head are auxiliary design bases in contact with the ends of the piston lugs and the piston pin sleeve. The hole of the large connecting rod head is made according to the tolerances of IT6–IT7 qualities, which for most non-removable connecting rods is 0.03–0.07 mm. The end surfaces of the large connecting rod head are made according to the tolerances of IT10–IT11 qualities, which for most non-removable connecting rods is 0.15–0.20 mm. The accuracy of the holes of the small connecting rod head is not so high and is limited by IT10–IT11 qualities and the accuracy of the ends of the small connecting rod head IT11–IT12, respectively. Precision requirements for oil channel holes are limited to IT14 tolerances. The deviation of the center-to-center distances should not exceed 0.05–0.1 mm, depending on the dimensions of the connecting rod. The axes of the holes of the large and small connecting rod heads must be parallel with a tolerance of 0.05:100 to 0.15:100, and their ends must be perpendicular to the axes of these holes in the range from 0.1:100 to 0.3:100. The roughness of the upper hole in the large head of the connecting rod should be Ra = 0.32–0.63 μm, and in the small head—Ra = 1.6–3.2 μm. The roughness of the ends of both heads is in the range of Ra = 1.6–3.2 μm. The hardness of the connecting rods is regulated by their official purpose and is within the HRC 50–55.

2.3. Manufacturing Process

Given the technological capabilities of up-to-date multi-purpose machines and design approaches successfully used in developing fixtures for fork-type parts, it is proposed for machine parts such as one-piece connection rods (Figure 1c). This allows carrying out one replacement by combining four drilling-milling-boring operations of a typical manufacturing process into one—at the CNC multi-purpose machining center. For parts such as one-piece clip connection rods (Figure 1d), it is proposed to carry out machining for one institution by combining five operations of a typical manufacturing process into one at the CNC multi-purpose machining center. This approach allows you to intensify the production process, clearly shown in Figure 2.

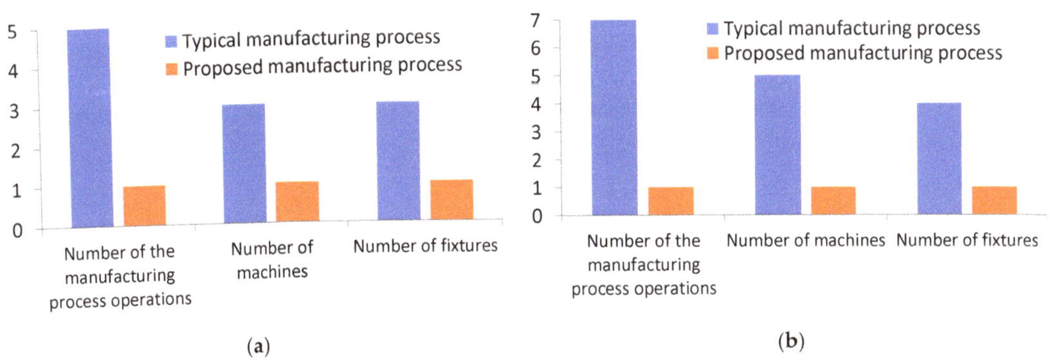

Figure 2. Diagram comparing manufacturing process indicators: (**a**) One-piece connection rods; (**b**) One-piece clip connection rods.

For one-piece connection rods, it was possible to reduce the manufacturing process by four machining operations, the number of machines from three units to a single CNC multi-purpose machine, and the number of special fixtures from three units to one flexible fixture. For one-piece clip rods, it is possible to reduce the manufacturing process by six machining operations, the number of machines from five units to the CNC multi-purpose machine, and the number of special fixtures from four units to one flexible fixture.

When using a modern machine with automated positioning of the part, pre-machining of base surfaces is not required. This fact simplifies the fixture design and reduces the manufacturing process by one operation compared to the manufacturing process.

The proposed manufacturing process of machining all surfaces of the one-piece connection rod that requires drilling, milling, or boring can be implemented in the machining center with CNC, and vertical and horizontal layouts in six positions.

2.4. Fixture Design

Despite various machines containing connecting rods in a global market, the design of one-piece connecting rods is almost indistinguishable. Their difference can only be in changing the type-size of the configuration of machined surfaces and their characteristics (accuracy, roughness) depending on the purpose of the machine/unit, so it is advisable to develop a flexible fixture that installs one-piece connection rods in a specific range of sizes and shapes. This fixture should ensure full tool accessibility and allow multi-axis machining of all surfaces needed in a single step. Simultaneously, the standard dimensions and design parameters of non-removable connecting rods and one-piece clip rods are relatively similar, and several types of connecting rods can be processed in one fixture. The use of modern progressive machines allows us to depart from the principle of the orientation of the workpiece in the fixture and requires only a secure fastening to ensure the invariability of the position of the workpiece during machining.

As a result, a flexible fixture was developed for the installation of one-piece connection rods of various sizes in the range of 140–200 mm in length, 35–50 mm in width, and 17–24 mm in height, which is carried out by adjusting the screw mechanisms that change the distance between the locating-clamping elements (Figure 3) Machined surfaces are shown in Figure 3 by letters A, B, C, D, E.

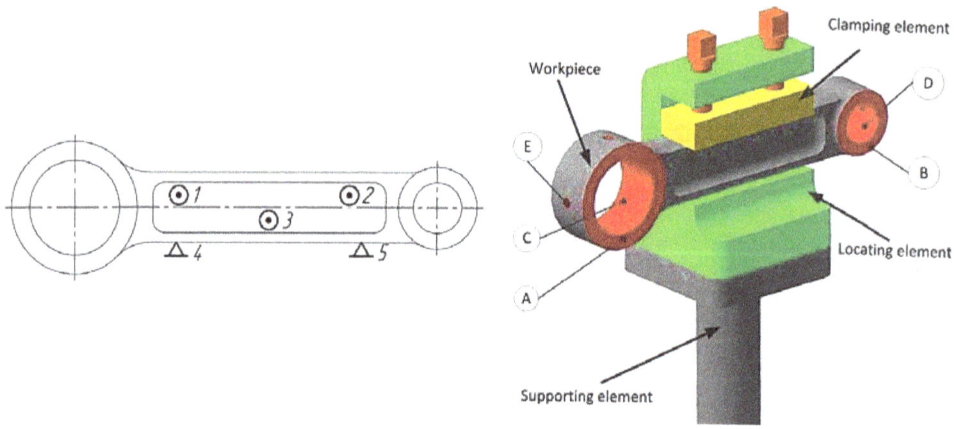

(a) (b)

Figure 3. Flexible fixture for machining a one-piece connecting rod with machined surfaces in a CNC multi-purpose operation: (**a**) Locating chart; (**b**) 3D visualization.

This fixture is modular, i.e., it can be installed due to standardized mounting and clamping elements in the base modules or a three-jaw chuck or machine vise with prismatic jaws. The proposed fixture has a relatively simple design with a minimum of parts and connections and, therefore, is expected to have high rigidity for machining with the maximum allowable cutting modes. This technical solution will allow you to perform machining on a three-coordinate milling machine with CNC, equipped with only a two-coordinate rotating table. The only condition is the presence of a system of a preliminary determination of the position of the workpiece relative to the machine's coordinate system.

3. Results

3.1. Stress-Strain Analysis of the System "Fixture–Workpiece"

At the initial stage of the study, the proposed design of the fixture was tested for a non-removable connecting rod of a simpler design that does not contain a terminal connection. Since this concept was developed for the first time, to minimize time from the potential risks of unsatisfactory modeling results, it was performed for the design of the connecting-rod, which contains fewer machining transitions and, accordingly, the number of operations for numerical simulation of these transitions.

Since the principle of operation of the proposed fixture is based on the need to ensure absolute reliability of fastening, to determine the possibility of achieving an accuracy of size, shape, and relative position of surfaces during machining, it is necessary to perform a stress-strain analysis of the system "fixture–workpiece". The displacement of the "fixture–workpiece" elements under the action of external loads (clamping forces, cutting forces, and cutting moments) in the proposed fixture for one-piece connection rods is determined. The fixture's strength was determined by obtaining values of equivalent stresses. The contact interaction models between the workpiece and functional fixture elements are considered, and stress concentrators are identified. The maximum values of equivalent stresses determined in the ANSYS Workbench software according to von Mises equivalent strength were compared with the permissible value for the specific material from which the fixture parts are made and the material of the machining workpiece.

The fixture model considers the Coulomb friction coefficient of 0.1 between the contact surfaces of the fixture, which have approximately the same roughness Ra = 1.6 μm.

The boundary conditions are presented in Figure 4 and Table 1.

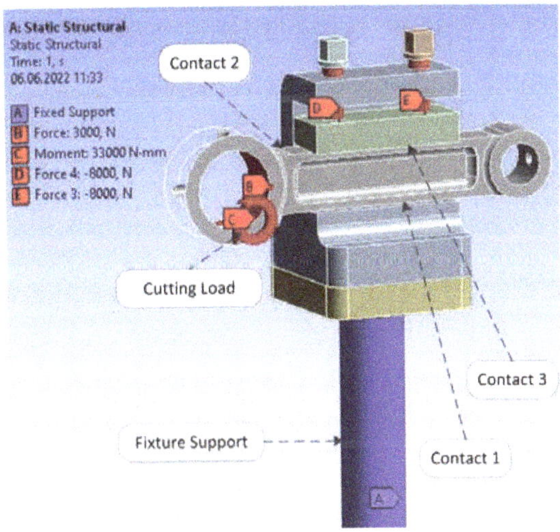

Figure 4. Boundary conditions and contact layers for the fixture.

In the process of modeling, the forces and torques of cutting were applied to each of the machined surfaces in turn, as the technological capabilities of the machine can perform only one-tool machining. The value of forces at different machining steps varied between 0.51–2.94 kN and torques 45–142 N·m depending on the method and stage of machining. Simultaneously, on a pressure plate from screws, the constant clamping force 8.0 kN applied to each of them acted.

Table 1. Boundary conditions for numerical simulation of the system "fixture–workpiece".

Reference Surface/Fixing Type	Parameters of the Bonding Groups			
	Contact	Contact Surfaces	Types of the Contact Surfaces	Friction Coefficient
The cylindrical surface of the fixture/fixture support	1	The lower plane of the support/the lower surface of the connecting rod neck	smooth/non-machining	0.2
	2	the end surface of the support/the end plane of the connecting rod neck	smooth/non-machining	0.2
	3	the surface of the pressure plate/the upper surface of the connecting rod neck	grooving/non-machining	0.7

The material properties of the fixture and the workpiece are summarized in Table 2.

Table 2. Mechanical properties of the materials for the workpiece and fixture elements.

Material (DIN Standard)	Young's Modulus, GPa	Poisson's Ratio	Density, kg/m^3	Tensile Strength, GPa	The Ultimate Strength of the Compression, GPa	Yield Strength, GPa
Structural alloy steel 40Cr	200	0.3	7850	0.960	0.960	0.765
Structural steel C45 (after heat treatment)	200	0.3	7850	0.950	0.950	0.726

To evaluate the results of numerical modeling and theoretical performance of the structure, we compare them with the results of similar indicators obtained in previous studies [32] and the allowable values of the permissible parameters for the connecting rod from the average values of its dimensions: length–170 mm, width–40 mm, and height–20 mm.

These boundary conditions allowed simulating the process under static loads. The values of the maximum equivalent stress according to the von Mises hypothesis and the maximum displacements of the elements and surfaces of the system "fixture–workpiece" are presented in Table 3.

Table 3. Numerical simulation results.

Surface (Figure 3)	Manufacturing Step with a Maximum Loading	Maximum Displacement, mm		Permissible Values of Displacements, mm	Maximum Equivalent Stress, MPa		Permissible Values of Stresses, MPa
		Fixture with Incomplete Locating	Fixture with Complete Locating		Fixture with Incomplete Locating	Fixture with Complete Locating	
A	Milling	0.039	0.048	0.15	297	315	726
B	Milling	0.082	0.095	0.20	419	427	726
C	Drilling	0.092	0.107	0.25	329	361	726
D	Drilling	0.085	0.097	0.20	314	338	726
E	Drilling	0.006	0.006	0.25	112	94	726

The results show that the proposed fixture design provides the specified accuracy. The corresponding indicators do not exceed the values for a flexible fixture with a complete workpiece locating.

3.2. *Eigenfrequencies of the System "Fixture–Workpiece"*

Vibrations inevitably occur in the technological system during machining. Technological parameters of machining modes cause them (e.g., cutting depth, feed, speed, dimensions, number of tools, etc.). Eigenfrequencies for these oscillations depend on the design parameters and the material of parts.

The eigenfrequencies have been determined using the built-in module "Modal Analysis" of the ANSYS Workbench software to prevent the resonant modes during the machining of one-piece connection rods in the developed fixture. The corresponding comparison with the machining frequencies at all steps of drilling-milling-boring operations for the studied fixture has been performed.

The analysis results determine the required detuning from the resonance (Table 4).

Table 4. The results of fixture eigenfrequencies investigation.

Fixture	Eigenfrequency, Hz			The Maximum Frequency of the Machining, Hz	Manufacturing Step with the Maximum Frequency of the Cutting Process
	1st	2nd	3rd		
Incomplete locating	1729	1817	2231	100	Drilling of the hole (diameter 4 mm, speed 3000 rpm)
Complete locating	1425	1427	2073		

The mode shape examples for the first two eigenfrequencies are shown in Figure 5.

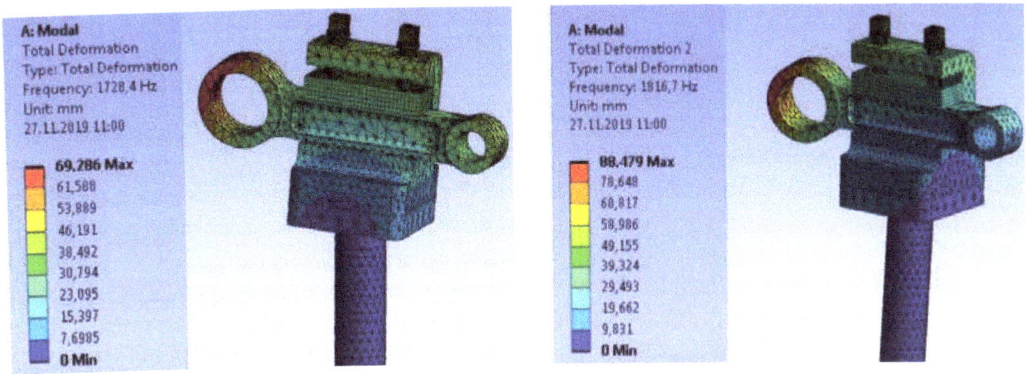

(**a**) (**b**)

Figure 5. An example of the proposed fixture's oscillation frequencies: (**a**) The 1st eigenfrequency; (**b**) The 2nd eigenfrequency.

Remarkably, at the first eigenfrequency, oscillations occur in the XY plane and the second one in the YZ plane. Since the first eigenfrequency significantly exceeds the operating frequency of the cutting process (1729 Hz >> 100 Hz), resonance does not occur.

Nevertheless, after comparing the values for the complete and incomplete locating fixture, it should be noted that the oscillation frequencies for the incomplete locating fixture are higher, which can be explained by fewer parts and connections between them, respectively.

3.3. *Forced Oscillations of the System "Fixture–Workpiece"*

To check the oscillation amplitudes of of the system "fixture–workpiece", the amplitude of dynamic forces and torques during cutting was chosen as 20% of the nominal values. Therefore, the amplitude of cutting forces at the machining steps of one-piece connection rod surfaces varied from 101 N to 587 N, and the amplitude of cutting torques was in the range of 9.0–28.4 N·m.

Previous studies have shown that the range of operating frequencies of the cutting process when machining one-piece connection rods does not exceed 100 Hz. This fact allows limiting the range 0–100 Hz on the frequency response charts when determining the amount of displacement.

To establish the possibility of fixture operation in a wider range and review the amplitude-frequency characteristics of fixture, it was decided to set the range 0–2 kHz, which corresponds to 30,000 rpm for single-blade tools, and 15,000 rpm–for two-blade tools such as drills. This may be relevant for the treatment of small diameter holes that are intended for lubrication. Therefore, to increase the efficiency of processing such holes, it is possible to use tools made of hard alloy that allow high cutting speeds and, accordingly, to ensure optimal operating conditions of these tools, it is necessary to provide them with high speeds. Therefore, the study of the amplitude of oscillations in the range 0–2 kHz is relevant information for further research.

The numerical simulation of forced oscillations for the system "fixture–workpiece" allowed for obtaining the amplitudes of displacements at points in the machining zone during the cutting process (Table 5).

Table 5. The results of the dynamic analysis.

Surface (Figure 3)	Manufacturing Step with a Maximum Loading	The Resulting Force at the Transition, N	Force's Amplitude, N	Torque, N·m	Torque's Amplitude, N·m	Maximum Displacement for the Fixture, μm	
						Incomplete Locating	Complete Locating
A	Milling	895	179	68	13.6	4.8	5.1
B	Milling	1070	214	83	16.6	6.4	6.6
C	Drilling (Ø 29.5 mm)	2937	587	142	28.4	9.6	10.3
D	Drilling (Ø 15.5 mm)	1706	341	107	21.4	7.1	7.3
E	Drilling	505	101	45	9.0	2.1	2.1

An example of the frequency response when drilling a hole with a diameter of 29.5 mm in the large head of the one-piece connecting rod is shown in Figure 6.

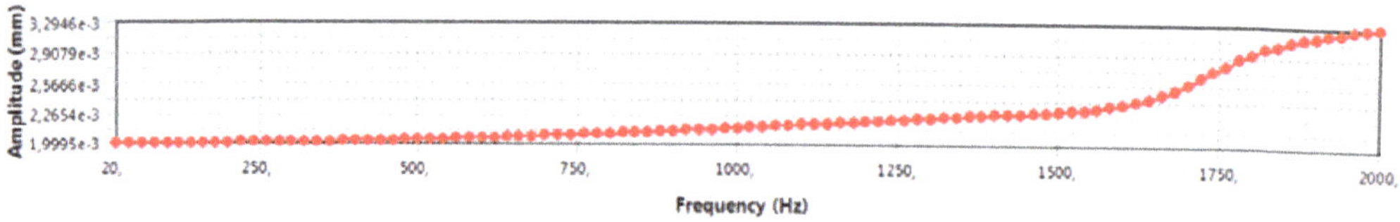

Figure 6. The amplitude-frequency response of the fixture with incomplete locating when machining the surface of a one-piece connection rod.

The amplitudes of oscillations that occur at all machining steps are less than the tolerances on the corresponding machined surfaces. Therefore, the needed accuracy is achieved.

For determining the reserve of the proposed fixture, the dynamic response has been evaluated for the machining step of drilling a hole with a diameter of 29.5 mm in the large head of the connecting rod (Table 6).

Table 6. The results for the calculation of the dynamic stiffness of the fixture.

Fixture	Manufacturing Step	The Amplitude of the Dynamic Component of the Cutting Force, N	The Maximum Amplitude of Displacements, μm
Incomplete locating	Drilling a hole (Ø 29.5 mm)	587	9.6
Complete locating		587	10.3

4. Discussion

Thus, the theoretically hypothesis of achieving the accuracy parameters of machining in a fixture with incomplete locating of workpiece on the example of the developed design of a flexible fixture installing one-piece connection rods for multi-nomenclature production is proved. Thus, the proposed approach to the fixture design based on the reliable clamping of the workpiece and the use of modern CNC machine tools with the functions of automatic determination of the coordinates of the workpiece has shown its effectiveness.

According to the results of the numerical simulation of the stress-strain state of the proposed design of a flexible fixture for one-piece connection rods, it is established that the values of the accuracy parameters specified by the designer during machining will be provided. After all, the maximum displacement of the fixture elements during machining of a one-piece connection rod was 0.092 mm when drilling and 0.082 mm when milling, not exceeding the tolerances according to the requirements of the drawing 0.25 mm and

0.2 mm, respectively, for this connecting rod design. The displacement data are also elastic, as the values of the maximum stresses at the most loaded step do not exceed the strength limits of the material of the fixture parts and the workpiece material.

The developed fixture design also meets the strength conditions. The safety factor for this system at maximum stresses of 419 MPa equals 1.8. As a result of comparing fixtures with complete and incomplete locations, it was found that the displacements and stresses are smaller by 7% and 9%, respectively, in the fixture with an incomplete location. This fact is explained by the smaller number of parts and connections in such a fixture.

The results of harmonic analysis of the proposed design indicate the high rigidity of the developed technical solution and the lack of resonance due to design shortcomings of the fixture because the first eigenfrequency of the fixture is more than 14 times higher than the maximum frequency of the cutting process. Analysis of the dynamic state of the elements of the system "fixture–workpiece" of the proposed fixture showed that the amplitudes of oscillations that occur during removal of the allowance during cutting in places of machined surfaces do not exceed tolerances for machining at the appropriate steps. The maximum value of the oscillation amplitude for the operating step of drilling a hole with a diameter of 29.5 mm in the large head of the connecting rod is 0.096 mm.

Analyses of free and forced oscillations also confirm the higher performance of the proposed fixture design for connecting rods machining. Overall, it was established that the proposed fixture with an incomplete location for one-piece connection rods at the theoretical stage provides the necessary indicators of machining accuracy.

Overall, the proposed methodology continues recent strategies in the development of flexible fixtures [33], increases the reliability of the automated flexible fixture systems for mass production [34], improves methods for flexible fixture design in the automotive industry [35], extends recent developments in reconfigurable fixture design [36], and improves simulation approaches in fixture design [37].

Further research will evaluate the magnitudes of displacements under static and dynamic loads during machining, considering time-varying dynamic components of cutting forces and torques.

5. Conclusions

1. The efficiency of the developed flexible fixture for machining of rods with incomplete basing based on comparison of the maximum stresses, movements, and frequencies of oscillations with their admissible values for a rod of the concrete size is theoretically proved. Actual performance will be tested in further experimental studies.
2. The proposed approach to the machining of connecting rods with incomplete location allowed for reducing the number of technological operations by 4 and 6, the number of units of equipment by 2 and 4, and the number of special fixtures by 2 and 3, depending on the type of one-piece connecting rods.
3. A comprehensive approach to the numerical modeling of the stress-strain state and free and forced oscillations of the "workpiece-fixture" system is proposed. The reliability of the proposed fixture is justified in terms of strength, rigidity, and vibration safety.
4. It is established that the values of displacements, frequencies of oscillations, and amplitudes of oscillations in the fixture with incomplete locating are 3–7% less than in fixtures with full locating, which was developed and studied earlier.

Author Contributions: Conceptualization, V.I.; methodology, I.D.; software, F.B. and J.T.; validation, M.K., A.E. and J.T.; formal analysis, I.P.; investigation, I.D.; resources, F.B.; data curation, V.I.; writing—original draft preparation, I.D. and A.E.; writing—review and editing, I.P.; visualization, J.T.; supervision, M.K.; project administration, V.I. and I.P.; funding acquisition, F.B. and M.K. All authors have read and agreed to the published version of the manuscript.

Funding: This research was funded by the research project "Improvement of the Production Planning by Implementation of the Computer-Aided Fixture Design System" within the Joint Ukrainian–Slovak

R&D Projects for the period of 2022–2023 granted by the Slovak Research and Development Agency (SK-UA-21-0060) and the Ministry of Education and Science of Ukraine (State reg. no. 0122U002657). The Research was supported by project VEGA 1/0051/20.

Institutional Review Board Statement: Not applicable.

Informed Consent Statement: Not applicable.

Data Availability Statement: The data presented in this study are available on request from the corresponding author.

Acknowledgments: The scientific results have been partially obtained within the research project "Fulfillment of tasks of the perspective plan of development of a scientific direction "Technical sciences" Sumy State University," ordered by the Ministry of Education and Science of Ukraine (state reg. no. 0121U112684). The research was partially supported by the Research and Educational Center for Industrial Engineering (Sumy State University) and the International Association for Technological Development and Innovations.

Conflicts of Interest: The authors declare no conflict of interest.

References

1. Gothwal, S.; Raj, T. Different aspects in design and development of flexible fixtures: Review and future directions. *Int. J. Serv. Oper. Manag.* **2017**, *26*, 386–410. [CrossRef]
2. Denysenko, Y.; Dynnyk, O.; Yashyna, T.; Malovana, N.; Zaloga, V. Implementation of CALS-Technologies in Quality Management of Product Life Cycle Processes. In Proceedings of the Advances in Design, Simulation and Manufacturing, DSMIE-2018, Lecture Notes in Mechanical, Engineering. Sumy, Ukraine, 12–15 June 2018; Ivanov, V., Rong, Y., Trojanowska, J., Venus, J., Liaposhchenko, O., Zajac, J., Pavlenko, I., Edl, M., Perakovic, D., Eds.; Springer: Cham, Switzerland, 2019; pp. 3–12. [CrossRef]
3. Karabegovic, I.; Karabegovic, E.; Mahmic, M.; Husak, E. The role of smart sensors in production processes and the implementation of Industry 4.0. *J. Eng. Sci.* **2019**, *6*, B8–B13. [CrossRef]
4. Akimov, O.V.; Marchenko, A.P.; Alyokhin, V.I.; Soloshenko, V.; Shinsky, O.Y.; Klymenko, S.I.; Kostyk, K.O. Computer engineering and design of cast parts for internal combustion engine crankcase. *J. Eng. Sci.* **2019**, *6*, E24–E30. [CrossRef]
5. Osadchiy, I.; Kryvoruchko, D.; Kolesnyk, V.; Hatala, M.; Duplak, J.; Mital, D. Development of integrated technology of FRP gear manufacturing. *Manuf. Technol.* **2016**, *16*, 574–578. [CrossRef]
6. Gleeson, F.; Coughlan, P.; Goodman, L.; Newell, A.; Hargaden, V. Improving manufacturing productivity by combining cognitive engineering and lean-six sigma methods. *Procedia CIRP* **2019**, *81*, 641–646. [CrossRef]
7. Kondakov, A. Productivity as measure of multiproduct metal processing production efficiency. *Procedia Eng.* **2016**, *150*, 987–991. [CrossRef]
8. Shanmugasundar, G.; Dharanidharan, M.; Vishwa, D.; Sanjeev Kumar, A.P. Design, analysis and topology optimization of connecting rod. *Mater. Today Proc.* **2021**, *46*, 3430–3438. [CrossRef]
9. Xiao, J.; Qin, W.; Xu, S. Interference and thickness design of bushing of connecting rod small end for Anti-loosening. *Eng. Fail. Anal.* **2021**, *127*, 105556. [CrossRef]
10. Basavaraj, Y.; Joshi, R.; Raghavendra Setty, R.; Samiullah, M.; Museb, M.; Tayab, M.; Banu, H. FEA of NX-11 unigraphics modeled connecting rod using different materials. *Mater. Today Proc.* **2021**, *46*, 2807–2813. [CrossRef]
11. Seyedzavvar, M.; Seyedzavvar, M. Design of high duty diesel engine connecting rod based on finite element analysis. *J. Braz. Soc. Mech. Sci. Eng.* **2018**, *40*, 59. [CrossRef]
12. Muhammad, A.; Ali, M.; Shanono, I. Design optimization of a diesel connecting rod. *Mater. Today Proc.* **2020**, *22*, 1600–1609. [CrossRef]
13. Chao, J. Fretting-fatigue induced failure of a connecting rod. *Eng. Fail. Anal.* **2019**, *96*, 186–201. [CrossRef]
14. Strozzi, A.; Baldini, A.; Giacopini, M.; Bertocchi, E.; Mantovani, S. A repertoire of failures in connecting rods for internal combustion engines, and indications on traditional and advanced design methods. *Eng. Fail. Anal.* **2016**, *60*, 20–39. [CrossRef]
15. Jia, D.; Li, Y.; Deng, X.; Lei, J.; Deng, W. Design research on forging mark of connecting rod. *Eng. Fail. Anal.* **2020**, *113*, 104537. [CrossRef]
16. Balamurugan, R.; Kirubagharan, R.; Ramesh, C. Implementation of lean tools and techniques in a connecting rod manufacturing industry. *Mater. Today Proc.* **2020**, *33*, 3108–3113. [CrossRef]
17. Liu, X.; Liu, J.; Zhou, H.; Ni, Z. Chapter 12–Digital twin driven process design evaluation. In *Digital Twin Driven Smart Design*; Tao, F., Liu, A., Hu, T., Nee, A.Y.C., Eds.; Academic Press: Amsterdam, The Netherlands, 2020; pp. 309–332. [CrossRef]
18. Machining Fixture for Connecting Rod. Available online: https://www.indiamart.com/proddetail/machining-fixture-for-connecting-rod-14363410788.html (accessed on 3 May 2022).
19. USA-Made Racer Elite Connecting Rods Deliver Factory-Level Performance. Available online: https://racerxonline.com/2021/09/15/usa-made-racer-elite-connecting-rods-deliver-factory-level-performance-advertiser-wiseco (accessed on 3 May 2022).

20. Kar, A. Connecting Rod Manufacturing. *Report Number E341076-2018*; Czech Technical University in Prague, 2019. Available online: https://www.researchgate.net/publication/330183972_Connecting_Rod_Manufacturing (accessed on 20 March 2022).
21. Raijada, S.; Dudhatra, A. Design of fixture of connecting rod for boring operation. *Int. J. Sci. Res. Dev.* **2014**, *2*, 538–540. Available online: http://ijsrd.com/Article.php?manuscript=IJSRDV2I9266 (accessed on 20 March 2022).
22. Cui, C.; Han, Z.; Yu, T. Design of mechanical fixture based on computer CAD research. *J. Phys. Conf. Ser.* **2021**, *2023*, 012028. [CrossRef]
23. Raja, S.; Shaikh, M.B.N.; Majeed, M.; Varshney, A.; Samad, A. Design, modelling, fabrication, and testing of vertical milling machine fixture for friction stir welding operation. In *Advancement in Materials Processing Technology*; Springer Proceedings in Materials; Springer: Singapore, 2022; pp. 95–106. [CrossRef]
24. Litrop, A.; Klemenc, J.; Nagode, M.; Šeruga, D. Design of a shear test fixture system and optimization of a metal shear specimen for cyclic loading. *J. Test. Eval.* **2022**, *50*, JTE20210681. [CrossRef]
25. Lu, Y.; Zhang, X.; Li, M.; Zhao, M. Robust optimization design of fixture locating scheme based on constrained multi-objective optimization algorithm. *Acta Armamentarii* **2022**, *43*, 686–693. [CrossRef]
26. Corrado, A.; Polini, W. Tolerance analysis tools for fixture design: A comparison. *Procedia CIRP* **2020**, *92*, 112–117. [CrossRef]
27. Parvaz, H.; Nategh, M. Development of locating system design module for freeform workpieces in computer-aided fixture design platform. *Comput.-Aided Des.* **2018**, *104*, 1–14. [CrossRef]
28. Wu, D.; Wang, H.; Huang, L.; Wang, Y. Computer aided machining fixture design algorithm and software based on case learning for near-net-shaped jet engine blade. *J. Manuf. Process.* **2021**, *69*, 545–555. [CrossRef]
29. Luo, G.; Wang, X.; Yan, X. Geometric Theorems and Application of the DOF Analysis of the Workpiece Based on the Constraint Normal Line. *Adv. Mater. Sci. Eng.* **2021**, *2021*, 1014708. [CrossRef]
30. Kamble, V.; Mathew, A. Brief review of methodologies for creation of cohesive fixture design. *Mater. Today Proc.* **2020**, *22*, 3353–3363. [CrossRef]
31. Singh, D. *Fundamentals of Manufacturing Engineering*, 3rd ed.; Routledge: London, UK, 2018.
32. Ivanov, V.; Dehtiarov, I.; Zaloga, V.; Kosov, I.; Savchuk, V. Increasing Productivity of Connecting Rods Machining. In *Advances in Design, Simulation and Manufacturing III*; Lecture Notes in Mechanical, Engineering; Ivanov, V., Trojanowska, J., Pavlenko, I., Zajac, J., Peraković, D., Eds.; Springer: Cham, Switzerland, 2020; pp. 264–275. [CrossRef]
33. Bakker, O.J.; Papastathis, T.N.; Ratchev, S.M.; Popov, A.A. Recent research on flexible fixtures for manufacturing processes. *Recent Pat. Mech. Eng.* **2013**, *6*, 107–121. [CrossRef]
34. Illidge, A.; Bright, G. An automated flexible fixture system for mass customization. *S. Afr. J. Ind. Eng.* **2018**, *29*, 21–34. [CrossRef]
35. Yu, K.; Wang, S.; Wang, Y.; Yang, Z. A flexible fixture design method research for similar automotive body parts of different automobiles. *Adv. Mech. Eng.* **2018**, *10*, 1687814018761272. [CrossRef]
36. Daniyan, I.A.; Adeodu, A.O.; Oladapo, B.I.; Daniyan, O.L.; Ajetomobi, O.R. Development of a reconfigurable fixture for low weight machining operations. *Cogent Eng.* **2018**, *6*, 1579455. [CrossRef]
37. Vidoni, M.; Sarkar, S.; Farivar, H.; Sims, J.; Jordan, M.; Münter, D. Simulation of fixture hardening for tool and process design optimization. *HTM-J. Heat Treat. Mater.* **2022**, *76*, 356–369. [CrossRef]

machines

Article

Measurement and Modelling of a Cycloidal Gearbox in Actuator with Permanent Magnet Synchronous Machine

Viktor Šlapák [1], Jozef Ivan [1], Karol Kyslan [1], Matúš Hric [2], František Ďurovský [1,*], Dušan Paulišin [2] and Marek Kočiško [3]

1 Department of Electrical Engineering and Mechatronics, Faculty of Electrical Engineering and Informatics, Technical University of Košice, Letná 9, 042 00 Košice, Slovakia; viktor.slapak@tuke.sk (V.Š); jozef.ivan@tuke.sk (J.V.); karol.kyslan@tuke.sk (K.K.)
2 SPINEA Technologies Ltd., Volgogradská 4782/13, 080 01 Prešov, Slovakia; matus.hric@spinea-technologies.com (M.H.); dusan.paulisin@spinea-technologies.com (D.P.)
3 Faculty of Manufacturing Technologies of the Technical University of Košice with the Seat in Prešov, Bayerova 1, 080 01 Prešov, Slovakia; marek.kocisko@tuke.sk
* Correspondence: frantisek.durovsky@tuke.sk; Tel.: +421-55-602-2267

Abstract: Compact geared servo drives are a common part of modern industrial automation; thus, their proper modelling is a necessary part for the application and control design. The presented paper focuses on the mathematical model of the cycloidal gearbox, which is used in the compact actuator with a permanent magnet synchronous motor. A measurement procedure to obtain the necessary gearbox parameters is presented along with its mathematical model. A new approach was used to model the stiction and nonlinear gearbox friction behaviour in all four quadrants. A simulation of the actuator with the modelled gearbox is described and its results are compared with the real system measurement. Obtained results show a high match between simulation and experimental results and confirm the correctness of the simulation model.

Keywords: cycloidal gearbox; friction; actuator; servomotor; permanent magnet synchronous machine

Citation: Šlapák, V.; Ivan, J.; Kyslan, K.; Hric, M.; Ďurovský, F.; Paulišin, D.; Kočiško, M. Measurement and Modelling of a Cycloidal Gearbox in Actuator with Permanent Magnet Synchronous Machine. *Machines* **2022**, *10*, 344. https://doi.org/10.3390/machines10050344

Academic Editor: Domenico Mundo

Received: 30 March 2022
Accepted: 5 May 2022
Published: 7 May 2022

1. Introduction

Modern industry is based on automation and robotic technology, creating very complex systems with a large number of drives and mechanical gearings. Therefore, it is desirable to integrate more machine parts into single replaceable units. A great example is the integration of mechanical gears, an electric motor, a position sensor, and, optionally, a brake into a single electric actuator. It is nothing unusual to even combine multiple actuators into a more complex modular positioning unit or to attach more sensors to the actuator or its output shaft for some special applications.

In precise industry machines, such as industrial robots, positioning tables, CNC machines, or some medical devices, a compact solution is needed, as the mounting space is often very limited (i.e., the actuator should be placed directly into the robotic joint). Altogether, this compact solution should deliver high-output torque within a low-speed operation, high positioning accuracy, and high torsional stiffness, and should withstand high tilting torques. Three different gearbox types can be used in such a compact design to meet all of the foregoing requirements: namely, the planetary, harmonic, or cycloidal gearbox.

The planetary gearbox is a mechanically simple, reliable, and compact solution with a low manufacturing price. It is capable of delivering high torques and of withstanding tilting stress. However, compared to the harmonic and cycloidal gearboxes, and assuming a similar gear ratio, it is significantly bulkier and provides much less precision regarding the output position. Moreover, there is a higher backlash, compared to other types of precise gearboxes.

The harmonic gearbox is very simple from the construction point of view, as it consists of three main parts—a wave generator, a flexible spline, and a circular spline. It is very compact and precise; however, due to the technologically demanding manufacturing process of the flexible spline, it is more expensive than the planetary gearbox. However, the planetary gearbox requires an optional additional output bearing to withstand the tilting stress, which takes up some more mounting space and increases the overall price of the gearbox.

Compared to the previous two types of gearboxes, the mechanical construction of the cycloidal gearbox is much more complex, due to the use of an eccentric shaft with cycloidal wheels, needles, transformation elements, bearings, and other mechanical components. This mechanical complexity increases the overall price but delivers a high torsional stiffness, high positioning precision, and, due to the implemented axial and radial bearing system, the gearbox offers high load capacity, which includes axial forces and tilting moment. On the other hand, these properties are accompanied by a highly nonlinear friction characteristic. Regardless of the gearbox type, the sizing of the compact geared actuator and its parts, such as the electric motor or brake, requires knowledge of the gearbox characteristics.

The efficiency and friction curve are fundamental characteristics of the gearbox. Friction has a significant impact on the proper sizing of the electric motor and affects the control of the actuator. The implementation of friction compensation in the control structure improves the dynamics and positioning accuracy of the actuator. Moreover, friction decreases the motor torque needed for dynamic braking; thus, faster braking can be achieved. This is very important in collaborative robotics or automated guided vehicle propulsion. The model of the gearbox, considering the friction characteristic, can be used in the precise output torque control. Various approaches for the modelling of the efficiency and friction of gearboxes can be found in the following literature:

- for planetary gears: [1–9];
- for cycloidal gears: [10–14];
- for harmonic gears: [15,16].

The analytical model of a planetary gearbox is presented in [1], and the relationship between the power transmitted through the gearbox and the speed ratio has been found. The proposed method enables the evaluation of efficiency with conventional programming languages. The optimisation methods to achieve the best possible efficiency of a planetary gearbox were applied in [2], resulting in a mathematical model with eight design variables and eighteen constraint conditions. The experimental results in [3] show that the spin power loss of a planetary gearbox increased with the increasing number of planets, and that it is not viable to change the number of planets to improve the efficiency of the gearbox. The simple Simulink model was presented in [5], with the estimation of viscous friction. The finite element method was applied to investigate the level of stress and deformation of a planetary gearbox in [6]. The aim was to improve the efficiency of the drive; however, an analytical model of the gearbox was not provided.

An efficiency study of a cycloidal gearbox for robotic applications can be found in [10], where it was shown that manufacturing deficiencies increase the stress and losses in particular mechanical spots of the gearbox. An efficiency adjustment of cycloidal drives based on the varying circumference of cycloids was presented in [11]. The developed model predicted a large range of efficiencies across various torque ratios and roller constraints. The quantitative analysis of meshing power losses in the cycloidal gearbox was provided in [12], where the formula for gear meshing efficiency was proposed for a load-dependent type of losses. The improvement in the efficiency for cycloidal gearbox and its comparison with other types of gearboxes was shown in [13]. Various methods of analytical efficiency calculation, based on theoretical mathematical models of cycloidal gearboxes, and the determination of power losses and friction due to contact of gearbox-specific elements, were presented in [14]. The modelling of the harmonic drive was presented in [15]. The resulting model reproduced dynamic responses of the gearbox with acceptable precision. The parameters of the model were determined by an analysis of the harmonic-drive system; however, the model performance was highly sensitive to transmission properties. The novel approach to modelling harmonic drive transmission can be found in [16], where the proposed model captures the nonlinear stiffness and hysteresis behaviour of the harmonic gearbox.

The complexity of the high-precision gearboxes caused by the nonlinear friction characteristic makes it very difficult to analytically derive an accurate model of the gearbox. In addition, it is almost impossible to construct two gearboxes with identical parameters; thus, every high-precision gear has unique properties. Therefore, the direct measurement of the gearbox efficiency by employing a torque transducer provides the most accurate information about the efficiency and performance of a specific gear.

In this paper, efficiency and friction have been studied for a PMSM machine and cycloidal gearbox integrated into a single actuator. Compared to previous studies based on the experimental measurement-based modelling [10,16,17], the model of the gearbox was developed from the measured dataset of speeds and torques. The efficiency of the actuator is calculated for all four quadrants of the torque–speed area, and the nonlinear friction characteristic is modelled by a unique exponential function. The complex gearbox friction model, incorporating the friction characteristic and the stiction, is introduced and experimentally verified.

2. Experimental Setup for the Measurement of Actuator Characteristics

The analysed actuator with the cycloidal gear (denoted as DUT—drive under test; parameters can be found in Appendix A) was measured on a test bench as shown in Figure 1. The DUT was loaded by another electric drive (denoted as LM—loading machine). A planetary gearbox was used on the output of the LM, so the LM did not have to be unnecessarily large and the machine with the lower rated torque could be employed. This reduced the cost of the experimental setup and reduced the space requirements.

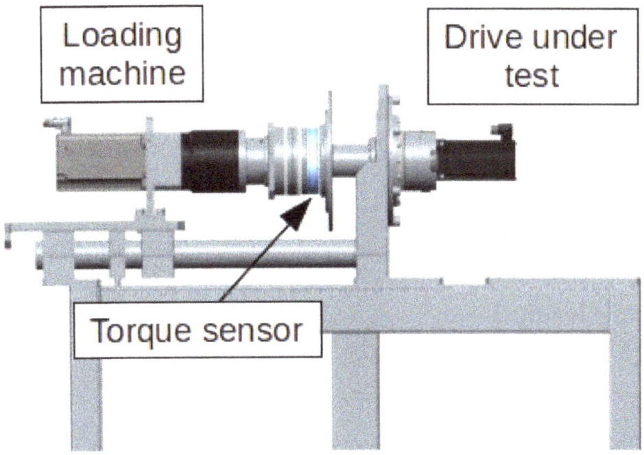

Figure 1. Experimental setup for the measurements and tests.

Both drives, DUT and LM, were fed by industrial SIEMENS Sinamics S120 servo drives with a single SIMOTION control unit. The actual torque on the common output shaft was measured by the torque transducer Magtrol TF213. The measurement range was 500 Nm with 0.1% accuracy. The sensor was mounted directly on the DUT output flange, and on the other side, it was connected via flexible coupling to the LMs planetary gearbox output shaft. The electric output of the sensor was connected to and evaluated by the SIMOTION control system.

The motor torque cannot be measured directly, as the motor is an inseparable part of the compact actuator. Moreover, in the final application, it is not desirable to use a torque sensor either. Therefore, the motor torque was evaluated by the servo drive from the motor current. Nevertheless, considering the use of the mathematical model in the actuator control, it is advantageous to model the relationship between the actuator output torque and the motor torque, which is calculated by the motor control system. Here, it is not important to analyse the efficiency of the motor and gearbox separately; thus, the efficiency of the motor is included in the gearbox efficiency.

To sum up, the measured variables were:

- the actual motor speed of DUT;
- the actual motor torque of DUT (derived from the motor current as the internal value of the servo drive);
- the actual torque on the output shaft.

The measurement process was based on an incremental step-change of load torque from negative to positive values during the constant speed operation of DUT, where each step lasted for 5 s. This step-change of load torque was repeated for various speed values, from −2000 rpm to 2000 rpm, with a 100 rpm step. A similar testing profile was used in [10]. The applied testing profile is shown in Figure 2 with a 100 ms sampling time; the figure displays 1000 s from more then 3 h. Using this approach, all speed–torque quadrants were covered with approximately 50 samples, taken in each operating point.

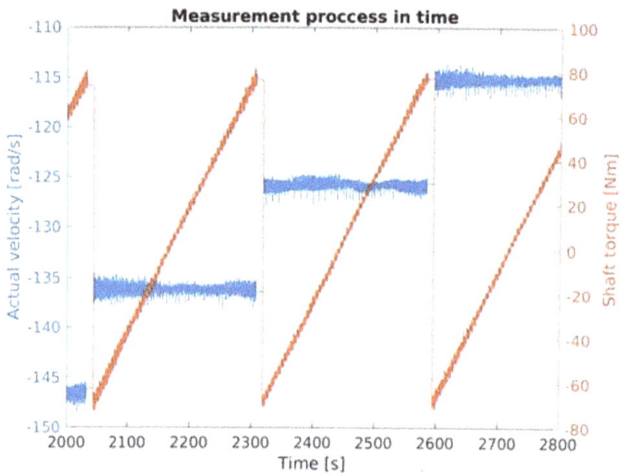

Figure 2. A testing profile for the measurement of the gearbox friction characteristic.

The range of the measurement parameters was determined by the operating range of DUT. The measured data for each torque–speed combination were averaged to obtain a single value of speed and torque for each operating point.

It is important to mention that the friction depends significantly on the actual temperature of the gearbox. However, the actual temperature inside the gearbox is difficult to monitor; thus, only the temperature on the gearbox housing is measured usually. The housing temperature can be, as such, considered to be only an informative value due to the thermal inertia of the gearbox. The actual temperature of the gearbox components inside the housing can vary significantly, compared to the housing temperature. Therefore, the actuator was pre-heated to a steady operating temperature to prevent the temperature influence on the friction measurement. The temperature of the actuator can be considered constant during the measurement.

3. Stiction

Stiction is the phenomenon in which friction limits the movement of the rotor until sufficient torque is applied [18]. After that, the rotor starts to rotate and the friction drops below a static friction value level. Then, with the speed increase, the other friction components rise to the friction characteristic, as described in Sections 4 and 5.

The stiction measurement is performed using a slow torque up–down ramp; the observed values are the motor torque and the rotor speed (Figure 3). The stiction T_{st} is determined as a value of torque in the instant in which the rotor starts its movement. To obtain an accurate value of the stiction, several points were measured and averaged.

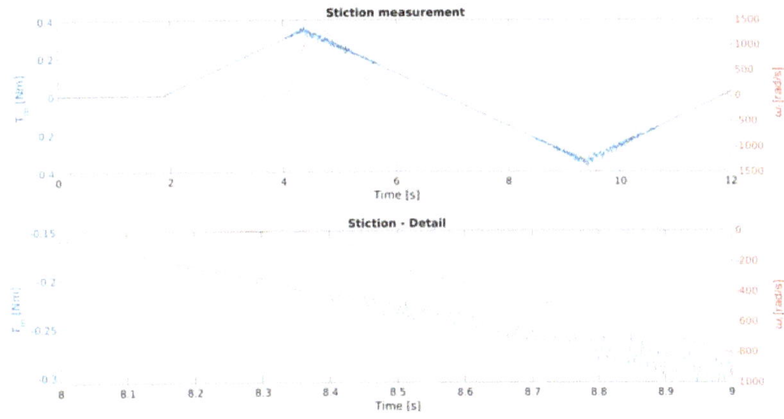

Figure 3. A stiction measurement.

4. Gearbox Efficiency

The gearbox efficiency, in terms of torque conversion, is defined as:

$$T_{out} = T_{in}\, i\, \eta, \tag{1}$$

where T_{out} is the output torque, T_{in} is the gearbox input torque, i is the gearbox ratio, and η is the gearbox efficiency.

Assuming the gearbox operation in steady temperature, the gearbox efficiency depends on the actual speed and torque:

$$\eta = f(T_{in}, \omega_r), \tag{2}$$

where ω_r is the input angular speed.

Using the measured and subsequently averaged data, the actuator efficiency is calculated as follows:

$$\eta = \frac{T_{out}}{i\, T_{in}} \times 100\%. \tag{3}$$

Note that Equation (3) was used only to calculate the efficiency from measured values. In practice, it is advantageous to relate the efficiency to the input torque (as defined in Equation (2) and later in Equation (9)), as it is easier to measure (the information may be read from the servo drive and no additional sensor is needed in the application). Therefore, all measurements results are related to input torque and speed.

A graphical representation of the measured data is in Figure 4. Red dots represent the efficiency in a negative speed direction, and blue dots stand for a positive direction.

The measured efficiency, as is depicted in Figure 4, varies notably in each quadrant. As it was expected, the efficiency is above 100% in generator mode, as Equation (3) has been used for the motor as well as the generator mode.

The efficiency approaches negative or positive infinity for motor torque values close to 0 Nm in all four quadrants. The main reason for this is the static friction (or stiction), as the gearbox itself can hold a certain load without any motor torque or even with low motor torque in the same direction.

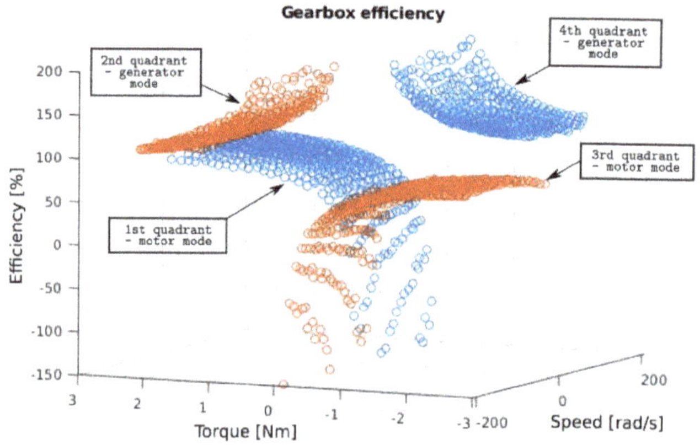

Figure 4. Four-quadrant visualisation of efficiency points of the gearbox, obtained by experimental measurement.

A mathematical description of the measured efficiency is needed to build up a mathematical model for further usage in simulations and system control. However, as can be seen, there are large discontinuities present, i.e., the efficiency during the speed transition from the positive to negative direction within the same load can suddenly increase from 60% to 120%. Therefore, the mathematical approximation of the efficiency should be conducted separately for each quadrant. Nevertheless, using such an approximation in simulation or in drive control will cause undesired peaks during the transients between quadrants, as the efficiency value may suddenly change by more than tens of percents in the worst case. Thus, another approach to model the efficiency with the same dataset will be described in the next section.

5. Friction Characteristic

The main factor affecting the overall mechanical efficiency is the friction in the mechanics, which causes general mechanical losses T_{loss}. These mechanical losses are, in the most simple way, modelled only by viscous friction:

$$T_{loss} = B\omega_r, \tag{4}$$

where B is the coefficient of a viscous friction. However, especially in cycloidal gearboxes, the friction coefficient is not constant. Instead, friction depends on actual angular velocity, load torque, and temperature. When assuming a constant temperature, friction losses can be defined as a nonlinear friction characteristic, as a function of angular velocity and input torque:

$$T_{loss} = \omega_r f_f(T_{in}, \omega_r), \tag{5}$$

where $f_f(T_{in}, \omega_r)$ is the friction characteristic.

Using this approach, the original mechanical equation (Equation (1)) can be expressed as follows:

$$T_{out} = T_{in}\,\eta\, i = (T_{in} - T_{loss})\, i = (T_{in} - \omega_r f_f(T_{in}, \omega_r))\, i. \tag{6}$$

The friction characteristic can be determined, using the same measured data and Equation (6), as:

$$f_{fm}(T_{in}, \omega_r) = \frac{(T_{in} - T_{out}/i)}{\omega_r}. \tag{7}$$

The determined friction characteristic is depicted in Figure 5 from the input (i.e., motor) shaft perspective, where red dots are from the negative speed region (e.g., 2nd and 3rd quadrant) and blue dots belong to the positive speed region (e.g., 1st and 4th quadrant).

As can be seen, the friction characteristic does not contain large discontinuities, as was the case in Figure 4 for efficiency. Thus, it is sufficient to divide characteristics into two regions only—one each for positive and negative speed. This is needed for further data processing and mathematical approximations.

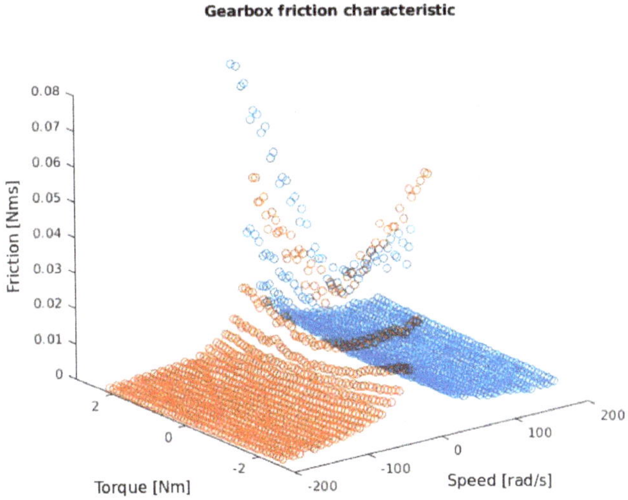

Figure 5. Four-quadrant visualisation of friction points of the gearbox, obtained by experimental measurement (input shaft perspective).

At first, the measured friction was approximated for the points with the same load by the following exponential function:

$$f_f(T_{in} = const, \omega_r) = a\ \exp(b\ \omega_r) + c\ \exp(d\ \omega_r), \quad (8)$$

Next, function parameters a, b, c, d for every torque value were compared and plotted as a dependency on the torque. This way, the possible approximation of each parameter could be observable. It was apparently visible that parameter a showed a quadrature dependency in relation to input torque, that parameter b showed a linear dependency, and that parameters c, d were almost constant over every torque value. Therefore, they were fitted using the MATLAB fit function. As a result, the following form of the approximation function was found:

$$f_{fj}(T_{in}, \omega_r) = (p_{j1}\ T_{in}^2 + p_{j2}\ T_{in} + p_{j3})\ \exp((p_{j4}\ T_{in} + p_{j5})\omega_r) + p_{j6}\ \exp(p_{j7}\omega_r), \quad (9)$$

where index $j = 1$ stands for the positive speed interval, index $j = 2$ stands for the negative speed interval, and p_{jx} are the coefficients of substitute functions, as described above and listed in Appendix A, Table A3. The graphical representation of the friction characteristic approximation is shown in Figure 6, where red circles represent measured values.

The function given by Equation (9) approximates the measured values, with the R-square value equal to 0.988 in this case. The equation used in the R-square evaluation is presented in Equation (10) with the sampled data and the calculated friction function for the same input data.

$$R^2 = 1 - \frac{\sum_{k=1}^{n}(f_{fm}(T_{ink}, \omega_{rk}) - f_f(T_{ink}, \omega_{rk}))^2}{\sum_{k=1}^{n}(f_{fm}(T_{ink}, \omega_{rk}) - \bar{f}_{fm})^2}, \quad (10)$$

where n is the number of measured values and $\bar{f}_{fm}$ is the average value of the measured friction.

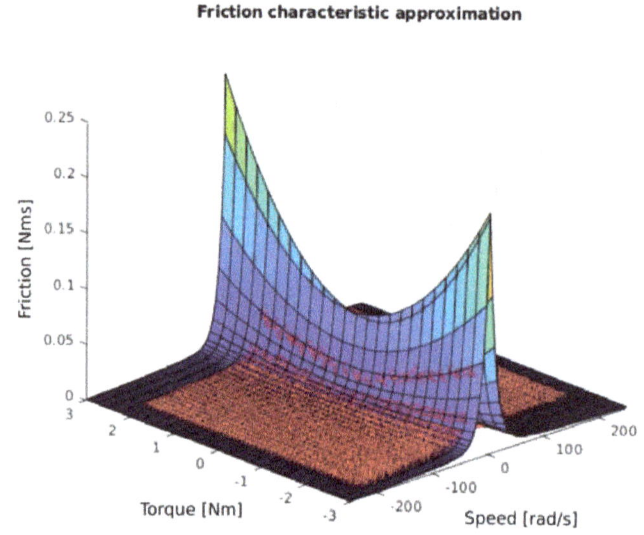

Figure 6. A comparison of the measured and approximated friction characteristics of the actuator.

6. Simulation Model of the Servoactuator

The simulation model of the measured servoactuator consists of the electrical and the mechanical part, including the gearbox equations. The electrical part of the servoactuator is described by the following model in dq-coordinates:

$$\frac{\mathrm{d}i_d}{\mathrm{d}t} = \frac{1}{L_d}(-Ri_d + pL_q i_q \omega_r + u_d), \tag{11}$$

$$\frac{\mathrm{d}i_q}{\mathrm{d}t} = \frac{1}{L_q}(-Ri_q - pL_d i_d \omega_r - \frac{2k_t}{3}\omega_r + u_q), \tag{12}$$

$$T_e = k_t i_q, \tag{13}$$

where i_d and i_q are direct (field-generating) and quadrature (torque-generating) currents, R is the phase resistance, L_d and L_q are direct and quadrature inductances, k_t is the motor torque constant, p is the number of pole pairs, and u_d and u_q are direct and quadrature voltages, respectively.

The mechanical part with the gearbox is described as follows:

$$\frac{\mathrm{d}\omega_r}{\mathrm{d}t} = \frac{1}{J}(T_e - T_{loss} - \frac{T_L}{i}), \tag{14}$$

$$T_{loss} = \begin{cases} \operatorname{sgn} T_{st}, & \text{if } |f_f(T_{in}, \omega_r)| < T_{st} \\ \omega_r \, f_f(T_{in}, \omega_r), & \textit{otherwise} \end{cases} \tag{15}$$

$$\omega_r = \begin{cases} 0, & \text{if } |T_e| < T_{st} \text{ and } |\omega_r| < \omega_{st} \\ \int \mathrm{d}\omega_r \, \mathrm{d}t, & \textit{otherwise} \end{cases} \tag{16}$$

$$\omega_{out} = \frac{\omega_r}{i}, \tag{17}$$

where J is the system inertia calculated to the input side of the gearbox (e.g., motor side), T_L is the load torque, T_{st} is the stiction, ω_{st} is the stiction breakaway speed at which the stiction affects the motor, and ω_{out} is the output angular velocity. The stiction breakaway speed has a small value, selected by the trial-and-error method, to avoid current chattering at standstill.

Here, Equations (15) and (16) are divided into two cases to model the stiction effect, which is present when accelerating from zero speed and during speed direction transition. Firstly, the input torque T_e must prevail over the stiction to start the movement. Then, when the rotor starts to rotate, friction function $f_f(T_{in}, \omega_r)$ takes place.

The mathematical model was simulated with the parameters listed in Appendix A, and the simulation results are shown in Figure 7. The simulation was performed for the complex gearbox model as well as for the simple friction model, according to Equation (4).

The movement starts at the time $t = 1$ s by the torque-generating current peak, followed by a little drop. This is caused by the stiction described in Section 3. Next, the torque-generating current rises according to the acceleration and gearbox friction characteristics.

In transition, when the speed changes its direction (at the time $t = 8$ s), the stiction effect is observable again. This causes a short steady-state phase with the zero speed, until the torque rises to a higher value, along with the stiction. The simple friction model given by Equation (4) significantly differs in transition, which can be observed in the waveform of the torque-generating current and the speed direction change.

At the end of the movement, in standstill operation with zero speed, the torque-generating current remains at a level which is not sufficient to move the rotor. This phenomenon is a consequence of the stiction, as the speed drops to zero at the moment when the motor torque drops under stiction.

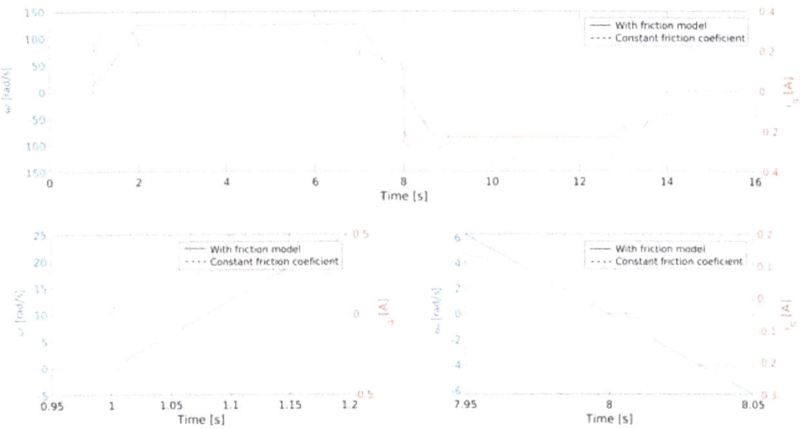

Figure 7. A simulation of a simple movement of the actuator without load, where the angular velocity of the motor is blue and the torque-generating current is red. Continuous lines represent the simulation with the presented friction model, whereas dashed lines represent the simulation with the basic friction modelling. The upper plot shows the whole movement, the bottom-left plot shows the movement start in detail, and the bottom-right plot shows the stiction effect during speed transition.

7. Experimental Results

The same movement as in Figure 7 was performed by the actuator on the test bench. Results are compared with the simulation in Figure 8. Again, the torque-generating current and the angular velocity of the motor was observed. No load was applied during the experiment.

It can be concluded that simulated waveforms match the responses of the real system. The differences in the transients are caused by inequalities of the characteristics over the mechanical revolution of the gearbox, as all mechanical parts are manufactured with some mechanical tolerance, which affects the mutual mechanical contact of internal gearbox components.

The model of stiction, causing current peaks, a small steady phase during speed direction change, and a non-zero current at standstill, are reproduced fairly. This can be observed in the detailed waveforms provided by Figure 9.

A comparison of the mathematical model and the real measurement for another speed value is shown in Figure 10. Here, the results are very close to each other as well.

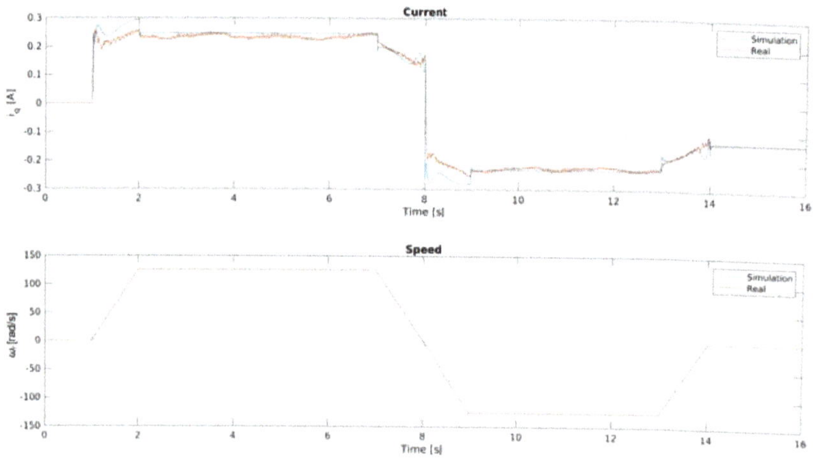

Figure 8. A comparison of the simulation and the real measurement of the actuator movement without load. The upper plot shows the waveforms of torque-generating currents; the bottom plot shows the angular velocity.

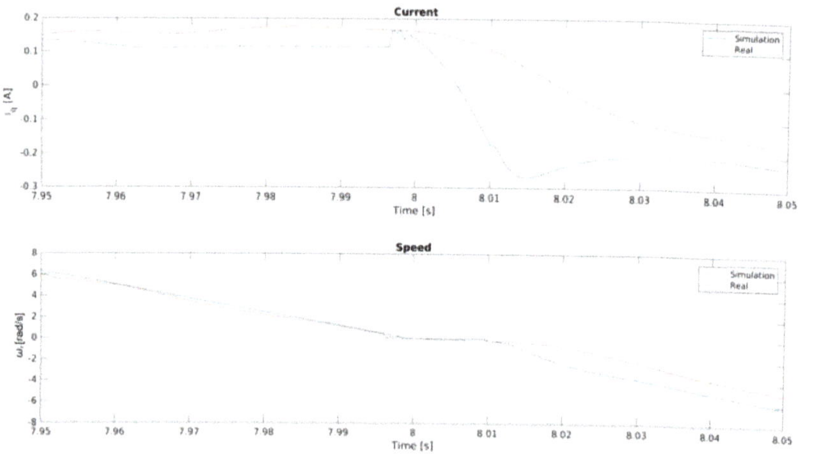

Figure 9. A detail of the transition from positive to negative speed during actuator movement.

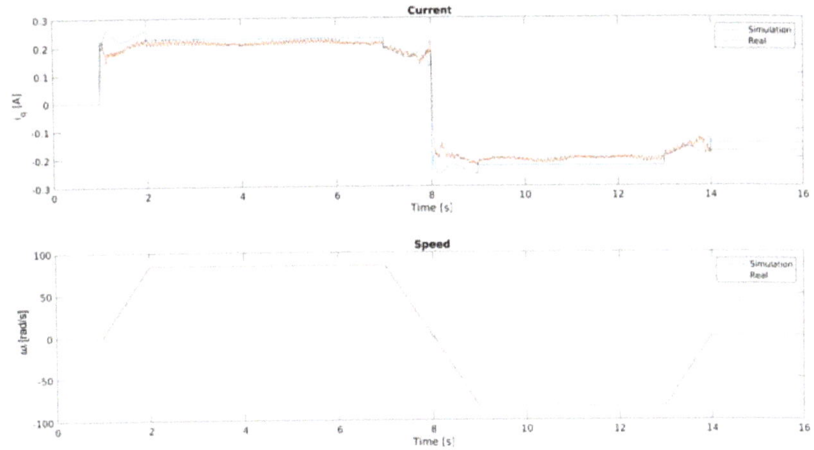

Figure 10. A comparison of the simulation and the real measurement of the actuator movement without load, with 800 rpm as the speed setpoint.

8. Conclusions

The presented mathematical model of the gearbox was focused on modelling the friction in the cycloidal gearbox, attached to the permanent magnet synchronous motor in the compact actuator. The friction model is based on the new approach, considering the stiction phenomena and the nonlinear friction characteristic.

The mathematical model was verified by a comparison of the simulation with the real system—an actuator with a cycloidal gearbox. The results confirm that the designed mathematical model of gearbox has a high level of accuracy.

The model can be used in actuator or motor control to compensate for the friction, which can further improve the dynamic behaviour of the controlled system. Moreover, the model shows various aspects (e.g., current peaks during the motor speed-up or non-zero current at standstill) which have to be considered in the actuator and the whole system design. Finally, the model finds usage in the various model-based control techniques as well.

Although the accuracy of the model in the terms of friction is very high, the model does not cover the torsional stiffness and backlash, which are important for modelling positioning accuracy. However, the presented model can be used as the basis or as an input part of a more complex mechanical model of the cycloidal gearbox.

The disadvantage of the presented mathematical model is the fact that it is based on experimental data; thus, there is no rule for calculating parameters in the friction model. However, such a rule would be hard to establish. The main reason for this is that there can be significant differences between every gearbox produced. In reality, each cycloidal gearbox has its unique performance, thanks to the mechanical complexity of the gearbox and a lot of parameters affecting the gearbox performance during its production. Therefore, the authors suppose, to obtain general model parameters, a statistically significant amount of measurement and the determination of the parameters for a given product line should be performed and statistically evaluated.

Author Contributions: Conceptualisation, V.Š. and J.I.; methodology, V.Š.; software, J.I.; validation, M.H., D.P., and M.K.; formal analysis, K.K.; investigation, V.Š. and J.I.; resources, J.I. and K.K.; data curation, J.I.; writing—original draft preparation, V.Š.; writing—review and editing, K.K.; visualisation, J.I.; supervision, M.H. and F.Ď.; project administration, F.Ď. and M.K.; funding acquisition, M.H. and D.P. All authors have read and agreed to the published version of the manuscript.

Funding: This work was supported by the Slovak Research and Development Agency, under contract no. APVV-16-0355. This work was also supported by the Scientific Grant Agency of the Ministry of Education of the Slovak Republic, under the project VEGA 1/0493/19.

Data Availability Statement: The data presented in this study are available on request from the corresponding author.

Conflicts of Interest: The authors declare no conflict of interest.

Appendix A

The parameters of the measured actuator are summarised in Table A1. The controllers' settings are given in Table A2. The parameters determined from measurements and used in the simulation are summarised in Tables A3 and A4.

Table A1. Parameters of the measured actuator.

Parameter	Symbol	Value	Unit
Rated voltage	U_{dc}	560	V
Rated current	I_n	2.81	A
Phase resistance	R	2.61	Ω
Direct inductance	L_d	18.5	mH
Quadrature inductance	L_q	18.5	mH
Torque constant	k_t	1.25	NmA^{-1}
Pole pairs	p	5	–
Gearbox ratio [1]	i	−32	–

[1] The measured gearbox inverse of the direction of the rotation, which is represented by the negative sign before the gear ratio.

Table A2. Controllers' settings.

Parameter	Symbol	Value	Unit
Current loop—Proportional gain	K_i	30	VA^{-1}
Current loop—Integral reset time	T_i	1.1	ms
Speed loop—Proportional gain	K_ω	0.15	$Nms.rad^{-1}$
Speed loop—Integral reset time	T_ω	25	ms

Table A3. Friction characteristic model parameters.

Parameter	Value
p_{11}	0.017
p_{12}	0.0065
p_{13}	0.0704
p_{14}	−0.0038
p_{15}	−0.0968
p_{16}	0.0085
p_{17}	−0.0078
p_{21}	0.014
p_{22}	−0.0035
p_{23}	0.0713
p_{24}	0
p_{25}	0.1004
p_{26}	0.0094
p_{27}	0.0092

Table A4. Stiction model parameters.

Parameter	Symbol	Value	Unit
Stiction breakaway speed	ω_{st}	0.5	s^{-1}
Stiction	T_{st}	0.2	Nm

References

1. Castillo, J.M.D. The analytical expression of the efficiency of planetary gear trains. *Mech. Mach. Theory* **2002**, *37*, 197–214. [CrossRef]
2. Qimin, X.; Qili, X. Study on Optimal Design of Planetary Gear Reducer Based on Particle Swarm Algorithm and Matlab. In Proceedings of the 2010 Sixth International Conference on Semantics, Knowledge and Grids, Beijing, China, 1–3 November 2010; pp. 391–394. [CrossRef]
3. Talbot, D.C.; Kahraman, A.; Singh, A. An Experimental Investigation of the Efficiency of Planetary Gear Sets. *J. Mech. Des.* **2012**, *134*, 021003. [CrossRef]
4. Yao, W.S.; Lin, C.Y. Design of Active Continuous Variable Transmission Control System with Planetary Gear. *Electronics* **2022**, *11*, 986. [CrossRef]
5. Kecskés, I.; Burkus, E.; Odry, P. Gear efficiency modeling in a simulation model of a DC gearmotor. In Proceedings of the 2018 IEEE 18th International Symposium on Computational Intelligence and Informatics (CINTI), Budapest, Hungary, 21–22 November 2018; pp. 000065–000070. [CrossRef]
6. Mohsine, A.; Boudi, E.M.; El Marjani, A. Modeling and structural analysis of planetary gear of a wind turbine. In Proceedings of the 2016 International Renewable and Sustainable Energy Conference (IRSEC), Marrakech, Morocco, 14–17 November 2016; pp. 462–466. [CrossRef]
7. Ngandu Kalala, G.; Chiementin, X.; Rasolofondraibe, L.; Boujelben, A.; Kilundu, B. Modeling Impulsive Ball Mill Forces Effects on the Dynamic Behavior of a Single-Stage Gearbox. *Machines* **2022**, *10*, 226. [CrossRef]
8. Pennestri, E.; Freudenstein, F. The Mechanical Efficiency of Epicyclic Gear Trains. *J. Mech. Des.* **1993**, *115*, 645–651. [CrossRef]
9. Pennestri, E.; Valentini, P.P. A Review of Formulas for the Mechanical Efficiency Analysis of Two Degrees-of-Freedom Epicyclic Gear Trains. *J. Mech. Des.* **2003**, *125*, 602–608. [CrossRef]
10. Farrell, L.C.; Holley, J.; Bluethmann, W.; O'Malley, M.K. Cycloidal Geartrain In-Use Efficiency Study. In Proceedings of the 42nd Mechanisms and Robotics Conference, Quebec City, QC, USA, 26–29 August 2018. [CrossRef]
11. Sensinger, J.W. Efficiency of High-Sensitivity Gear Trains, Such as Cycloid Drives. *J. Mech. Des.* **2013**, *135*, 071006. [CrossRef]
12. Wang, R.; Gao, F.; Lu, M.; Liu, T. Meshing Efficiency Analysis of Modified Cycloidal Gear Used in the RV Reducer. *Tribol. Trans.* **2019**, *62*, 337–349. [CrossRef]
13. Lee, K.; Hong, S.; Oh, J.H. Development of a Lightweight and High-efficiency Compact Cycloidal Reducer for Legged Robots. *Int. J. Precis. Eng. Manuf.* **2019**, *21*, 415–425. [CrossRef]
14. Matejic, M.; Blagojevic, M.; Cofaru, I.I.; Kostic, N.; Petrovic, N.; Marjanovic, N. Determining efficiency of cycloid reducers using different calculation methods. In *MATEC Web of Conferences, Proceedings of the 9th International Conference on Manufacturing Science and Education—MSE 2019 "Trends in New Industrial Revolution", Sibiu, Romania, 5–7 June 2019*; EDP Sciences: Ulys, France, 2019; Volume 290. [CrossRef]
15. Tuttle, T.; Seering, W. A nonlinear model of a harmonic drive gear transmission. *IEEE Trans. Robot. Autom.* **1996**, *12*, 368–374. [CrossRef]
16. Zhang, H.; Ahmad, S.; Liu, G. Modeling of Torsional Compliance and Hysteresis Behaviors in Harmonic Drives. *IEEE/ASME Trans. Mechatron.* **2015**, *20*, 178–185. [CrossRef]
17. Wang, C.; Wang, S.R.; Wang, G.Q. A method for calculating gear meshing efficiency by measured data from gear test machine. *Measurement* **2018**, *119*, 97–101. [CrossRef]
18. Lee, D.H.; Ahn, J.W. Dual Speed Control Scheme of Servo Drive System for a Nonlinear Friction Compensation. *IEEE Trans. Power Electron.* **2008**, *23*, 959–965. [CrossRef]

machines MDPI

Article

Design of Electromagnetic Control of the Needle Gripping Mechanism

Jiří Komárek * and Vojtěch Klogner

Department of Textile Machine Design, Faculty of Mechanical Engineering, Technical University of Liberec, 461 17 Liberec, Czech Republic; vojtech.klogner12@gmail.com
* Correspondence: jiri.komarek@tul.cz; Tel.: +420-485-353-418

Abstract: This paper deals with the modification of the mechanical system of the needle bar. The purpose of this work is to reduce the vibration and noise of the sewing machine for creating a decorative stitch. A special floating needle is used to sew this stitch, in which two mechanical systems of needle bars handover through the sewn material, so that a perfect imitation of a hand stitch is created. The original system, which controls the release of the needle at the handover location by abruptly stopping the needle bar control element, could be replaced by a new system that uses magnetic force to release the needle. In addition to the usual design procedure, numerical simulations of the attractive force of the electromagnet are also used in the design of a suitable electromagnet. At the same time, an electrical circuit is also designed to allow the needle to be released and gripped quickly. The advantages of the new system lie not only in reducing vibrations and the associated increase in the operation speed of the machine, but also in making it easier for the machine to switch to possible automated or semi-automated production.

Keywords: sewing machine; needle bar; floating needle; electromagnet; electromagnetic simulation; noise reduction

Citation: Komárek, J.; Klogner, V. Design of Electromagnetic Control of the Needle Gripping Mechanism. *Machines* **2022**, *10*, 309. https://doi.org/10.3390/machines10050309

Academic Editor: Med Amine Laribi

Received: 31 March 2022
Accepted: 23 April 2022
Published: 26 April 2022

1. Introduction

The subject of the research is a special sewing machine, which is used to sew a decorative stitch. To do this, it uses a so-called floating needle, which has a tip on both sides and an eye in the middle. This needle passes through the sewn material back and forth with each stitch, thus imitating the hand stitch very well. In this process, two mechanical systems of needle bar (hereinafter also referred to as the needle bar), located one above and the other below the machine's worktop, transmit the needle. The release of the needle during its transfer is controlled on the needle bar by a mechanical stop. Excessive vibration and noise occur when the needle bar control element hits the stop [1]. Modifications to the existing mechanism to reduce noise have been discussed before and the results have been published in [2,3].

The principle of imitation of a hand stitch, which is used by the investigated sewing machine, has been known for a long time. The first mention of a needle with spikes at both ends appeared in 1755. At the time, it was a proposal by the German inventor Charles F. Weisenthal, who wanted to replace low-productivity hand sewing with machine sewing [4,5]. Long after the discovery of the sewing machine with a chain stitch, and later also with a lock stitch [6], Weisenthal's needle again found application in hand-stitch imitation machines. The principle of making a hand stitch was invented by Jessie Langsdorf in 1936. His patent, issued in 1937, was used by the Naftali brothers in collaboration with the AMF (American Machine and Foundry Company, Brooklyn, New York, NY, USA) in the manufacture of tie-making machines [7,8]. In the following years, the system was gradually improved and found its use in decorative sewing machines and button sewing machines. Since then, the sewing machine has undergone several modifications, but this system is still unsatisfactory.

Modern sewing machines must meet many requirements, such as quiet operation, minimal vibration [9], long life of the mechanisms used and easy operation of the machine. Great emphasis is also placed on reducing the time of the sewing process, which has a direct effect on increasing sewing productivity [10]. New work is also based on these assumptions, which deals with the design of a completely new system that does not require a mechanical bond to release the needle. This will allow the noise generated by the release of the needle on the original system to be completely eliminated. The motivation for this research is, on the one hand, the great pressure of the consumer industry to reduce production costs and, on the other hand, the need to protect the working operator of the sewing machine from higher noise levels [11–13].

The mechanical system of the needle bar performs a rectilinear reciprocating movement, which is realized on the original machine by a cam mechanism. According to previous studies, it has been found that replacing an existing needle bar drive with a controlled servo drive will also significantly affect vibration reduction, including meeting the assumption that the operating speed of the machine will increase. The design of a new needle attachment system should also contribute to this. In addition, replacing the original needle bar drive with a mechatronic system should facilitate the transition to automated production [14–17]. Thus, there is a noticeable support for automation in the field of the clothing industry, where there is an effort to at least partially automate the process of manufacturing clothing. It is quite certain that if the sewing process is to remain reliable after increasing the operating speed, the mechanical system of the needle bar needs to be adjusted.

2. The Current Method of Gripping the Needle

The section of the mechanical system of the needle bar is shown in Figure 1. The needle bar performs a rectilinear reciprocating motion. This is ensured by the carrier 12, which is connected to the needle bar shell 4 and drives the needle bar according to the required stroke, which is generated by the cam mechanism on the original sewing machine.

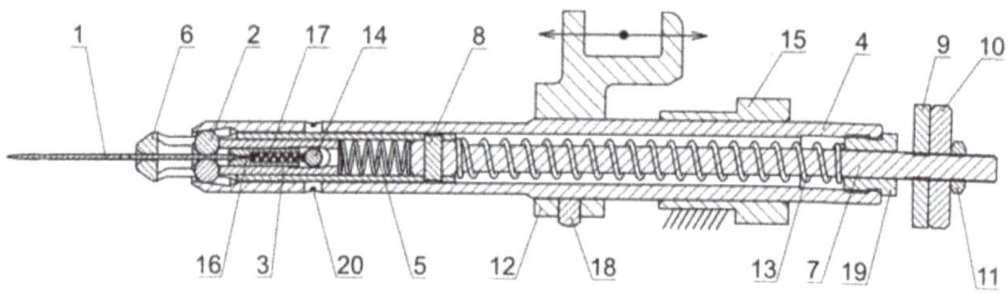

Figure 1. The current method of gripping the needle—gripping the needle.

The floating needle 1 is held in the needle bar by gripping system 2. The system consists of a pair of balls which are housed in a roller 3. Figure 1 shows the mechanical system of the needle bar holding the needle, where the balls are pressed against the conical surface by means of springs 5, and 13, which causes them to be pinched.

The release of the needle is controlled by the impact of the control element (which consists of parts 6, 7, 8, 9, 10, and 11) on the stopper 15, located on the machine frame. The release occurs during the needle bar movement to the needle handover location. This condition is shown in Figure 2.

The original mechanical system of the needle bar is designed for lower speed, and therefore at higher speed the inner parts of the needle bar oscillate, which is described in detail in [18,19].

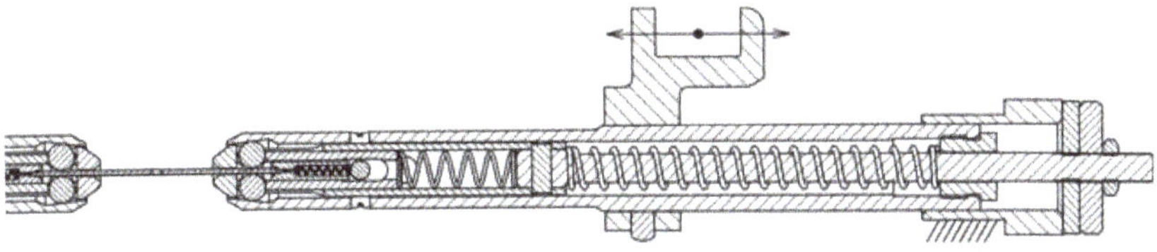

Figure 2. The current method of gripping the needle—releasing the needle.

3. Design of a New System

3.1. Requirements for a New Mechanism

During the sewing process, forces act on the needle which affect the gripping part of the needle bar. A measuring aid was designed to determine these forces. It was a modified stitch plate, which was equipped with strain gauges. The measurement of the forces acting on the needle during sewing on the investigated sewing machine is described in detail in [20]. This measurement determined the forces with which the needle should be held during the sewing process. The needle must be held with a minimum force of 25 N when piercing. The force must be at least 15 N during drawing through the sewn material. The needle must be flexible during transfer. The weight of the needle bar should not increase significantly so as not to negatively affect the required torque of the drive motor.

One of the solutions to simplify the mechanical system and at the same time eliminate the problematic impact of the control element into the stop is to use the magnetic force to attach the needle using an electromagnet [21]. The general advantage of electromagnetic control is that it allows the power supply parameters to be changed independently, regardless of the other mechanisms of the sewing machine, because they are not connected by any mechanical coupling. Another advantage is that the electromagnets can be overloaded for a short time, which allows their more efficient use. A relatively small stroke, approximately 2 mm, is required to operate the existing needle gripping system. For such small strokes, the use of electromagnets is a suitable choice due to their attractive force characteristics [22].

There are a number of arrangements, or constructions, of the electromagnet. The most basic electromagnetic arrangement is the front electromagnet, which is shown in Figure 3. It excels in high strength in the attracted state. However, as the air gap increases, its strength decreases very sharply.

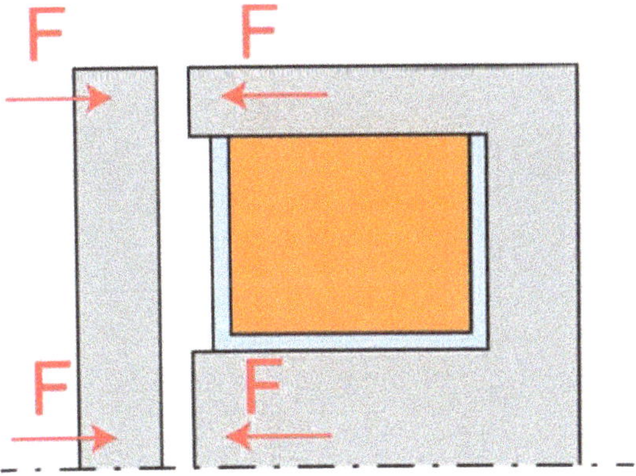

Figure 3. Front electromagnet.

Another variant is the so-called solenoid, where the armature is pulled by an electromagnet. This solution is characterized by a more gradual decrease of the attractive force. It is therefore suitable for applications in which it is necessary to apply force at different anchor positions. The electromagnetic solenoid can exist in two variants, pushing and pulling, according to the direction of the derived arm force. The pushing arrangement shown in Figure 4 has a flatter force characteristic, i.e., in the position of the zero air gap, it has a smaller attractive force, and on the contrary, in the position of the maximum air gap, it has a greater attractive force. The pulling arrangement is shown in Figure 5.

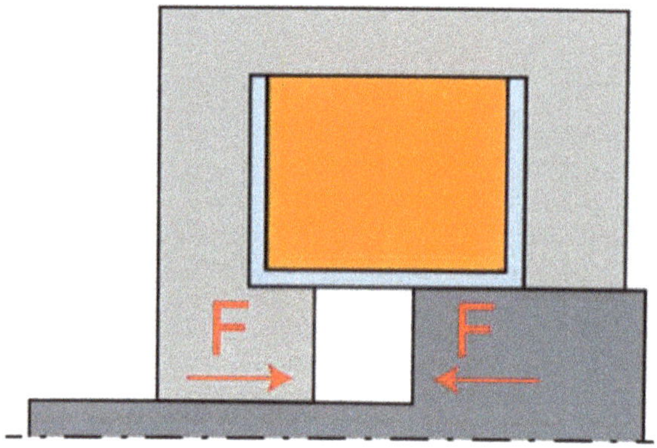

Figure 4. Push solenoid.

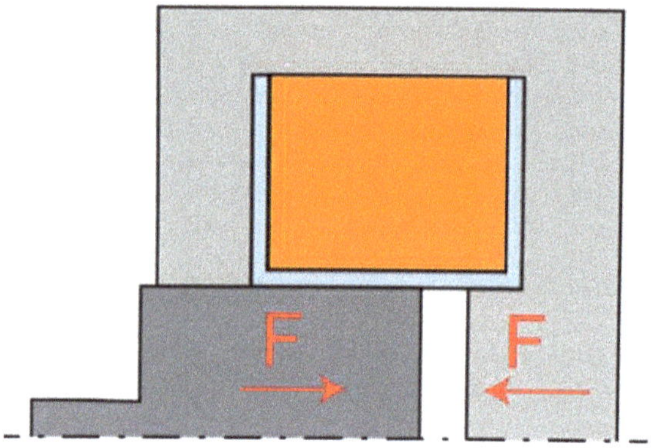

Figure 5. Pull solenoid.

In general, the magnetic force depends on the perpendicular cross-sections of the surfaces that make up the magnetic circuit, i.e., the larger diameter armature has a greater attractive force than the smaller diameter armature while maintaining the coil parameters.

Another possibility is to use the needle itself as the armature of the electromagnet and, unlike the previous two variants, to omit the ball gripping mechanism, as shown in Figure 6. However, due to the very small needle diameter, approximately 1 mm, the attractive force would be very small. This would not be a problem when piercing, but without a self-holding ball gripping mechanism, the needle could be easily pulled out of the gripper. This variant would therefore only be suitable for sewing delicate fabrics where there are no large frictional forces during drawing.

Two solenoid layout options allow us to design two variants of the mechanism. In the first, when the electromagnet is not energized, the spring closes the mechanism and keeps the needle gripped. In the second, when the electromagnet is not energized, the spring opens the mechanism. From the point of view of the functionality of the mechanism, it would be more appropriate to use the variant where the needle remains gripped and does not fall out spontaneously when the power supply is disconnected. However, it is necessary to perform a force analysis to select a more suitable variant for the dynamic characteristics of the needle bar.

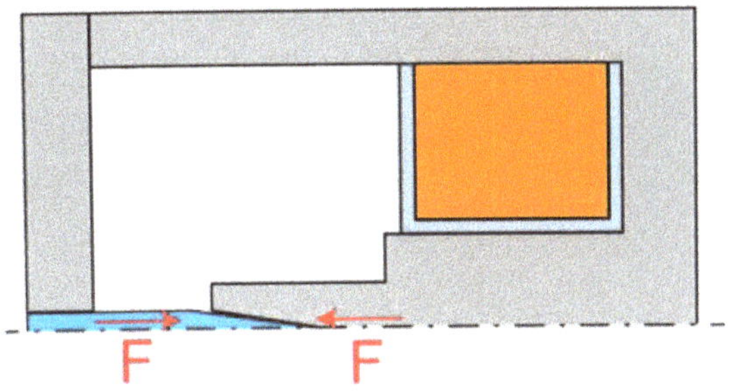

Figure 6. The needle itself as the armature of the electromagnet.

For an initial consideration of electromagnet design, we can specify which forces act on the control element. Inasmuch as we require the control element to move at the same speed and acceleration as the needle bar shell most of the time, we can define the required force that will act on the control element. Because the acceleration of the needle bar is precisely defined by the stroke dependence, we can determine the course of this inertial force relatively accurately.

Another force present is the force exerted by the spring, which returns the mechanism to its initial position. We assume that the spring has a linear stiffness characteristic. In the graph in Figure 7, it should have a general force function in the transition area, but for simplicity it is shown as linear with respect to the rotation of the control cam [23]. The last significant force is the attractive force of the solenoid. The force characteristic of the solenoid is hyperbolic depending on the actual size of the working air gap.

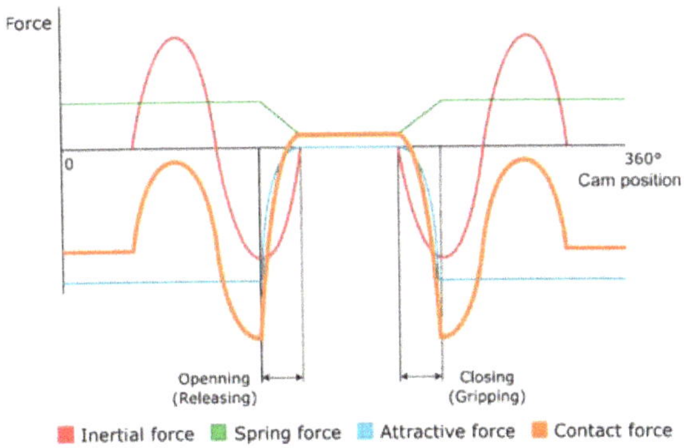

Figure 7. The course of forces on the mechanism with the push solenoid.

We can also show the contact force between the control element and the needle bar shell. If we want the control element to be in a clearly defined position, the contact force must be negative for the closed state and positive for the open state. Contact force can be zero with no contact. During needle release, this force determines the acceleration of the control element relative to the needle bar shell. This phenomenon is very important as it determines the time for which the needle will be released or attached. When analysing the force characteristic, the time at which the gripping mechanism is closed or opened is important. For the purpose of simplification, in this case, this time is chosen in the sinusoidal acceleration amplitude. If we plot all the forces in the graph in Figure 8 as a function of the cam rotation, we can observe several important findings.

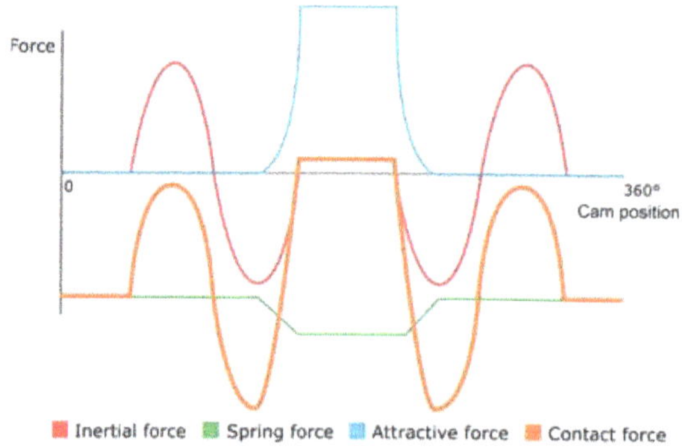

Figure 8. The course of forces on the mechanism with the pull solenoid.

In the case of a solenoid in the pushing arrangement, its applied force in the attracted state must be greater than the sum of the force exerted by the spring and the amplitude of the inertial force so as not to disengage the control element from the needle bar shell. In other words, the total force acting on the control element must be less than zero in the interval when the required needle gripping and greater than zero in the interval when the needle is released.

If we compare the variant of the push and pull solenoid, we find that there are a number of differences. The fundamental difference is that the pull solenoid only exerts force when the mechanism is in the open state. This means that in order to compensate for the accelerating force when moving the needle bar, there is only a spring, which must therefore exert several times more force than in the case of the push solenoid. Assuming that the working stroke is the same for both springs, this spring would certainly have larger installation dimensions and greater weight. The second problem is that the pull solenoid has the least force at the largest air gap, that is, at the exact moment when we need to open the gripper quickly.

For the following reasons, the variant with the push solenoid was chosen for the solution. The acceleration of the needle bar only occurs when the needle is gripped, i.e., at the moment when the push solenoid exerts the greatest force. In addition, due to the shape of the armature, the push solenoid exerts more force than the pull solenoid at the moment of the largest air gap. Nevertheless, there is one disadvantage. When the power supply is disconnected, the needle is released by the return spring.

Figure 9 schematically shows the new needle bar design. The needle bar shell 1 must be made of several pieces in order to be able to assemble the mechanism. The needle 2 is held by balls 3 housed in a roller 4. The roller is controlled by a part 9, which is screwed onto the armature of the electromagnet 11. The armature is returned to its initial position by a spring 8. The electromagnet itself consists of a coil 13 and a frame 12. The movement of the armature is damped by a rubber pad 10. To seat the needle, there is a roller 5, which is returned by the spring 6 to the basic position. The spring is supported on a pin 7, which is fixed in the needle bar shell.

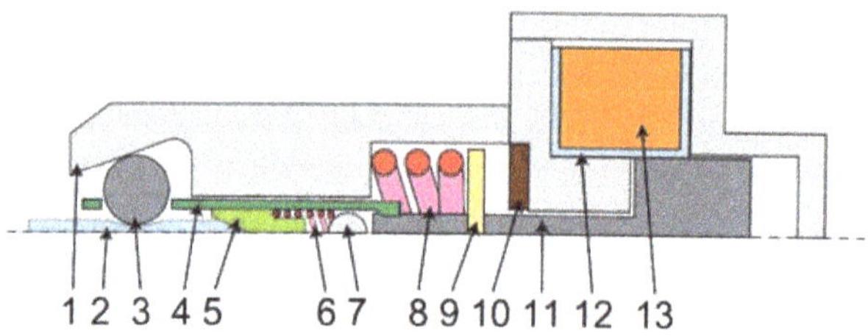

Figure 9. Design of the new mechanical system of the needle bar.

3.2. Winding Dimensioning

The design of the electromagnet winding is a relatively complex issue that solves the magnetic fluxes around the coil, the heat dissipation from the coil space to the surroundings also plays a big role here. The usual procedure in the design of electromagnets is the use of empirical relationships, which are based on simplifying assumptions, especially discretization of a continuous problem into several material blocks. Here, the methodology of Professor Cigánek is used [24].

The calculation is based on the empirical equation of the attractive force F of the electromagnet:

$$F = \pi 10^5 B^2 d^2 \varepsilon^2 (1 + \nu) \tag{1}$$

including the induction in the centre of the coil B, the armature diameter d, the magnetic flux deflection factor ε and the flux increase factor ν.

For an approximate calculation of the armature diameter, we can use this equation:

$$d = 0.115 \sqrt[5]{\frac{F}{1+\nu}\delta^2} \tag{2}$$

in which only the required attractive force F occurs at the required air gap δ.

For a more accurate calculation, it is then necessary to use this equation:

$$d = \sqrt[5]{\frac{20.2 \times 10^5 \rho k_s^2 (1+\beta) F \delta^2}{\xi \beta \varepsilon^2 \lambda (1+2\beta)(1+\nu)\alpha \Delta\vartheta_m}} \tag{3}$$

in which there are other auxiliary factors and selected parameters. Specifically, the ratio of winding thickness to armature diameter β, length ratio λ to armature diameter, coil winding factor ξ (i.e., how much geometric space of the coil is occupied by the copper itself), conductor resistivity at operating temperature ρ, magnetic circuit saturation factor k_s, heat conduction coefficient α, and finally the temperature difference between the operating temperature and the environment ϑ_m.

Figure 10 shows the course of the coil weight (green-yellow area) and the armature weight (purple area) depending on the parameters β and λ. It is possible to see the opposite trends of both dependencies, so it is necessary to select values that will compromise the requirements for the weight of the armature and coil.

The calculation of the factors ν and ε can be performed using the following relations, although these relations contain an average d, which is currently unknown. It is either possible to use an approximate relation to calculate the mean d, or to calculate these three equations iteratively until the deviation is acceptably small. In practice, the average d is chosen as an integer for production reasons, so accuracy does not matter much at this stage.

$$\nu = \frac{2\delta^2}{d^3 \varepsilon^2 \ln(1+2\beta)} \tag{4}$$

$$\varepsilon = 1 + \frac{\delta}{d} - \left(\frac{\delta}{d}\right)^2 \tag{5}$$

This relation applies to the cross section of conductor S:

$$S = \frac{4480}{U\varepsilon} k_s (1+0.75) \rho \delta \sqrt{\frac{F}{1+\nu}} \tag{6}$$

The number of turns of coil N is calculated as follows:

$$N = \frac{\xi}{S} \beta \lambda d^2 \tag{7}$$

The approximate value of the induction in the centre of the coil B can be expressed by the relation:

$$B = \sqrt{\frac{F}{\pi 10^5 d^2 \varepsilon^2 (1+\nu)}} \tag{8}$$

The last equation needed for the calculation is the steady state equation of the flowing current I:

$$I = \frac{10^7 B\delta k_s}{4\pi N} \tag{9}$$

The result of these empirical equations are the calculated values of armature diameter d (respectively radius $r1$), coil length L and coil diameter D (respectively radius R), coil conductor diameter d_C and current I flowing through the coil at a given voltage. Other values that need to be selected are related to the design of the electromagnet itself, especially the air gap between the components.

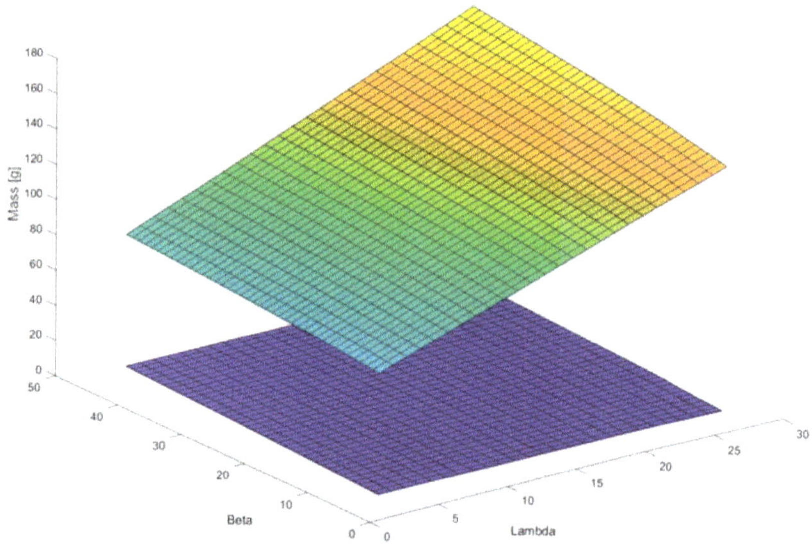

Figure 10. Graphical representation of the dependence of the coil and armature weights on the dimensional ratios.

The calculation performed should be taken as a guide only, as it applies to the pull solenoid, not the push solenoid, which has a different design and different magnetic conditions. Nevertheless, it provides enough input information to create an electromagnetic simulation that calculates the attractive force much more accurately.

Table 1 shows the reference parameters. Table 2 shows the main calculation parameters of the coil.

In Figure 11, it is possible to see the proposed dimensions that are needed to create a simulation model of the attractive force; the values of the dimensions are entered in Table 3. The table contains the construction dimensions of the electromagnet, which are based on the technological possibilities of electromagnet production and the construction arrangement of the needle bar. The basic parameters obtained on the basis of the calculation given in Section 3.2 were respected during the design.

Table 1. Reference parameters.

Parameter	Symbol	Value
Attractive force at maximum stroke	F	6 N
Air gab at maximum stroke	δ	2 mm
Coil voltage	U	3 V
Coil winding factor	ξ	0.7
Magnetic circuit saturation factor	k_s	1.3
Flux increase factor	ν	1.41×10^{-5}
Magnetic flux deflection factor	ε	1
Length ratio to armature diameter	λ	2
Ratio of winding thickness to armature diameter	β	0.4

Table 2. Calculated parameters.

Parameter	Symbol	Value
Armature diameter	d	12 mm
Coil diameter	D	25.6 mm
Coil length	L	28.8 mm
Coil conductor diameter	d_C	0.6 mm
Magnetic induction in the centre of the coil	B	0.36 T
Number of turns of coil	N	400
Static current	I	1.9 A

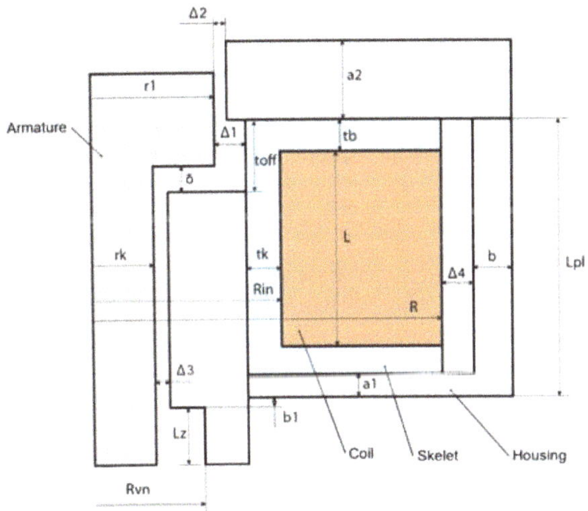

Figure 11. Electromagnet dimensions.

Table 3. Electromagnet dimensions.

Parameter	Symbol	Value
Air gap	δ	0–2 mm
Internal air gap 1	Δ1	1 mm
Internal air gap 2	Δ2	0.1 mm
Internal air gap 3	Δ3	0.1 mm
External air gap	Δ4	0.1 mm
Shaft radius	rk	2 mm
Armature radius	r1	6 mm
Coil inside radius	Rin	8 mm
Coil outside radius	R	12.8 mm

Table 3. *Cont.*

Parameter	Symbol	Value
Skeletal thickness radial	tk	1 mm
Skeletal thickness axial	tb	1 mm
Coil length	L	28.8 mm
Housing thickness 1	a1	7 mm

3.3. Simulation of the Attractive Force of an Electromagnet

As already mentioned, the empirical design serves only to approximate the dimensions and is, today, no longer sufficient due to its low accuracy. At present, it is possible to use numerical electromagnetic simulation based on the finite element method. There are several pieces of software that deal with this issue; the program FEMM 4.2 [25,26] was chosen for this work, because it has an integrated interface to the Matlab environment, which continues to work with data.

The computational model is static, i.e., the entire magnetic circuit is in a steady state. This is advantageous for simulating the end positions in which the electromagnet remains. In this case, however, we are also interested in the state between the end positions, which can be achieved by a quasi-static problem, i.e., by dividing into discrete time steps in which the circuit is static. This introduces a certain inaccuracy into the model, but this procedure is sufficient to determine the attractive force characteristic.

As a result of the electromagnetic simulation, we are particularly interested in the characteristics of the attractive force of the electromagnet, i.e., its dependence on the size of the air gap. The second important datum is the total inductance of the electromagnet, which is important for the design of the electrical circuit of the coil supply.

In the simulation, we discretize the whole problem into a so-called finite element network; the software can create this network itself based on the defined geometry and parameters, such as the maximum element size. The network element type is set to a triangle, the minimum angle of the triangle is set to 30°, and the required accuracy is set to 1×10^{-8}.

The boundary condition is set to a free environment whose radius is 120 mm, as can be seen in Figure 12.

Furthermore, it is necessary to define the materials of individual bodies because during magnetization, it is necessary to know the relative permeability and hysteresis B-H curve. The coil is defined by the resistivity of the conductor, the cross section, the number of turns and the current flowing.

The material of the core, hence the metal casing of the coil, must be chosen with regard to its magnetic properties. In general, relative permeability affects the magnetic flux, so the greater the relative permeability, the less resistance of the material to the passing magnetic flux. The shape of the hysteresis B-H curve affects the magnetic losses or the inductance of the entire electromagnet.

Magnetically soft steels are used as core materials, i.e., steels which, after the elimination of the magnetic field, do not remain magnetized, so that their attractive force is close to zero. These steels consist of either almost pure iron or, conversely, a very high alloy content. For comparison, two steels were selected that have all of their parameters defined in the FEMM 4.20 program, namely AISI 1010, which has an iron content of 99.5% and a steel called Supermalloy, which contains 5% iron, 79% nickel and 5% molybdenum.

The FEMM simulation program can be controlled using commands in the Matlab script, which allows us to iterate between the individual variants very quickly, as everything can be automated. The quasi-static state is achieved by gradually shifting the geometry of the anchor, i.e., by reducing the air gap. At each state, the attractive force is calculated, which is equal to the volume integral of the Lorenz force.

$$F_{lorenz} = \int J \times B dV \tag{10}$$

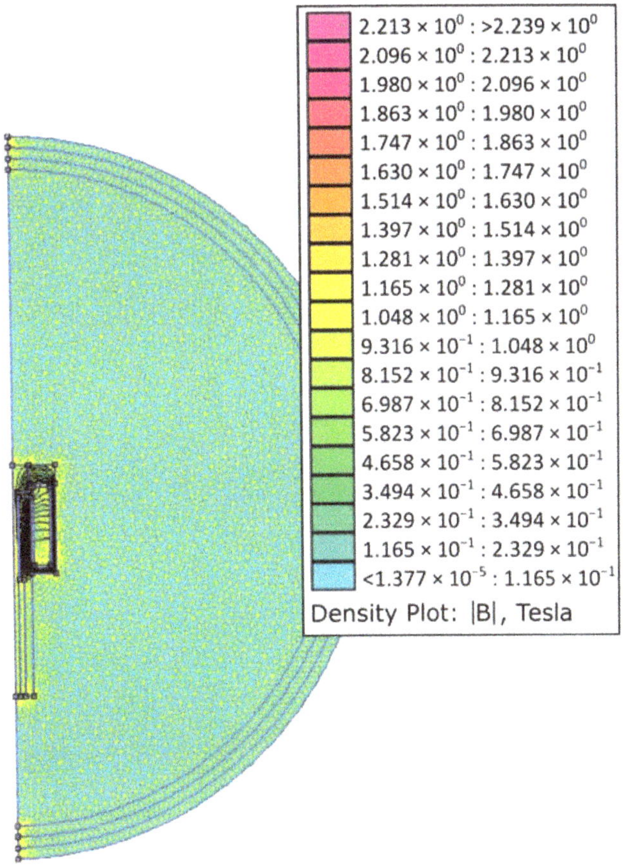

Figure 12. Simulation of the attractive force of an electromagnet: the boundary condition.

Figure 13 on the left shows the model in the FEMM 4.20 program with the finite element network and defined materials; the right part of the model, after simulation in the postprocessor, shows the magnetic flux lines and colour rendering of the magnetic field intensity, where purple represents the highest intensity. The carcass material, which is in fact plastic, is replaced by air because they have similar magnetic properties.

Figure 14 shows the simulation results for the initial state when the air gap is zero. Figure 15 shows the simulation results for the maximum air gap condition.

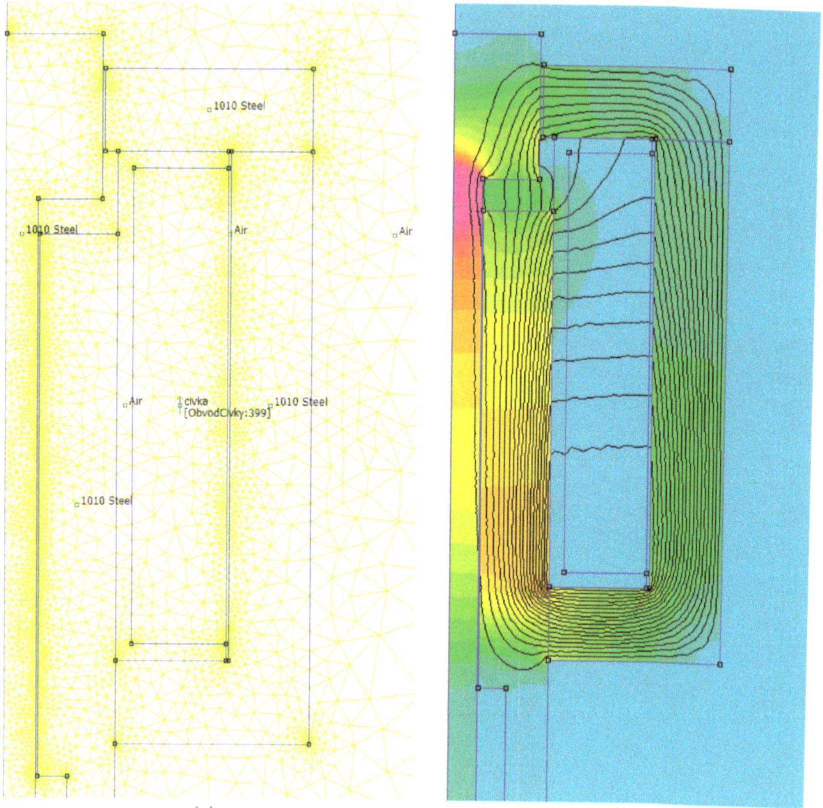

(**a**) (**b**)

Figure 13. Simulation of the attractive force of an electromagnet: (**a**) mathematical model with created simulation network; (**b**) simulation result with shown magnetic flux lines and magnetic field intensity.

As already mentioned, the main output of the simulation is the characteristic of the attractive force. Figure 16 shows the characteristics for the two materials mentioned above. It can be seen that the AISI 1010 steel has a more linear course than the Supermaloy steel, which has a sharp break in its air gap area of 0.5 mm but a 50% greater attractive force in the attracted state. The core material AISI 1010 or DIN 1.0413 was chosen for its design, due to greater material availability, lower price and also because the electromagnet from this material has a lower inductance of 0.033 H than the electromagnet from Supermaloy, of 0.068 H.

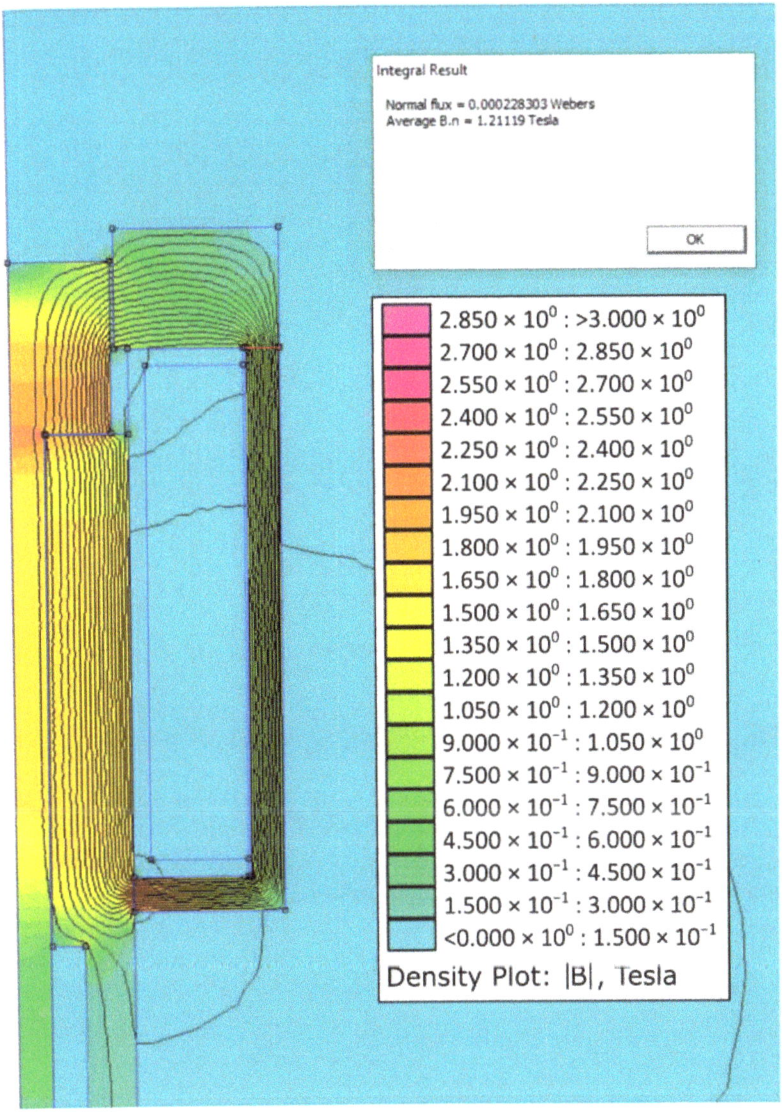

Figure 14. Distribution of magnetic flux and magnetic field intensity for minimum air gap.

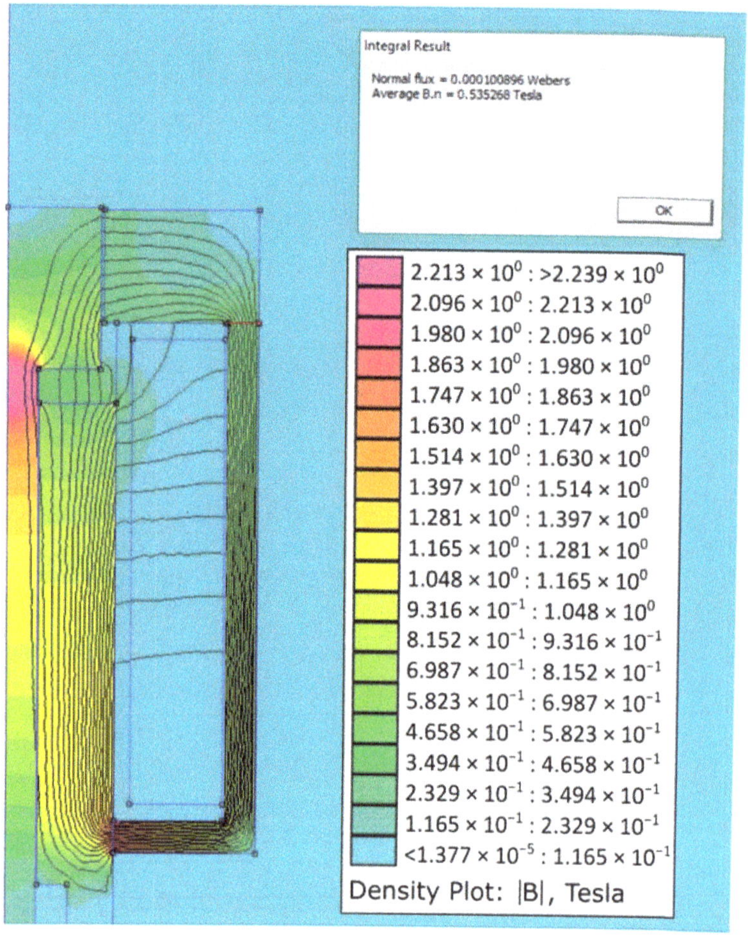

Figure 15. Distribution of magnetic flux and magnetic field intensity for maximum air gap.

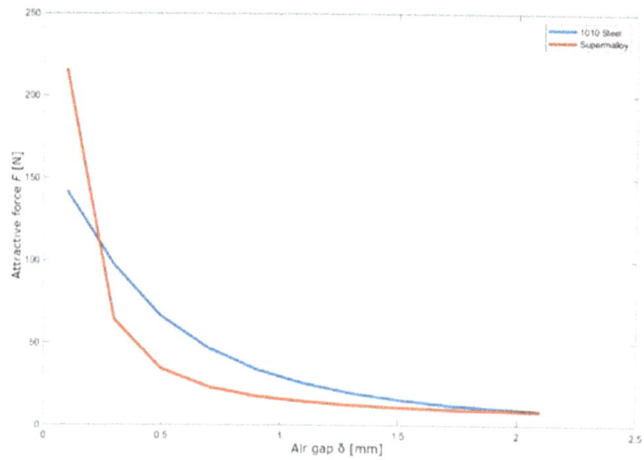

Figure 16. Courses of the electromagnet attractive force obtained from the simulation.

3.4. Coil Electrical Circuit Design

Because the electromagnet operates with relatively high dynamics, the transients of the coil are important, especially the dependence of the current flowing through the magnetic field on time and voltage. In general, the higher the voltage, the faster the saturation of the magnetic circuit, but if the time of increased voltage is too long, the coil heats up excessively. Therefore, it is necessary to determine how long it takes for the circuit to saturate.

From the previous simulation we know the value of the inductance of the electromagnet, i.e., the inductance of the coil with the inserted soft core, so we can use the following equation to describe the current and voltage:

$$U = L\frac{dI}{dt} + RI \tag{11}$$

Assuming that the resistivity and inductance are constant, the equation can be adjusted to these forms by time derivation:

$$\dot{I} + \frac{RI}{L} - \frac{U}{L} = 0 \tag{12}$$

$$\dot{I} = \frac{U - RI}{L} \tag{13}$$

The differential equation [27] can then be solved using a homogeneous solution:

$$I(t) = C_1 + C_2 e^{-\left(\frac{R}{L}\right)t} \tag{14}$$

There is defining initial and final coil saturation conditions. At the beginning, there is zero current; at infinity, the current approaches the constant value given by Ohm's law:

$$I = \frac{U}{R} \tag{15}$$

$$I(0) = 0 = C_1 + C_2 \times 1 \tag{16}$$

$$C_1 = -C_2 \tag{17}$$

$$\lim_{(t\to\infty)} I = \frac{U_{stat}}{R} = -C_2 + C_2 e^{-\left(\frac{R}{L}\right)T} = C_2\left(-1 + e^{-\left(\frac{R}{L}\right)\infty}\right) \tag{18}$$

$$C_2 = \frac{U_\infty}{R} \tag{19}$$

From which, the constants C_1 and C_2 can be calculated and substituted into the solution:

$$I(t) = \frac{U_{stat}}{R}\left(1 - e^{-\left(\frac{R}{L}\right)t}\right) \tag{20}$$

In this case, we are interested in how long the current value approaches the asymptote $\frac{U_{stat}}{R}$. The value of the approximation can be defined, for example, as:

$$C_{95} = 0.95 \tag{21}$$

$$I_{95} = C_{95} I_\infty = C_{95}\frac{U_{stat}}{R} \tag{22}$$

so we are looking for a current that has a value of 95% of the constant current value.

The equation is, therefore:

$$C_{95}\frac{U_{stat}}{R} = \frac{U_{stat}}{R}\left(1 - e^{-\left(\frac{R}{L}\right)T}\right) \tag{23}$$

where T is the time sought when the coil will be sufficiently saturated:

$$T = -\frac{L}{R}\ln(1 - C_{95}) \tag{24}$$

These equations apply to the case when the voltage value is equal to the static voltage value at which the coil is rated. However, for faster saturation, it is possible to increase the voltage value after a while. After that, however, the equations compiled above no longer apply, as this is a discontinuous problem. We can partially approach the desired value of the saturation time if we adjust the current equation to the following form:

$$\frac{U_{\text{stat}}}{R} = \frac{U}{R}\left(1 - e^{-\left(\frac{R}{L}\right)T}\right) \tag{25}$$

$$T = -\frac{L}{R}\ln\left(\frac{U - U_{\text{stat}}}{U}\right) \tag{26}$$

when the value of the asymptote approximation remains the same, but the course of the exponential curve is adjusted to the new voltage value. The asymptote approximation value is no longer needed because the new curve intersects the static current value.

Instead of an analytical calculation, it is also possible to calculate the original differential equation in the Matlab program, with the current set point.

To describe the state of discharge of the coil, we can use a slightly modified equation, where we will again be interested in the time in which a certain current value is reached; in this case, when it will be equal to zero. Here, too, the principle can be used to use a higher voltage than the nominal one for the transient to speed up the process. However, due to the polarity of the previous charge, it is now necessary to connect a negative voltage that is a multiple of the nominal static voltage of the coil:

$$I(t) = \frac{U_{\text{stat}}}{R}\left(e^{-\left(\frac{R}{L}\right)t}\right) \tag{27}$$

All of the above waveforms can be seen in the graph in Figure 17. The current functions during charging and discharging the coil have the same shape, only mirrored and offset along the horizontal axis. It can be seen that at a higher charging voltage $2U_{ref}$, a faster charge will occur in time t_{n2} than at a nominal voltage t_{n1}. The situation is the same for discharging, only due to the initial conditions, it is necessary to move the discharge at higher voltages along the horizontal axis so that at time $t = 0$, the current value is equal to the static value of the coil current. The graph also plots the values of sufficient approximation to the current asymptote, one for charging and the other for discharging the coil.

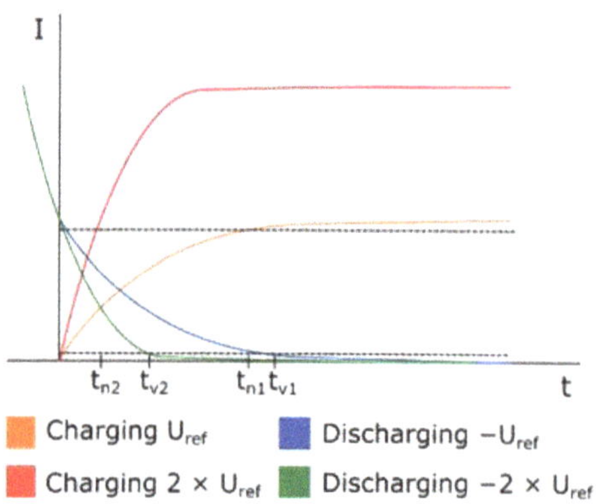

Figure 17. Courses of current during charging and discharging at different voltages.

The resulting selected voltage profile can be similar to Figure 18; this is a solution where the electromagnet is in a pushing arrangement. The voltage changes abruptly depending on the position of the cam; in the mechanism gripped state, the current is reduced, as not all the force that would be generated by the electromagnet is needed. The reduction in the voltage will bring a smaller value of the generated waste heat in the conductor. To release the needle, it is necessary to demagnetize the entire magnetic circuit, so a negative voltage is applied during demagnetization. Thereafter, the coil is de-energized until just before the point of gripping the needle, when a voltage multiplier is applied, which results in a sharp increase in current. At this point, it is possible to briefly exceed the rated current of the coil to create a value stronger than the rated force of the electromagnet. Then, the stress is reduced again to the initial value; between these two states, it is possible to insert another intermediate stage to achieve the desired characteristics of the attractive force.

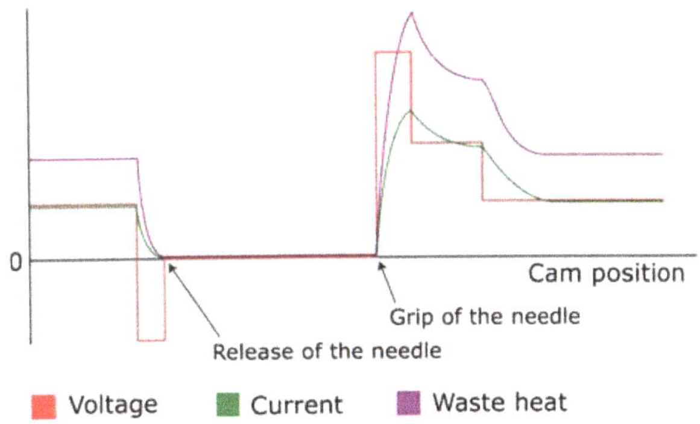

Figure 18. Courses of physical quantities of the coil.

The ideal source with a hard characteristic, i.e., with a constant voltage at different loads, is chosen as the voltage source. The voltage of the source is externally controlled so that the optimal behaviour of the whole needle bar system is achieved. It can be assumed that in the real case, the source will not have the ideal designed characteristics and will contain delays and instabilities, which will bring a certain amount of deviation into the system, which must be taken into account.

4. Discussion

The current design of the mechanical system of the needle bar of the special sewing machine imitating a hand stitch makes it impossible to increase the operating speed of the sewing machine due to the increase in vibration and noise. Scientific work has been done in the past focused on needle bar innovation to reduce these problems. This work deals with the design of a completely new principle of needle gripping and releasing control using an electromagnet located in the body of the needle bar.

In the first part, the type of electromagnet was chosen together with the design of its parameters and dimensions according to the typically used design procedure. The result of the preliminary design is a push solenoid, which has a working stroke of 2 mm and an attractive force at a maximum air gap of 6 N, supplied by a controlled direct current of 3 V, but with the possibility of overload to 24 V as the armature moves.

In the next part, the computational model of the electromagnet was built in the finite element program FEMM 4.20 and the characteristics of the attractive force, depending on the size of the air gap or the position of the armature in the working range, were determined. Since the empirical design of the solenoid parameters is valid for the pull solenoid, the calculated forces in the FEMM 4.20 program differed due to the different design; specifically for push solenoids, the force is greater at maximum air gap, which was confirmed by calculation of 8.2 N.

It is clear from the design of the electromagnet that this could be the right way to create an alternative means to control the needle grip and release in the needle bar. However, for the correct design of the electromagnet and the theoretical verification of its functionality, it will be necessary to create several more numerical computational models that describe the various aspects of the needle bar operation. The plan is to create, in parallel with the above-mentioned models, a complex model for the Matlab Simulink environment. This model should describe the mechanical parts of the needle bar and the bonds between them, the kinematic excitation of the needle bar shell, the electrical circuit that supplies the solenoid, and the solenoid itself, which is located between the shell and the core. It will be possible to apply a number of optimization tasks to this complex model. For example, the design of return spring parameters that will exert sufficient force to meet the required dynamics but will not unnecessarily load the mechanism with excessive force in the expanded state. It will also be necessary to design a suitable power supply for the electromagnet, as the gripping of the needle must be realized in the order of milliseconds. It will be necessary to design the coil overvoltage from nominal 3 V to 24 V in order to saturate the magnetic circuit faster. When demagnetizing, a negative voltage of −24 V will be used for a faster collapse of the magnetic field and, thus, the release of the needle. It will therefore be necessary to determine the durations of this overvoltage or the application of a negative voltage in order to avoid overloading the coil or creating the opposite magnetic field.

The proposed system could most likely lead to a reduction in vibrations and noise. The basic assumption is that the vibrations caused by the original control of the needle release, which uses the impact of the control element on the stopper, will be completely eliminated, because this very significant source of vibrations will be replaced by a newly designed magnetic system. That is, this current source of vibration will be completely eliminated. Nevertheless, of course, we do not expect the vibrations of the magnetic control of needle attachment to be zero, but they will be significantly lower. Evidence will be provided when the research is in the phase when a functional model of the developed magnetic system is produced; then, the measurement of actual noise values will be performed and compared with the original system.

This innovation could also allow the machine to increase operating speed and, thus, productivity. Due to the use of a higher charging voltage, the time required to attach and release the needle can be reduced. However, it is also possible to design an electromagnetic circuit with respect to higher speeds of the needle bar.

There is also a noticeable support for automation in the field of the clothing industry. The new needle bar, together with the previously designed needle bar servo drive system, could facilitate the transition of the sewing machine to automated production.

Moreover, the structure is also interesting for intelligent sewing machines that are under development. Intelligent sewing machines have the ability to automatically adapt to changes in the technological processes of sewing clothes. Therefore, the presented needle attachment structure could be one of the solutions when needles are changed on other sewing machines due to the sewing of very different materials from which clothes are made; these are frequent cases in modern clothing production processes.

In this way, this research activity responds to the pressure of the consumer industry to reduce production costs and, at the same time, seeks to support the automation of the sewing process and also to ensure greater safety for machine operators by reducing noise levels.

Author Contributions: Conceptualization, J.K. and V.K.; methodology, J.K. and V.K.; software, V.K.; validation, V.K.; formal analysis, V.K.; investigation, J.K. and V.K.; resources, J.K. and V.K.; data curation, J.K. and V.K.; writing—original draft preparation, V.K.; writing—review and editing, J.K.; visualization, J.K.; supervision, J.K.; project administration, J.K.; funding acquisition, J.K. All authors have read and agreed to the published version of the manuscript.

Funding: This research was funded by the Student Grant Competition of the Technical University of Liberec under the project No. SGS-2022-5046.

Conflicts of Interest: The authors declare no conflict of interest. The funders had no role in the design of the study, in the collection, analyses, or interpretation of data, in the writing of the manuscript, or in the decision to publish the results.

References

1. Beran, J.; Bílek, M.; Komárek, J.; Němeček, P. Experimental analysis of the sewing machine. *MATEC Web Conf.* **2016**, *89*, 01002. [CrossRef]
2. Pejchar, K.; Beran, J. Optimization of the Needle Bar Mechanism. In Proceedings of the X. International Scientific Conference Transfer 2009, Trenčín, Slovakia, 17–18 November 2009; pp. 44–49, ISBN 978-80-8075-414-3.
3. Komárek, J.; Beran, J.; Lima, M.; Machado, J.; Silva, J. Finding the Optimal Setting of the Sewing Needle Transfer Mechanism Using Simulation Software. In Proceedings of the TRS 2012, the 41st Textile Research Symposium, Guimaraes, Portugal, 12–14 September 2012; pp. 401–404, ISBN 978-972-8063-67-2.
4. Barborák, O. *Oděvnictví II, Šijacie Stroje v Odevnej Výrobe*; Trenčianská univerzita A. Dubčeka v Trenčíně: Trenčín, Slovakia, 2002; ISBN 80-88914-77-9.
5. Fiddlebase: Website about the History of Sewing Machines. Available online: https://www.fiddlebase.com/the-invention-of-the-sewing-machine-by-grace-rogers-cooper/ (accessed on 31 March 2022).
6. Clifford, L.B. The Sewing Machine Master Guide: From Basic to Expert; Blodget Publishing LLC: 2013; ISBN 0990022722, 978-0990022725. Available online: https://www.amazon.com/Sewing-Machine-Master-Guide-Expert/dp/0990022722 (accessed on 18 April 2022).
7. Langner, S. Improvements in or Relating to Double-Pointed Needles Adapted for Use in Sewing Machines Having Alternately Operating Needle Bars. U.S. Patent 496,859, 1938.
8. Johnson, W.C. Sewing Machine Needle Bar Mechanism. U.S. Patent US2402251A, 18 June 1946.
9. Šidlof, P.; Votrubec, V. Vibrations Reduction of Industrial Sewing Machines. In *Vibration Problems ICOVP 2011*; Springer Proceedings in Physics; Springer: Berlin/Heidelberg, Germany, 2011; Volume 139. [CrossRef]
10. Asif, A.; Hasan, M. Lean Manufacturing for Improving Productivity at Sewing Section in Apparel Industry: An Empirical Study. *J. Text. Eng.* **2018**, *8*, 1–9. [CrossRef]
11. Tomozei, C.; Pipa, A.; Irimia, O.; Panainte-Lehadus, M.; Nedeff, F. Measuring noise level in the textile industry. *J. Eng. Stud. Res.* **2019**, *24*, 43–49. [CrossRef]
12. Kirin, S.; Lauš, K. A study of noise levels in the sewing process. *Saf. J. Saf. Work. Organ. Living Environ.* **2011**, *53*, 243–250. Available online: https://hrcak.srce.hr/71905 (accessed on 18 April 2022).
13. Zhang, Z.; Liu, X.; Huang, C.; Pan, D. Noise source identification for industrial sewing machines based on non-linear partial least squares regression model. *Proc. Inst. Mech. Eng. Part C J. Mech. Eng. Sci.* **2016**, *230*, 220–225. [CrossRef]
14. Nayak, R.; Padhye, R. *Automation in Garment Manufacturing*; Woodhead Publishing: Duxford, UK, 2018; pp. 1–290.
15. Jindal, H.; Kaur, S. Robotics and Automation in Textile Industry. *Int. J. Sci. Res. Sci. Eng. Technol.* **2021**, *8*, 40–45. [CrossRef]
16. Salahuddin, M.; Lee, Y.A. Automation with Robotics in Garment Manufacturing. In *Leading Edge Technologies in Fashion Innovation. Palgrave Studies in Practice: Global Fashion Brand Management*; Palgrave Macmillan: London, UK, 2022. [CrossRef]
17. Lee, S.; Rho, S.; Lim, D.; Jeong, W. A Basic Study on Establishing the Automatic Sewing Process According to Textile Properties. *Processes* **2021**, *9*, 1206. [CrossRef]
18. Komárek, J. Dynamic model of the mechanical system of the needle bar. In Proceedings of the XII International Conference on the Theory of Machines and Mechanisms, Liberec, Czech Republic, 6–8 September 2016; pp. 323–329, ISBN 978-3-319-44086-6.
19. Komárek, J.; Nag, R. Verification of Mathematical Model of Mechanical System of Needle Bar using Laser Doppler Vibrometer, In Proceedings of the International Conference on the Theory of Machines and Mechanisms, Liberec, Czech Republic, 7–9 September 2021; Advances in Mechanism Design III, TMM 2020; Mechanisms and Machine Science; Springer: Berlin/Heidelberg, Germany, 2020; Volume 85, pp. 304–311. [CrossRef]
20. Krpeš, P. Design of the Needle Exchange System of Sewing Machine DECO 2000. Diploma Thesis, Technical University of Liberec, Liberec, Czech Republic, 2009.
21. Asadi Khanouki, M.; Sedaghati, R.; Hemmatian, M. Multidisciplinary Design Optimization of a Novel Sandwich Beam-Based Adaptive Tuned Vibration Absorber Featuring Magnetorheological Elastomer. *Materials* **2020**, *13*, 2261. [CrossRef] [PubMed]
22. Lyshevski, S.E. *Electromechanical Systems, Electric Machines, and Applied Mechatronics*; CRC Press: Boca Raton, FL, USA, 1996.
23. Robert, N.L. *Cam Design and Manufacturing Handbook*; Industrial Press: New York, NY, USA, 2002; ISBN 0-8311-3122-5.
24. Cigánek, L. *Elektrické přístroje spínací, ochranné a řídící: Působení, provedení*; SNTL: Praha, Czech Republic, 1956.
25. Baltzis, K. The finite element method magnetics (FEMM) freeware package: May it serve as an educational tool in teaching electromagnetics? *Educ. Inf. Technol.* **2010**, *15*, 19–36. [CrossRef]
26. Baltzis, K. The FEMM Package: A Simple, Fast, and Accurate Open Source Electromagnetic Tool in Science and Engineering. *J. Eng. Sci. Technol. Rev.* **2008**, *1*, 83–89. [CrossRef]
27. Hairer, E.; Nørsett, S.; Wanner, G. *Solving Ordinary Differential Equations I, Nonstiff Problems*; Springer: Berlin, Germany, 1993; ISBN 978-3-540-56670-0.

machines

MDPI

Article

Assessing the Effect of Interimplant Distance and Angle on Different Impression Techniques

Berkman Albayrak [1,*], İsmail Hakkı Korkmaz [2], Alvin G. Wee [3], Cortino Sukotjo [4,*] and Funda Bayındır [5]

1 Department of Prosthodontics, Bahçeşehir University School of Dental Medicine, Istanbul 34349, Turkey
2 Department of Mechanical Engineering, Erzurum Technical University Faculty of Engineering and Architecture, Erzurum 25050, Turkey; ismail.korkmaz@erzurum.edu.tr
3 Division of Prosthodontics—Department of Restorative Sciences, University of Minnesota School of Dentistry, Minneapolis, MN 55455, USA; wee00008@umn.edu
4 Department of Restorative Dentistry, University of Illinois at Chicago College of Dentistry, Chicago, IL 60612, USA
5 Department of Prosthodontics, Atatürk University Faculty of Dentistry, Erzurum 25240, Turkey; bayindirf@atauni.edu.tr
* Correspondence: berkman.albayrak@dent.bau.edu.tr (B.A.); csukotjo@uic.edu (C.S.); Tel.: +90-537-714-82-93 (B.A.); +1-(617)-272-55-12 (C.S.)

Abstract: We aimed to evaluate the trueness of digital and conventional impression techniques based on different angles and distances between implants and the deviation caused by the angle and distance parameters varying between implants. Eight implants were placed in a polyurethane edentulous mandibular model at different angles and distances. After obtaining a 3-dimensional (3D) reference model by using an optical scanner, the model was scanned with three intraoral scanners: Cerec Omnicam (DO), Trios 3 (DT), and Carestream 3500 (DC). Then, the master casts obtained from the conventional impressions (C) were also digitized, and all impression data were imported into reverse engineering software to be compared with the 3D reference model. Distance and angle measurements between adjacent implants were performed, and the data were analyzed with ANOVA–Tukey and Kruskal Wallis tests. The significance level was accepted as $p < 0.05$. While DT and C groups gave the best results for high interimplant distances, the trueness of intraoral scanners was found to be superior to the conventional method between closer implants. At higher angulations, the angular trueness of C group was found to be significantly lower. At short distances, digital groups showed superiority, and the trueness of conventional impression decreased with higher angulations.

Keywords: angle; digital implant impression; interimplant distance; intraoral scanner; trueness

Citation: Albayrak, B.; Korkmaz, İ.H.; Wee, A.G.; Sukotjo, C.; Bayındır, F. Assessing the Effect of Interimplant Distance and Angle on Different Impression Techniques. *Machines* **2022**, *10*, 293. https://doi.org/10.3390/machines10050293

Academic Editors: Marek Kočiško and Martin Pollák

Received: 30 March 2022
Accepted: 18 April 2022
Published: 21 April 2022

1. Introduction

One of the most important criteria for achieving long-term success in implant-supported restorations is to provide a passive fit [1–5]. However, it is not possible to obtain a perfect passive fit due to the dimensional changes that may occur during the impression and subsequent prosthetic production stages [6]. Hence, the initial step in order to obtain the optimum passive fit is to make an accurate conventional or digital impression [7] and transfer the spatial positions of the implants in the bone to the working models where the prosthetic design will be carried out.

Implant impressions use various components. Therefore, the inconsistencies in the junction between the impression copings, the implant, and the analogue can significantly affect the accuracy [8,9]. The amount of these deviations also increases exponentially in cases of multiple implants rather than single implants [10].

In cases with severe alveolar ridge resorption, considering some reasons such as time, cost, patient's age, health condition, and desire, tilted implant placement in non-ideal distances instead of combined treatments such as bone augmentation and sinus lift operation constitute a serious treatment alternative [11–13] However, in such cases, implant impression may become complicated, and the trueness of the impression might be compromised [14]. Although splinting of impression copings increases the accuracy, in full arch cases when the angulations exceed 20°, it has been shown that the non-splinted open tray impression is better than the splinted technique [15,16].

The effect of the angulation on the implant impressions was mostly investigated in two implants [17], and the effect of the angulation between each implant was not evaluated by examining the general deviation according to the knowledge of the authors in the complete arch studies performed [2,18–23]. In conventional impressions, deformation can occur even when using polyvinyl siloxane (PVS), an impression material with a high elastic memory, during the removal of the tray from the tilted implants [24,25]. Theoretically, digital implant impressions should not be affected from the angulations, as there is no material deformation or the movement of the impression coping within the impression. However, there are also some studies in the literature that show that digital impression accuracy decreases at angles above 15 to 30 degrees [21,26].

It has been shown that in cases of total and partial edentulism, the accuracy of the impression can vary, taking into account the increased number of implants and the increased distances between the implants [27]. In partial edentulism cases, linear and angular dimensional changes are lower compared to total edentulism, when conventional impression techniques are used [28]. With the increase in the scanned arch distance in the digital impression technique, it becomes difficult to obtain sufficient reference points between the point clouds. As it is not possible to combine the images obtained separately, optical noise and distortions occur in the 3-dimensional (3D) image, or because the software cannot use the surface matching algorithm, it cuts out some of the scanned areas [29]. Therefore, in theory, digital implant impressions are negatively affected by the increase in distance between the scan bodies. In particular, intraoral scanners that work on the principle of combining photos can stick different scan bodies on top of each other if they lose reference points that will continue scanning [30].

The aims of this study are to evaluate the trueness of digital and conventional impression techniques at different angles and distances between implants and the deviation caused by the angle and distance parameters varying between implants. The first hypothesis of this study is as follows: while the conventional method gives better results in areas where the distance between implants is large, the digital method is superior for areas with angulation. The second hypothesis is that the deviation will be higher in the regions where the distance and angulation between the adjacent implants are also high.

2. Materials and Methods

Eight implants (Dyna Dental Engineering BV, the Netherlands) (4.2 mm diameter and 11.5 mm length) were placed in a polyurethane edentulous mandible model (Promedicus, Poland) at different distances and angles. Implants were inserted to the sites of #47, #44, #43, #41, #33, #34, #35, and #37. The implants were placed at 40°, 0°, 20°, 0°, 15°, 0°, 0°, and 25° angles distally, respectively. All angles were arranged in the distal direction to mimic the distal angulation used in the production of cantilever segmented prosthesis [31].

First, in order to obtain a 3D reference model in which measurements will be made, Ti-base abutments (Dyna Dental Engineering BV, the Netherlands) were first placed on each of the implants. Eight abutment-level scan bodies (Dyna Dental Engineering BV, the Netherlands) were then seated on Ti-base abutments (Figure 1).

Afterwards, the model with the scan bodies was scanned using the Activity 885 Mark 2 Scanner (Smart Optics, Bochum, Germany) with an accuracy of 6 μm as in previous studies [20,32,33], and the 3D reference model was obtained by exporting the data in standard tesselation language (STL) format to be used in comparisons (Figure 2).

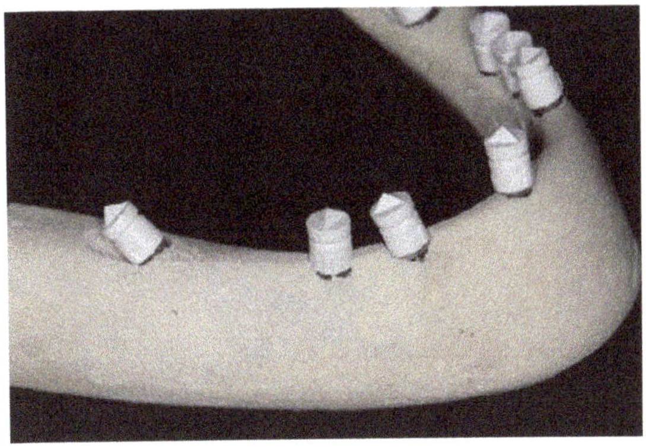

Figure 1. Working model with scan bodies attached to implants.

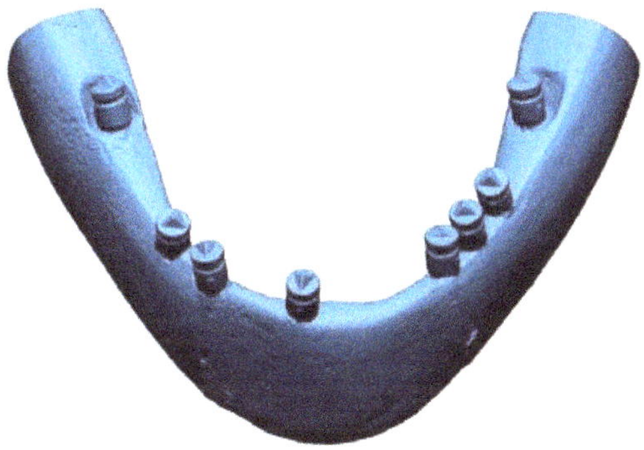

Figure 2. Reference 3D model obtained with optical scanner.

The study model was scanned 10 times with each scanner by a single operator before the final scans so that the user could adapt to each of the intraoral scanning systems. Then, starting from area #47 on the right posterior and continuing to the left posterior, scans were performed by following the scanning strategy recommended by the manufacturer of each scanner [34–36].

In the scans performed with the Cerec Omnicam (Dentsply Sirona, Bensheim, Germany), starting from the right posterior region (#47); occlusal, buccal, and lingual surfaces of each scan body were scanned, and the same procedure was continued until the left posterior region (#37). In Trios 3 (3-Shape, Copenhagen, Denmark) scans, scanning that started from the right posterior region occlusally was continued to the left posterior region, followed by scanning of the lingual and buccal surfaces. Unlike these, in Carestream 3500 scans, occlusal imaging of the scan bodies was completed, and then buccal and lingual surfaces were scanned. After each scan, the images were examined and the missing areas were scanned again; the digital impression phase was completed after the meshing procedure. Thus, a total of 30 STL files were obtained, including 10 from three different intraoral scanners.

For the conventional impression phase, eight open tray impression copings were attached to the implants after the removal of the scan bodies and Ti-base abutments. Before each impression, Kerr polyvinyl siloxane (PVS) tray adhesive (KaVo Dental GmbH, Bismarckring, Germany) was applied to the customized impression trays produced according to the model with 3-mm-space thickness. Then, the normal set Elite HD + putty soft and Elite HD + light body (Zhermack SpA, Rovigo, Italy) PVS were placed on the study model with the help of the tray to make single-step impression. After 10 min of setting time of the PVS, the trays were separated from the model and the analogs were attached to the impression copings. Soft tissue silicone (Zhermack Gingifast Elastic, Zhermack SpA, Rovigo, Italy) was applied to the areas where impression copings were located, and this silicone was allowed set. Afterwards, casting was performed using 150 g GC Fujirock type IV dental stone (GC Corporation, Tokyo, Japan) at a 1:5 powder–liquid ratio using a vacuum mixer. The gypsum was set (2 h) in accordance with the manufacturer's instructions, and in this phase, all models were kept at room temperature.

Ti-base abutments and scan bodies were placed on the cast models before the measurements were made. All cast models were digitized, and 3D models were obtained using a Straumann 7 Series laboratory scanner (Straumann Group, Basel, Switzerland). In total, 40 STL files were obtained, including 30 from the digital impression group and 10 from the conventional impression group (Figure 3).

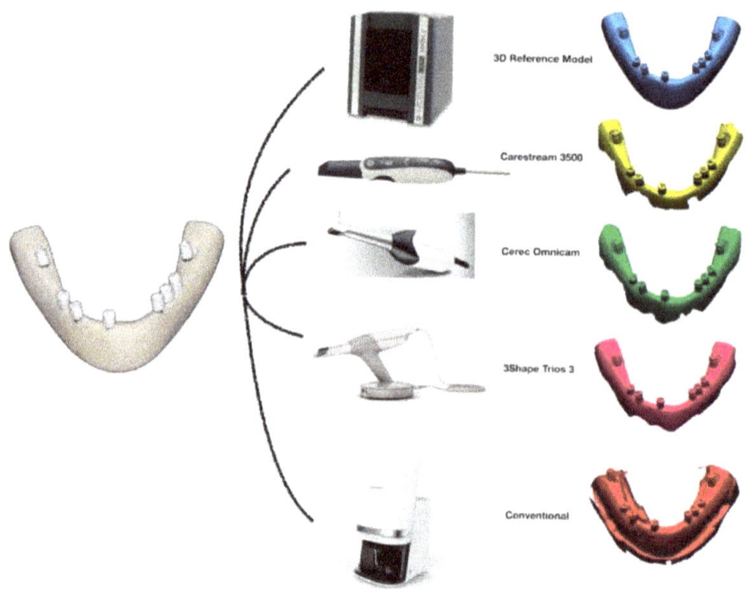

Figure 3. Entire impression groups.

During distance and angle measurements, first, all STL data obtained with the impressions were transferred to Rapidform (INUS Technology Inc., Seoul, South Korea) reverse engineering software. Reference points were determined on the scan bodies in the 3D images where measurements would be made. Two circles (Figure 4c) were formed in the upper and lower cylinders on the scan body, and their centers were determined to be 0.7 and 3.4 mm away from the base of the triangular pyramid with the "point" command (Figure 4a,b).

For all scan bodies in each scan, cartesian (x, y, z) coordinates of these points were exported from Rapidform in ".txt" format. The coordinates of the reference points of each scan body were determined by using midpoint of the centers of upper and lower circles. The distance between two reference points of $P_1(x_1, y_1, z_1)$ and P_2 (x_2, y_2, z_2) was calculated by using the following formula:

$$|P_1P_2| = \sqrt{(x_1 - x_2)^2 + (y_1 - y_2)^2 + (z_1 - z_2)^2}$$

(a)

(b)

Figure 4. *Cont.*

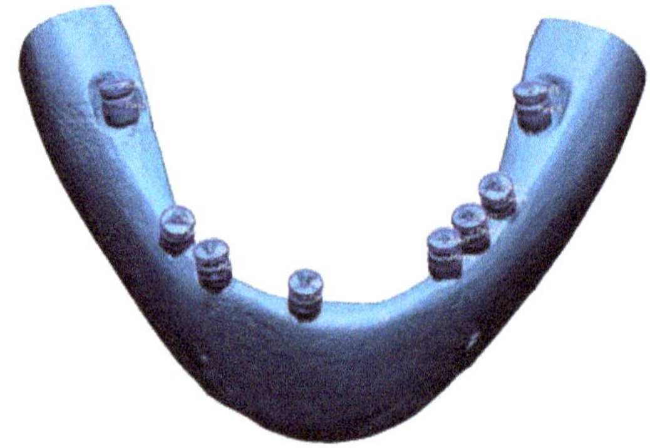

(c)

Figure 4. (**a**) First circle on the scan body, (**b**) Second circle on the scan body, (**c**) The centers of the two cross-section circles of all scan bodies.

Distance measurements were made by reference points (P) on scan bodies attached to implants, each located at a different distance. Accordingly, the distances between adjacent scan bodies were determined in both the reference model and the impression groups. (P_1-P_2, P_2-P_3, P_3-P_4, P_4-P_5, P_5-P_6, P_6-P_7, P_7-P_8).

Angle measurements were made with the following formula:

$$l_1 = \frac{x - x_1}{a_1} = \frac{y - y_1}{b_1} = \frac{z - z_1}{c_1};\ l_2 = \frac{x - x_2}{a_2} = \frac{y - y_2}{b_2} = \frac{z - z_2}{c_2}$$

$$\cos\varphi = \frac{\vec{s}_1 \vec{s}_2}{\left|\vec{s}_1\right| . \left|\vec{s}_2\right|} = \frac{a_1.a_2 + b_1.b_2 + c_1.c_2}{\sqrt{a_1^2 + b_1^2 + c_1^2}.\sqrt{a_2^2 + b_2^2 + c_2^2}}$$

In the angle measurements, the lines passing through the centers of the circles formed on the scan bodies were used. In the angle measurement formulation, the lines were found according to the points determined (l_1, l_2) and their direction vectors ($\vec{s}_1$, $\vec{s}_2$). A vector ($\vec{s}$) was defined for each scan body, taking into account the center points of scan body's drawn circles. In a Cartesian coordinate system, a vector with a component "x, y, z" can be calculated with this formula: ($\vec{s}$) = ai + bj + ck. The letters "a,b,c" are the coefficients that express the direction magnitude. Consequently, the angle between the reference scan body ($\vec{s}_1$) and the other scan body ($\vec{s}_2$) was defined in this way and it was calculated with the above formulation ($\vec{s}_1\vec{s}_2$, $\vec{s}_1\vec{s}_3$, $\vec{s}_1\vec{s}_4$, $\vec{s}_1\vec{s}_5$, $\vec{s}_1\vec{s}_6$, $\vec{s}_1\vec{s}_7$, $\vec{s}_1\vec{s}_8$).

In this study, starting from the right posterior region (#47), all eight scan bodies were numbered from P1 to P8. Distance (D) is expressed by adjacent scan body numbers: $D_{1\text{-}2}$, $D_{2\text{-}3}$, $D_{3\text{-}4}$, $D_{4\text{-}5}$, $D_{5\text{-}6}$, $D_{6\text{-}7}$, $D_{7\text{-}8}$. Angle (A) is also expressed by adjacent scan body numbers: $A_{1\text{-}2}$, $A_{2\text{-}3}$, $A_{3\text{-}4}$, $A_{4\text{-}5}$, $A_{5\text{-}6}$, $A_{6\text{-}7}$, $A_{7\text{-}8}$. By determining the deviation of the data of a total of four impression groups with respect to the reference model, comparisons were made between both the impression groups and different adjacent implants.

The distance and angle measurement data were analyzed with IBM SPSS Statistics 20.0 Release Notes program. First, the normality of the data was investigated with the Shapiro–Wilk test. The difference between the mean values of the variables with normal distribution was analyzed by one-way ANOVA, and Tukey's multiple comparison test was used to determine the differences between each group. The difference between the median values of the variables with non-normal distribution was analyzed by the Kruskal–Wallis test. First, the distance and angle deviations between the different adjacent implants of the impression groups were compared, and then the digital and conventional impression performances of the double-adjacent implants were evaluated. All statistical analyses were performed after taking absolute values of the data, and 0.05 was used as the level of significance.

3. Results

The first comparison was made between different impression groups: Cerec Omnicam (DO), Trios 3 (DT), Carestream 3500 (DC), and conventional (C).

3.1. Comparison of Distance Trueness

In the distance (D) parameter, there was no significant difference only in D_{1-2}, while a significant difference was found in all other distance deviations (Table 1). DT and C groups showed less deviation in the D_{4-5} and D_{7-8} regions, where the distance between the implants was greater than that in the other groups. In the regions where the distances between the implants were close (D_{2-3}, D_{3-4}, D_{5-6}, D_{6-7}) the conventional group showed the highest deviation, while the digital impression groups gave better results.

Table 1. Distance deviations of each impression group between adjacent implants.

	DO		DT		DC		C			
	Mean ± SD	Median (Min–Max)	Mean ± SD	Median (Min–Max)	Mean ± SD	Median (Min–Max)	Mean ± SD	Median (Min–Max)	Test Statistics	p
D_{1-2} *	0.35 ± 0.21	**0.32 (0.06–0.73)** ^{b}	0.5 ± 0.3	**0.45 (0.16–1.18)** ^{ab}	0.5 ± 0.43	**0.38 (0.15–1.54)** ^{b}	1.48 ± 0.78	**1.33 (0.27–3.02)** ^{a}	15,511	0.001
D_{2-3} *	0.25 ± 0.16	**0.22 (0.01–0.55)** ^{a}	0.44 ± 0.36	**0.32 (0.08–1.01)** ^{a}	0.73 ± 0.46	**0.73 (0.1–1.45)** ^{a}	1.72 ± 1.41	**1.43 (0.47–5.05)** ^{b}	17,118	0.001
D_{3-4} *	**0.42 ± 0.37** ^{a}	0.41 (0.03–1.24)	**0.37 ± 0.27** ^{a}	0.28 (0.03–0.74)	**0.73 ± 0.43** ^{ab}	0.63 (0.08–1.41)	**1.19 ± 0.83** ^{b}	1.12 (0.18–2.74)	5197	0.025
D_{4-5} *	1.18 ± 1.17	0.6 (0.03–3.03)	0.35 ± 0.28	0.22 (0.12–0.97)	0.51 ± 0.48	0.43 (0.08–1.52)	0.72 ± 0.74	0.57 (−0.19–2.26)	3773	0.287
D_{5-6} *	**0.44 ± 0.3** ^{a}	0.41 (0.07–0.99)	**0.81 ± 0.26** ^{ab}	0.82 (0.53–1.27)	**0.69 ± 0.37** ^{ab}	0.84 (0.07–1.11)	**1.39 ± 0.86** ^{b}	1.76 (0.19–2.7)	6255	0.011
D_{6-7} *	1.16 ± 0.25	**1.25 (0.77–1.49)** ^{b}	0.8 ± 0.25	**0.76 (0.44–1.16)** ^{b}	0.18 ± 0.19	**0.12 (0.01–0.53)** ^{a}	0.79 ± 0.6	**0.79 (0.09–1.62)** ^{b}	20,540	0.000
D_{7-8} *	**0.68 ± 0.36** ^{b}	0.7 (0.07–1.25)	**0.37 ± 0.3** ^{ab}	0.29 (0.01–0.89)	**0.32 ± 0.2** ^{a}	0.3 (0.01–0.65)	**0.39 ± 0.3** ^{ab}	0.28 (0.03–0.96)	3149	0.037

* **D** refers to distance deviations, and the numbers refer to adjacent implant pairs. The letters "a" and "b" indicate statistically significant difference at $p < 0.05$ within each row comparison.

3.2. Comparison of Angle Trueness

In the angle (A) parameter, it was determined that there was a significant difference between the groups for all angle deviations, except for A_{4-5}, which had an angulation of 15°. In the A_{1-2} region, where the angulation of 40° was the highest, C group deviated significantly more than in the DC and DO groups; in the A_{2-3} and A_{3-4} regions, which also had 20° angulation, the C group had the lowest performance at a significant level. In the parallel placed implant area, A_{6-7}, DC had significantly better results compared to other groups, but there was no significant difference between other digital and conventional impression groups (Table 2).

Table 2. Angle deviations of each impression group between adjacent implants.

	DO		DT		DC		C			
	Mean ± SD	Median (Min-Max)	Mean ± SD	Median (Min-Max)	Mean ± SD	Median (Min-Max)	Mean ± SD	Median (Min-Max)	Test Statistics	p
A_{1-2} *	0.17 ± 0.37	0.05 (0.01–1.23)	0.06 ± 0.02	0.07 (0.01–0.09)	0.05 ± 0.06	0.03 (0–0.19)	0.05 ± 0.03	0.05 (0.02–0.1)	4210	0.240
A_{2-3} *	0.01 ± 0	0 (0–0.01) ^{a}	0.01 ± 0.01	0.01 (0–0.03) ^{a}	0.01 ± 0.01	0.01 (0–0.03) ^{a}	0.13 ± 0.06	0.11 (0.06–0.27) ^{b}	25,948	0.000
A_{3-4} *	0.02 ± 0.01 ^{b}	0.02 (0–0.03)	0.02 ± 0.01 ^{b}	0.02 (0–0.03)	0.06 ± 0.03 ^{a}	0.05 (0.02–0.13)	0.08 ± 0.05 ^{a}	0.06 (0.01–0.16)	9410	0.001
A_{4-5} *	0.21 ± 0.4	0.05 (0–1.27) ^{ab}	0.03 ± 0.01	0.03 (0–0.04) ^{b}	0.23 ± 0.31	0.16 (0–1.11) ^{a}	0.06 ± 0.05	0.04 (0.01–0.15) ^{ab}	12,022	0.007
A_{5-6} *	0.03 ± 0.01 ^{a}	0.03 (0.01–0.04)	0.01 ± 0.01 ^{b}	0.01 (0–0.02)	0.03 ± 0.01 ^{a}	0.03 (0.02–0.03)	0.09 ± 0.08 ^{a}	0.07 (0–0.22)	8803	0.000
A_{6-7} *	0.01 ± 0.01 ^{a}	0.01 (0–0.02)	0.01 ± 0.01 ^{a}	0.01 (0–0.03)	0.01 ± 0 ^{a}	0.01 (0–0.02)	0.02 ± 0.01 ^{b}	0.02 (0–0.05)	7397	0.001
A_{7-8} *	0.04 ± 0.01	0.04 (0.01–0.06) ^{ab}	0.01 ± 0.01	0.01 (0–0.02) ^{b}	0.09 ± 0.02	0.09 (0.07–0.13) ^{ab}	0.02 ± 0.02	0.01 (0–0.07) ^{b}	27,985	0.000

* **A** refers to angle deviations, and the numbers refer to adjacent implant pairs. The letters "a" and "b" indicate statistically significant difference at $p < 0.05$ within each row comparison.

3.3. Effects of Distance and Angle Parameters on Different Impressions

Regardless of the impression groups, when the distance and angle deviations between the adjacent implants were compared, while a significant difference was detected between the groups in the deviation parameter ($p = 0.00$) (Figure 5), no significant difference was found between the groups in the angular deviation ($p = 0.067$) (Figure 6).

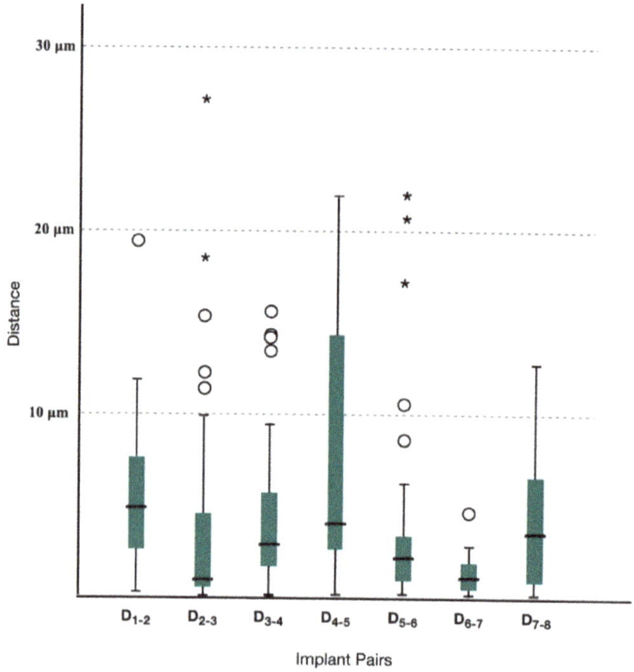

Figure 5. The distance deviations of implant pairs. * and ○ state outliers.

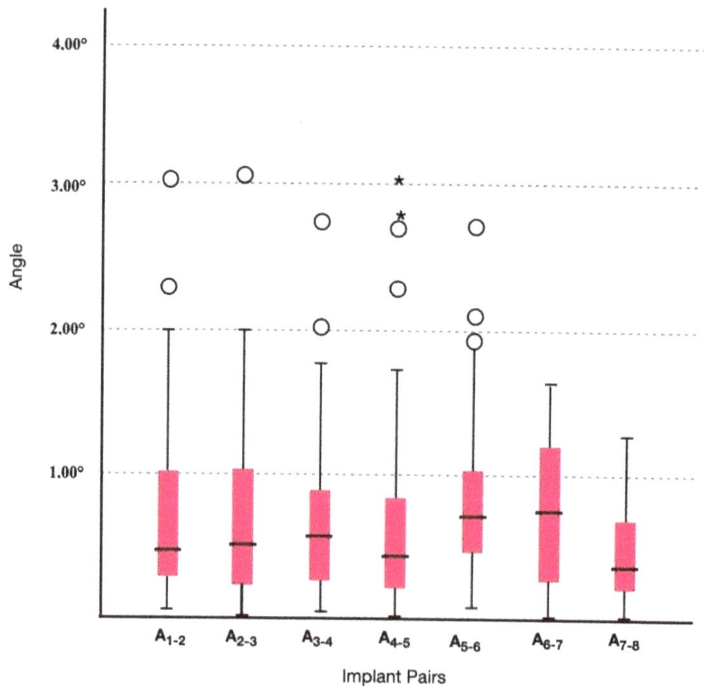

Figure 6. The angular deviations of implant pairs. * and ○ state outliers.

In $D_{6\text{-}7}$ and $D_{2\text{-}3}$, where the distance between implants was the lowest, the distance deviation was significantly the lowest, followed by $D_{5\text{-}6}$, which also did not have any edentulous area. These regions were followed by $D_{3\text{-}4}$ and $D_{7\text{-}8}$ with one missing tooth, while the $D_{4\text{-}5}$ and $D_{1\text{-}2}$ regions, where the body distance was more than one tooth, were the areas where deviation was detected the most (Table 3).

Table 3. Distance deviations between adjacent implants for entire impression groups.

	Mean ± SD	Median (Min–Max)
$D_{1\text{-}2}$	0.08 ± 0.19	0.05 (0–1.22) [d]
$D_{2\text{-}3}$	0.04 ± 0.06	0.01 (0–0.27) [ab]
$D_{3\text{-}4}$	0.04 ± 0.04	0.03 (0–0.16) [c]
$D_{4\text{-}5}$	0.13 ± 0.26	0.04 (0–1.27) [cd]
$D_{5\text{-}6}$	0.04 ± 0.05	0.02 (0–0.22) [b]
$D_{6\text{-}7}$	0.01 ± 0.01	0.01 (0–0.05) [a]
$D_{7\text{-}8}$	0.04 ± 0.04	0.04 (0–0.13) [c]
	Test Statistics	p
	52.508	0.000

The letters "a", "b", "c" and "d" indicate statistically significant difference at $p < 0.05$ within each column comparison.

4. Discussion

In the present study, the impressions of the scan bodies at different angles and distances were made with different techniques and different intraoral scanners, and the performances were compared. The deviation amounts of these angle and distance differences on the impressions were evaluated.

According to the results, the first hypothesis of the study that the conventional impression technique would give better results at points where the distance increased and digital impression technique would give better results at points where the angle increased was partially rejected. Intraoral scanners showed superiority over conventional impression in four regions where the distances between scan bodies were short, while Trios 3 was the group that showed the least deviation with the conventional group in two regions with increased distance. In addition, in areas with high angulation such as 40° and 20°, C was the group that showed the most deviation, as expected, while Cerec Omnicam showed the worst performance in a region with 25° angulation.

As with the effect of the edentulous space distance between implants on digital impression trueness [18,37–39], its effects on digital and conventional impressions have been studied in a comparative way in many studies [19,40]. As a result of the study conducted on two different models with six and eight implants, Tan et al. [19] stated that the reduction of the distance between the implants increased the digital impression accuracy, while the conventional impression technique was not affected by this parameter. Schmidt et al. [40] revealed that the deviation that occurs in extent increases with the length of the scanning distance. Likewise, Thanasrisuebwong et al. [37] stated that as the edentulous space between the two implants increased, the accuracy of the Trios 3 and Cerec Omnicam scanners decreased. In line with this research, while the largest deviations were observed in the D1-2, D4-5, and D7-8 regions in the digital impression groups, this situation was not observed in the conventional group in the present study.

The accuracy of the impressions made on angled implants has been examined in many studies [2,14,18,22,25,26,41] due to the problems experienced in removing the impression components away from the mouth and transferring them to the master cast. The fact that there is no material deformation and the impression components do not move in digital impressions makes the use of intraoral scanners for angled implants more ideal, theoretically [41]. In this study, digital groups performed better than the conventional in angulations such as 40° and 20°, which support this argument, while DC was the group with the highest deviation in the $A_{7\text{-}8}$ region with 25°. In a study conducted by Lin et al. [22], while the angulation between implants did not affect the conventional impression accuracy, it has been reported that it influences the digital impression, but the increase in angulation and deviation did not show a linear ratio. On the other hand, Gimenez et al. [26] stated that angulations had no effect on the digital impression accuracy as a result of the study they carried out on a full arch model including two implants with angulation of 30° and four parallel implants.

In the current study, the deviations between adjacent implants at different distances and angles were evaluated, and impressions with high trueness values were obtained in the $D_{2\text{-}3}$ and $D_{6\text{-}7}$ regions where the distance was short. In the $D_{1\text{-}2}$ and $D_{4\text{-}5}$ regions with two missing teeth in the pontic region, the highest deviation in the distance parameter was found in the impression groups. While these data support the second hypothesis of the study, no significant difference was detected in the angular deviations between adjacent implants. Therefore, the second hypothesis that the deviation would increase with advancing angulation and distance was also partially rejected.

Intraoral scanners must not lose the reference points they use during imaging so that they can take images of small areas and stitch them together to create all the data correctly [26]. In this study, various pauses were encountered in digital impressions, especially during the scanning of edentulous areas, and the impression was continued by returning and ensuring that the scanner caught the reference point. In these regions, it is predicted that the errors that may occur during the stitching of the newly acquired images with the previous ones by the software increase the deviation. On the contrary, in angled implants, it was observed that the system could easily distinguish the relevant scan body from the others and the scanning could be continued more easily. It is likely that this situation will enable the digital technique to outperform the conventional technique in angled implants.

Some researchers state that one of the factors that negatively affect digital measurement clarity is the positioning of the scanned implants at the chin curvature in the anterior region, rather than being on a straight line [42,43]. However, in the current study, similar deviation values were detected in posterior regions of $D_{1\text{-}2}$ and $D_{7\text{-}8}$, such as the $D_{4\text{-}5}$ region located at the curvature. The measurement technique can also affect the trueness values. In some studies, the study groups and the reference model were superimposed with the "best-fit alignment" technique [18,37], but some researchers advocate comparing distance and angle measurements between the reference model and other impression groups separately [21,44]. In the present study, the deviation values were obtained by comparing the distance and angle measurements done firstly on the reference model and then on the study groups.

Since this in vitro study was carried out on an artificial jaw model, it is not possible to reflect exactly the patient and mucosal movements that can be encountered on a digital impression. Therefore, it is important to evaluate the existing data together with future in vivo studies, in terms of clinical success to be achieved in implant impressions. In addition, the negative effects of oral fluids such as saliva and blood on the imaging of intraoral scanners could not be reflected. It will be useful to investigate the accuracy of intraoral scanners in long edentulous spans by adding components that will create a reference for the scanner.

5. Conclusions

Within the limitations of this in vitro study, it has been determined that digital impression gave better results in areas where the distance between the implants is shorter. With the increase in the distance, the Trios 3 scanner had the highest trueness with the conventional group. The use of Cerec Omnicam and Carestream 3500 scanners was reliable in areas where the distance was shorter between implants. In angular deviation, with the decrease in the performance of the conventional technique compared to the digital impression in the areas where the angle increases, a similar consistency was not found.

Author Contributions: Conceptualization, B.A., A.G.W., C.S. and F.B.; methodology, B.A., A.G.W.; İ.H.K., C.S. and F.B.; formal analysis, B.A., İ.H.K., C.S. and F.B.; investigation and data, B.A., İ.H.K., C.S. and F.B.; writing—original draft preparation, B.A. and İ.H.K.; writing—review and editing, B.A.,

A.G.W., C.S. and F.B.; supervision, A.G.W., C.S. and F.B.; project administration: C.S. and F.B. All authors have read and agreed to the published version of the manuscript.

Funding: This research received no external funding.

Data Availability Statement: Not applicable.

Acknowledgments: The authors thank the Dyna Dental Company for supplying all implant components and Zhermack Dental and GC Corporation for their support in terms of impression and cast materials. This study was partially supported by Atatürk University scientific research project. PRJ2015/434.

Conflicts of Interest: We wish to confirm that there are no known conflicts of interest associated with this publication and there has been no significant financial support for this work that could have influenced its outcome.

References

1. Moreno, A.; Gimenez, B.; Ozcan, M.; Pradies, G. A clinical protocol for intraoral digital impression of screw-retained CAD/CAM framework on multiple implants based on wavefront sampling technology. *Implant Dent.* **2013**, *22*, 320–325. [CrossRef] [PubMed]
2. Richi, M.W.; Kurtulmus-Yilmaz, S.; Ozan, O. Comparison of the accuracy of different impression procedures in case of multiple and angulated implants: Accuracy of impressions in multiple and angulated implants. *Head Face Med.* **2020**, *16*, 9. [CrossRef] [PubMed]
3. Bacchi, A.; Consani, R.L.; Mesquita, M.F.; Dos Santos, M.B. Effect of framework material and vertical misfit on stress distribution in implant-supported partial prosthesis under load application: 3-D finite element analysis. *Acta Odontol. Scand.* **2013**, *71*, 1243–1249. [CrossRef] [PubMed]
4. Hanif, A.; Qureshi, A. Complications in implant dentistry. *Eur. J. Dent.* **2017**, *11*, 135–140. [CrossRef]
5. Taylor, T.D.; Agar, J.R.; Vogiatzi, T. Implant prosthodontics: Current perspective and future directions. *Int. J. Oral Maxillofac. Implant.* **2000**, *15*, 66–75.
6. Abduo, J.; Bennani, V.; Waddell, N.; Lyons, K.; Swain, M. Assessing the fit of implant fixed prostheses: A critical review. *Int. J. Oral Maxillofac. Implant.* **2010**, *25*, 506–515.
7. Wee, A.G. Comparison of impression materials for direct multi-implant impressions. *J. Prosthet. Dent.* **2000**, *83*, 323–331. [CrossRef]
8. Kim, S.; Nicholls, J.I.; Han, C.H.; Lee, K.W. Displacement of implant components from impressions to definitive casts. *Int. J. Oral Maxillofac. Implant.* **2006**, *21*, 747–755.
9. Kwon, J.H.; Son, Y.H.; Han, C.H.; Kim, S. Accuracy of implant impressions without impression copings: A three-dimensional analysis. *J Prosthet. Dent.* **2011**, *105*, 367–373. [CrossRef]
10. Kim, K.R.; Seo, K.Y.; Kim, S. Conventional open-tray impression versus intraoral digital scan for implant-level complete-arch impression. *J. Prosthet. Dent.* **2019**, *122*, 543–549. [CrossRef]
11. Asawa, N.; Bulbule, N.; Kakade, D.; Shah, R. Angulated implants: An alternative to bone augmentation and sinus lift procedure: Systemic review. *J. Clinic Diag. Res.* **2015**, *9*, 10–13. [CrossRef] [PubMed]
12. Patel, M.P.; Anilkumar, S.; Chaukramath, R.; Gopalkrishnan, S. Rehabilitation of edentulous maxillary arch with implant- assisted fixed complete prosthesis using multi-unit straight and angulated abutments. *Eur. J. Prosthodont.* **2016**, *4*, 37–40. [CrossRef]
13. Oltra, D.P.; Marti, E.C.; Ali, J.A.; Diago, M.P. Rehabilitation of the atrophic maxilla with tilted implants: Review of literature. *J. Oral Implantol.* **2013**, *39*, 625–632. [CrossRef] [PubMed]
14. Gintaute, A.; Papatriantafyllou, N.; Aljehani, M.; Att, W. Accuracy of computerized and conventional impression-making procedures of straight and tilted dental implants. *Int. J. Esthet. Dent.* **2018**, *13*, 550–565.
15. Alikhasi, M.; Siadat, H.; Nasirpour, A.; Hasanzade, M. Three-dimensional accuracy of digital impression versus conventional method: Effect of implant angulation and connection type. *Int. J. Dent.* **2018**, *2018*, 3761750. [CrossRef]
16. Ribeiro, P.; Herrero-Climent, M.; Díaz-Castro, C.; Ríos-Santos, J.; Padrós, R.; Mur, J.; Falcão, C. Accuracy of implant casts generated with conventional and digital Impressions—An in vitro study. *Int. J. Environ. Res. Public Health* **2018**, *15*, 1599. [CrossRef]
17. Shim, J.S.; Ryu, J.J.; Shin, S.W.; Lee, J.Y. Effects of implant angulation and impression coping type on the dimensional accuracy of impressions. *Implant Dent.* **2015**, *24*, 726–729. [CrossRef]
18. Pereira, A.L.C.; De Freitas, R.F.C.P.; Campos, M.D.F.T.P.; Tôrres, A.C.S.P.; De Medeiros, A.K.B.; Carreiro, A.D.F.P. Trueness of a device for intraoral scanning to capture the angle and distance between implants in edentulous mandibular arches. *J. Prosthet. Dent.* **2021**, *in press*. [CrossRef]
19. Tan, M.Y.; Hui Xin Yee, S.; Wong, K.M.; Tan, Y.H.; Tan, K.B.C. Comparison of three-dimensional accuracy of digital and conventional implant impressions: Effect of interimplant distance in an edentulous arch. *Int. J. Oral Maxillofac. Implant.* **2019**, *34*, 366–380. [CrossRef]
20. Amin, S.; Weber, H.P.; Finkelman, M.; El Rafie, K.; Kudara, Y.; Papaspyridakos, P. Digital vs. conventional full-arch implant impressions: A comparative study. *Clin. Oral Implant. Res.* **2017**, *28*, 1360–1367. [CrossRef]

21. Gimenez, B.; Özcan, H.; Martínez-Rus, F.; Pradíes, L. Accuracy of a digital impression system based on active wavefront sampling technology for implants considering operator experience, implant angulation, and depth. *Clin. Implant. Dent. Relat. Res.* **2015**, *17*, e54–e64. [CrossRef] [PubMed]
22. Lin, W.S.; Harris, B.T.; Elathamna, E.N.; Abdel-Azim, T.; Morton, D. Effect of implant divergence on the accuracy of definitive casts created from traditional and digital implant-level impressions: An in vitro comparative study. *Int. J. Oral Maxillofac. Implant.* **2015**, *30*, 102–109. [CrossRef] [PubMed]
23. Gimenez, B.; Özcan, H.; Martínez-Rus, F.; Pradíes, L. Accuracy of a digital impression system based on active triangulation technology with blue light for implants: Effect of clinically relevant parameters. *Implant Dent.* **2015**, *24*, 498–504. [CrossRef] [PubMed]
24. Sorrentino, R.; Gherlone, E.F.; Calesini, G.; Zarone, F. Effect of implant angulation, connection length, and impression material on the dimensional accuracy of implant impressions: An in vitro comparative study. *Clin. Implant. Dent. Relat. Res.* **2010**, *12*, 63–76. [CrossRef]
25. Kurtulmuş-Yılmaz, S.; Ozan, O.; Özçelik, T.B.; Yağız, A. Digital evaluation of the accuracy of impression techniques and materials in angulated implants. *J. Dent.* **2014**, *42*, 1551–1559. [CrossRef]
26. Gimenez, B.; Özcan, M.; Martinez-Rus, F.; Pradies, G. Accuracy of a digital impression system based on parallel confocal laser technology for implants with consideration of operator experience and implant angulation and depth. *Int. J. Oral Maxillofac. Implant.* **2014**, *29*, 853–862. [CrossRef]
27. Ma, J.; Rubenstein, J.E. Complete arch implant impression technique. *J. Prosthet. Dent.* **2012**, *107*, 405–410. [CrossRef]
28. Flügge, T.; Van Der Meer, W.J.; Gonzalez, B.M.; Vach, K.; Wismeijer, D.; Wang, P. The accuracy of different dental impression techniques for implant-supported dental prostheses: A systematic review and meta-analysis. *Clin. Oral Implant. Res.* **2018**, *29*, 374–392. [CrossRef]
29. Gimenez, B.; Pradies, G.; Martinez-Rus, F.; Özcan, M. Accuracy of two digital implant impression systems based on confocal microscopy with variations in customized software and clinical parameters. *Int. J. Oral Maxillofac. Implant.* **2015**, *30*, 56–64. [CrossRef]
30. Flügge, T.; Att, W.; Metzger, M.; Nelson, K. A Novel method to evaluate precision of optical implant impressions with commercial scan bodies—An experimental approach. *J. Prosthodont.* **2017**, *26*, 34–41. [CrossRef]
31. Drago, C. Ratios of cantilever lengths and anterior-posterior spreads of definitive hybrid full-arch, screw-retained prostheses: Results of a clinical study. *J. Prosthodont.* **2018**, *27*, 402–408. [CrossRef] [PubMed]
32. Papaspyridakos, P.; Rajput, N.; Kudara, Y.; Weber, H.P. Digital workflow for fixed implant rehabilitation of an extremely atrophic edentulous mandible in three appointments. *J. Esthet. Restor. Dent.* **2017**, *29*, 178–188. [CrossRef] [PubMed]
33. Marghalani, A.; Weber, H.P.; Finkelman, M.; Kudara, Y.; El Rafie, K.; Papaspyridakos, P. Digital versus conventional implant impressions for partially edentulous arches: An evaluation of accuracy. *J. Prosthet. Dent.* **2018**, *119*, 574–579. [CrossRef] [PubMed]
34. CS 3500 Real-Time Full Arch Scan I Streamhealth Dental. Available online: https://www.youtube.com/watch?v=sMEykCIrBkI&t=31s (accessed on 29 November 2021).
35. Tutorial: CEREC Omnicam Scanning Technique. Available online: https://www.youtube.com/watch?v=E5hzbRNK7pk&t=95s (accessed on 29 November 2021).
36. 3Shape TRIOS Scan Strategy—Full Arch. Available online: https://www.youtube.com/watch?v=M_KbWcCianY&t=126s (accessed on 29 November 2021).
37. Thanasrisuebwong, P.; Kulchotirat, T.; Anunmana, C. Effects of inter-implant distance on the accuracy of intraoral scanner: An in vitro study. *J. Adv. Prosthodont.* **2021**, *13*, 107–116. [CrossRef] [PubMed]
38. Miyoshi, K.; Tanaka, S.; Yokoyama, S.; Sanda, M.; Baba, K. Effects of different types of intraoral scanners and scanning ranges on the precision of digital implant impressions in edentulous maxilla: An in vitro study. *Clin. Oral Implant. Res.* **2020**, *31*, 74–83. [CrossRef]
39. Mizumoto, R.M.; Alp, G.; Özcan, M.; Yilmaz, B. The effect of scanning the palate and scan body position on the accuracy of complete-arch implant scans. *Clin. Implant. Dent. Relat. Res.* **2019**, *21*, 987–994. [CrossRef]
40. Schmidt, A.; Billig, J.W.; Schlenz, M.A.; Wöstmann, B. The influence of using different types of scan bodies on the transfer accuracy of implant position: An in vitro study. *Int. J. Prosthodont.* **2021**, *34*, 254–260. [CrossRef]
41. Kim, R.J.Y.; Benic, G.I.; Park, J.M. Trueness of digital intraoral impression in reproducing multiple implant position. *PLoS ONE* **2019**, *14*, e0222070. [CrossRef]
42. Mangano, F.G.; Hauschild, U.; Veronesi, G.; Imburgia, M.; Mangano, C.; Admakin, O. Trueness and precision of 5 intraoral scanners in the impressions of single and multiple implants: A comparative in vitro study. *BMC Oral Health* **2019**, *19*, 101. [CrossRef]
43. Braian, M.; Wennerberg, A. Trueness and precision of 5 intraoral scanners for scanning edentulous and dentate complete-arch mandibular casts: A comparative in vitro study. *J. Prosthet. Dent.* **2019**, *122*, 129–136. [CrossRef]
44. Moura, R.V.; Kojima, A.N.; Saraceni, C.H.C.; Bassolli, L.; Balducci, I.; Özcan, M.; Mesquita, A.M.M. Evaluation of the accuracy of conventional and digital impression techniques for implant restorations. *J. Prosthodont.* **2019**, *28*, 530–535. [CrossRef] [PubMed]

machines

MDPI

Article

Research into the Impact of Spindle Speed and Feed Rate Changes on the Life of a Deep-Drilling Technology Tool

Martin Pollák [1,*], Marek Kočiško [1], Jaroslav Petrus [1], Sorin Dumitru Grozav [2] and Vasile Ceclan [2]

1 Faculty of Manufacturing Technologies with a Seat in Presov, Technical University of Kosice, Bayerova 1, 080 01 Presov, Slovakia; marek.kocisko@tuke.sk (M.K.); petrusj@centrum.sk (J.P.)
2 Faculty of Industrial Engineering, Robotics and Production Management, Technical University of Cluj-Napoca, Bulevardul Muncii nr. 103-105, 400641 Cluj-Napoca, Romania; sorin.grozav@tcm.utcluj.ro (S.D.G.); vasile.ceclan@tcm.utcluj.ro (V.C.)
* Correspondence: martin.pollak@tuke.sk; Tel.: +421-55-602-6460

Abstract: The drilling process in real production places ever-increasing demands on the length and accuracy of the holes made. The drilling of holes beyond a length-to-diameter ratio of 5–10 is called deep drilling. The aim of the research was to determine in detail the deep-drilling process input conditions, their impact on the stability of the cutting process and the degree to which the output requirements were achieved. The focus of the analysis was on how the monitored technological and physical impacts translate into achieving the required gun-drill life and the quality and dimensional accuracy of deep holes, as well as their overall impact on tool life. Based on the analysis, tests were conducted to verify the impact of individual parameters on tool life. The obtained results were then statistically evaluated and optimized. Drawing on the evaluated experimental results, solutions and procedures were proposed and implemented in the environment of a real operation. This research obtained the optimal values of the frequency of rotation and displacement to ensure maximum tool life while maintaining the efficiency of the production of drilled parts. At the same time, based on the research, a methodology and recommendations for deep-drilling technology were developed.

Keywords: gun drill tool; deep-drilling technology; optimization; tool life

Citation: Pollák, M.; Kočiško, M.; Petrus, J.; Grozav, S.D.; Ceclan, V. Research into the Impact of Spindle Speed and Feed Rate Changes on the Life of a Deep-Drilling Technology Tool. *Machines* **2022**, *10*, 268. https://doi.org/10.3390/machines10040268

Academic Editor: Davide Astolfi

Received: 8 March 2022
Accepted: 6 April 2022
Published: 8 April 2022

1. Introduction

The rapid development of new technologies and growing consumer demands mean that companies are trying to find new and more efficient ways of developing and researching their products and, thus, receive the retroactive economic benefits of production. Current technical progress and the application of specific solutions and innovations, thanks to which manufacturing companies can promptly improve established procedures and techniques, is expected to ensure high production efficiency. One such major technological operation is the drilling process [1].

The drilling of deep holes can be carried out in several ways: with a rotating workpiece or with a rotating tool, or with a tool rotating simultaneously with the rotating workpiece. Regardless of the drilling method, the basic principles of drilling and, moreover, the correct choice of cutting speed and feed rate, remain unchanged [2,3]. Successful chip formation and its removal from the cutting edge without damaging the tool and workpiece are, in principle, the most important factors influencing the quality of drilled holes [4]. Small-diameter holes are created using the technology of deep (gun-drill tool) drilling with the STS (single-tube system). An alternative to STS technology in cases where the drilling of a smaller production series is carried out and a specialized machine is not available is the ejector system [5,6]. Due to the state-of-the-art development of carbide monolithic tools, it is currently possible to make holes with smaller diameters with carbide screw augers featuring length-to-diameter ratios of up to 30 mm [7].

Both the design and the materials and coating used in deep-drilling technology have evolved [8]. The most recent solution is the design of gun-drill tool bodies with interchangeable cutting plates and interchangeable guide surfaces that provide the additional adjustment of dimensions using different washer widths [9]. Nevertheless, even with the use of the latest auger design, one of the most important conditions for maintaining a stable process is still relevant, namely the application of sufficient cutting fluid pressure to dissipate heat, to remove chips and to provide the cutting process with lubrication [10–12]. Other influences, such as the stability and accuracy of the machine, the properties of the machined material, the technological parameters and the use of a suitable tool also enter the process. All these factors affect the resulting process, which must, first and foremost, take into account the economic viability of production [13–15].

Several important authors from the worldwide scientific community address the issue of deep drilling. Hong J. et al. discuss a new method of data monitoring to estimate tool wear when making holes with the application of deep-drilling technology. The proposed monitoring method is a Gaussian regression process (GRS) based on a combination of force, torque and vibration signals that are extracted from a predefined segment. The results show that the accuracy of the tool wear estimate can be improved by the method proposed, which significantly outperforms other methods. Examples include linear regression and support vector regression [16]. In their publications, focusing on the force system and performance of the AlSi 1045 cutting steel drilling rig, Wang Y. et al. describe a drilling-force system created by integrating the forces on the main cutting edge by wear and outer diameter (OD) and mechanical equilibrium equations using a specific energy model, which consists of geometric parameters, machining parameters and material properties [17]. The researchers Zhang X. et al. are engaged in similar research and the development of high-quality augers with precisely controllable tool geometry. Through planned experiments, they experiment with the effects of changes in apex deviations on deviations in hole flatness. Through a series of FEM analyses and further measurements of the internal rolling of the machined holes, it was concluded that unbalanced cutting force components, applied to previously drilled holes due to inconsistent apex displacement, cause newly formed holes to deviate from the thin wall due to the greater material strain on the thin-walled side relative to the side with thick walls [18]. Studies of precision and surface roughness after deep drilling have been carried out by Kirsanov, S.V. and Babaev A.S., particularly on sintered carbide and water-resistant coating, which have higher wear resistance and ensure the lower surface roughness of drilled holes [19]. Based on their experimental results, Mann J.B. et al. conducted research in the field of drilling effects with overlapping controlled low-frequency modulation (MAM) and on the surface structures created after deep drilling [20]. With development of science and technology, the precision of the individual-component manufacturing process by deep-drilling technology increases together with the demand and constantly growing requirements for the tools and equipment needed for these operations. Therefore, Zhang K.L. researched the improvement in the accuracy and efficiency of these devices for real use in practice [21]. Schnabel D. et al. deal with deep-drilling processes resulting in small holes and openings, addressing especially the cooling and lubrication of the machined zone, which translates into longer tool life. In their publication, they model the movement of the cooling lubricant during the drilling process using the finite volume and the hydrodynamics of cooled particles. The results prove that changing the coolant close to the cut guarantees the desired cooling effect. The coolant is modeled by the smooth-pin hydrodynamics method and the end-point method is used to describe the points. This study also provides transient simulations of the hydrodynamics of the smooth particles, which show insufficient weight, and recommends replacing the cutting edges corresponding to the steady state and the results of the simulation of the final volume with a complete bore filled with coolant [22]. In their studies on point formation and wear behavior during the drilling process of titanium-coated forged steel S 48CS1V, Wang Y. et al. point out the complex wear of the hard carbide tool in the chip-forming tests performed on this steel. They claim that high alternative stress resulted in plastic strain, and that for atomic absorption, cold welding caused the carbide substrate grains to detach from the cobalt bond, thus causing the wear and tear of the tool [23]. In addition to turning and milling, drilling holes with augers is the most commonly used method in machining technology. A great number of machined holes is naturally reflected in the significant consumption of augers not only by individual companies, but also on a global scale. The choice of suitable types and auger design solutions therefore has a significant impact on the accuracy, efficiency and productivity of drilling, as well as on its economical use and consumption [24].

Drilling is a technological process resulting in circular cross-section holes. Their length depends on the design requirements and it is often necessary to drill holes with depths of up to five times their diameter [25]. This type of operation is referred to as deep drilling. Deep-drilling technology places greater emphasis on the cutting process than standard-hole (≤5D) drilling. To ensure a stable cutting process, as well as dimensional accuracy and quality standards, the following requirements must be met [26,27]:

(a) The design of the tool must ensure good guidance of the tool in the hole and the geometry of the cutting part of the tool must ensure the formation of chips that can be easily removed from the hole area.
(b) The tool must be sufficiently rigid to be able to drill efficiently in order to ensure that the drilling process runs smoothly.
(c) The design of the tool must allow smooth chip removal from the hole area to ensure the smooth running of the drilling process and to prevent damage to the tool.
(d) The technology must ensure a favorable relationship between cutting productivity and production economy through the appropriate choice of machine and other working conditions while adhering to all the required technical conditions for the accuracy of the drilled part.
(e) The tool design must provide the prescribed dimensional accuracy, geometric shape and surface quality.

Monolithic long screw augers STS22 1130 and STS 22 1150 or augers with internal cooling STS 22 1154, manufactured with cutting fluid supply conduits, are used for drilling deep holes. Because the auger grooves fill quickly with chips, the augers need to be pulled out of the hole often to cool and remove the chips. This type of drilling is called intermittent drilling [28]. To avoid auger overload, the work cycle for augers ranging between 10 to 30 mm in diameter is selected according to the following established rules: the first bore is drilled to a depth of 2.5 to 4D; for subsequent bores, the hole is deepened gradually by 2D, 1.5D, 1D; and every subsequent attempt deepens the bore by 0.5D. Drills with an automatic work cycle operate in such a way that the auger spindle returns to its initial position at rapid traverse upon reaching a certain depth. After the chips have been emptied, it returns to the bottom of the hole at rapid traverse and continues drilling [29,30].

Due to difficult working conditions, the cutting ratios are reduced compared to those in short-hole drilling. Taking these values as a basis, for a hole depth of L = 3 to 15D, the cutting speed is reduced to 0.9 or further down, to 0.4, while the feed rates are reduced to 0.3–0.6 times of the basic values.

Carbide augers work similarly to HSS. They are manufactured up to a length of 30xD and they always feature internal cooling, thanks to which they do not require an intermittent work cycle. A precondition is a pilot hole, which must meet the hole accuracy requirements, usually in IT8. The depth of the pilot hole is given by at least 2D. Compared to HSS drills, TK drills can be used for drilling at higher speeds. The feed rate for TK drills can be set min. 5 times higher and set spindle speed min. 3 times higher compared to HSS, which depends on the manufacturer's recommendations. The drills can also be cooled with oil mist (MQL), which is capable of ensuring good conditions for the stable cutting process [31].

The gun-drill tool is a single-wedge tool with internal flushing and external chip removal. It consists of three basic parts—the drill head, the drill tube, and the clamping shank. The tool has an outer straight V-shaped path and an inner conduit for the coolant and lubricant. The liquid flows through the tool axis, bypasses the cutting edge, carries away the emerging chips and washes them out of the hole through the external path. The drill head is typically carved from carbide, but in cases involving larger diameters, it can be made of steel with carbide elements (cutting edge and guide surfaces) bonded to them. The head has guide surfaces ground along the circumference, which are in contact with the drilled material during drilling. For larger diameters, replaceable elements (plates and guide surfaces) are widely used in current work. The forces acting on the tool during drilling strongly press the guide surfaces of the head against the wall of the drilled hole, so that the surface is slightly deformed, resulting in the high machining quality typical of all deep-drilling tools.

The aim of this research was to obtain the optimal values of the two main parameters influencing the tool life of deep-drilling technology, namely the feed rate and the tool rotation frequency. The first series of tests was performed to obtain the optimal spindle speed at a constant feed rate of 0.04 mm·min^{-1}. The second series of tests was focused on finding the optimal value of the feed rate while maintaining a constant spindle speed of 2600 min^{-1}.

2. Methodology

For the purposes of the experiment, the TBT gun drill 10.5 × 380 K15 shown in Figure 1 was used. It is a gun drill with a brazed carbide head, 10.5 mm in diameter and with a total length of 380 mm. The effective gun-drill length corresponds to the length of the hole on the test part, which is 280 mm. The carbide head type guide surfaces are of G design, complemented with a longitudinal groove on the main cutting edge, which helps to create chips. The carbide head is coated with TiAlN coating, which increases the auger's life. The smooth clamping shank is 20 mm in diameter and is designed for clamping in a hydraulic clamp. The selected gun-drill tool is clamped in a Schunk hydro clamp of TENDO ES type. Clamping is carried out directly in the connection cone of the clamp.

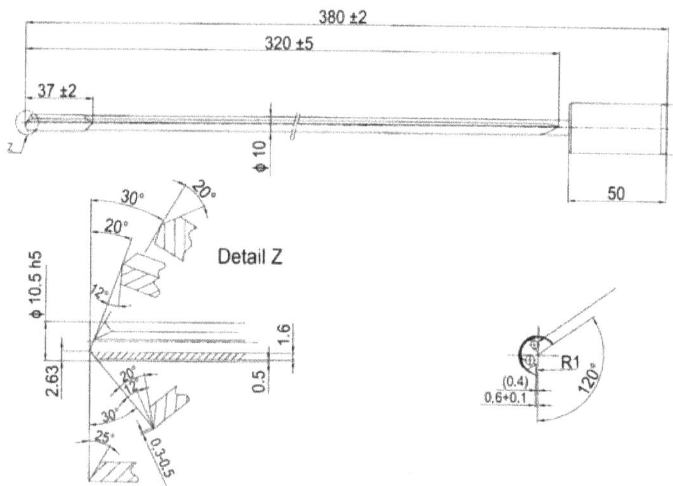

Figure 1. Gun-drill tool used in the experiment.

To present and analyze the input conditions, an Ishikawa diagram was constructed, as shown in Figure 2. The main axis of the diagram represents the main problem and the individual influences that affect the tool life during the deep-drilling process. Based on the aforementioned factors influencing tool life, the individual effects in this diagram were verified under real engineering operation. The main factors affecting tool life in deep-drilling are the tool, the workpiece material, the tool geometry, the machine, the clamping and the coolant.

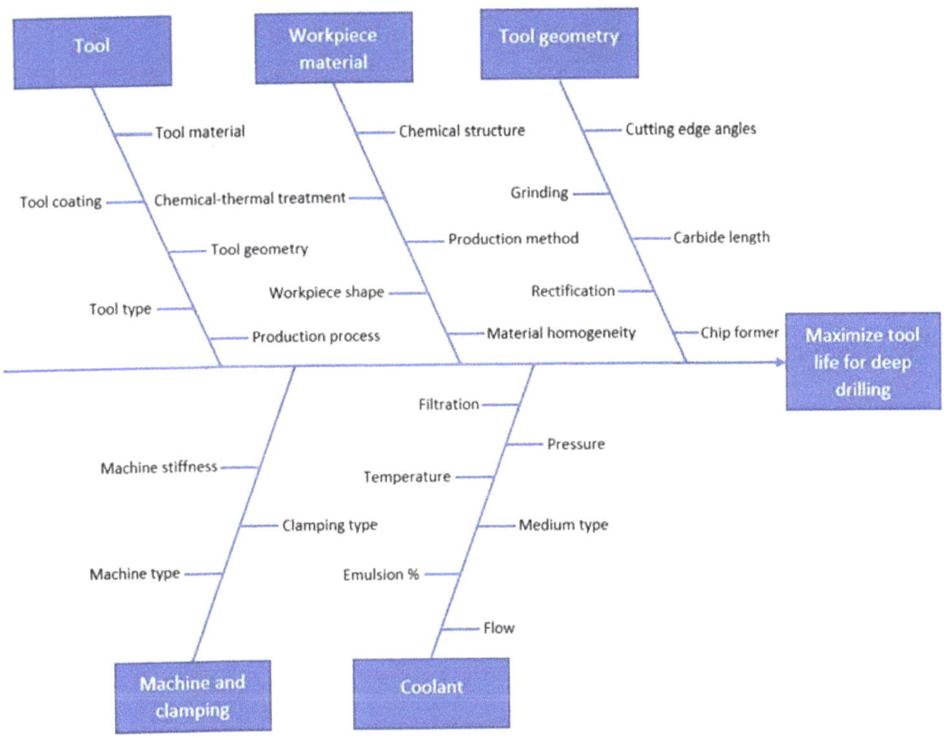

Figure 2. Ishikawa diagram.

Tests of the gun-drill tool life were performed in real production in a plant. The series of tests verified the technological and technical recommendations of the gun-drill manufacturer for a stable cutting process on a given part using deep-drilling technology.

2.1. Experimental Machinery

Testing was carried out under the automotive production conditions at the production line. The machine used for testing in the real production process was a thickness planer and boring center with a portal loader from CZ-Tech Celakovice of the ZAH720 type. Two heads were used on the ZAH720 machine, as shown in Figure 3; the right head was designed for VDI clamping and the other for Bolt-one. The tool holders used were from ALGRA. Electromechanical turrets of the TAN series are designed for installation on CNC lathes. They are used for clamping dovetail toolholders according to VDI 3425 for use in the automatic positioning of toolholders consisting of a solid base and a rotating head made of hardened ground steel. The cooling of the machining center tools is ensured by a system of devices, which consists of a low-pressure pump, a high-pressure pump, a filter, and a small-metal-particle magnetic filter. The coolant pump provides the pressure necessary to cool, lubricate and clean all chips off the tools. A wall-plus pump was used for these needs, supplying a coolant pressure of 18 bar at a flow rate of 10.5 L/min. Its power was sufficient for standard cooling of the turning tools and of the pilot auger. A Grundfos MTS high-pressure pump was used to cool the gun-drill tool.

Figure 3. ZAH720 thickness planer workspace.

2.2. *Parameters of the Test-Part Selection*

The transmission input shaft shown in Figure 4 was selected as a test part, with a lubrication hole 10.5 mm in diameter and 280 mm long. The deep-drilling tests were performed on the shaft of a six-speed DSG gearbox. Deep-hole drilling consists of drilling a 240 millimeter-long hole 10.5 mm in diameter. This was a lubrication hole, the diameter of which was also chosen with regard to the requirement of relieving the shaft. A vertical lubrication hole with a diameter of 3 mm, located at the end of the shaft, must be smoothened after drilling due to formation of sharp edges up to 10.5 mm in diameter. For this reason, the drilling method was adapted to drill a diameter of 10.5 mm to full extent before drilling a perpendicular hole with a diameter of 3 mm and again applying a drilling cycle to smoothen the D3 hole with the gun-drill. At this point, increased wear on the side edge of the cannon auger occurred, as this was an intermittent cut.

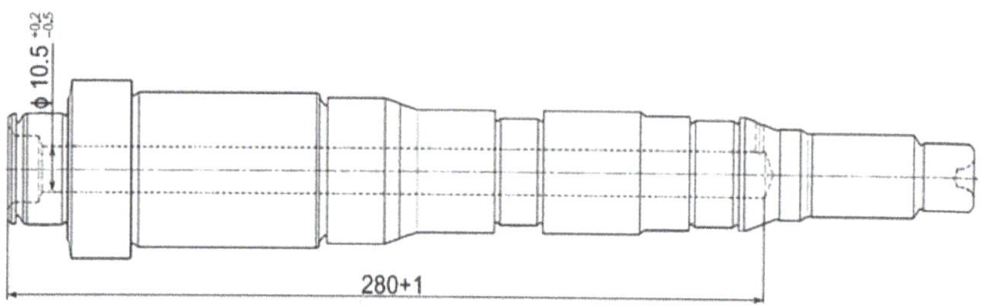

Figure 4. Transmission shaft of the part tested in the experiment.

The material used for the part selected to test the deep-drilling process complied with the company standard. With its composition and properties, the material resembled the 16MnCr5 and EN 10184. The chemical composition, according to the company standard, is presented in Table 1.

Table 1. Chemical composition of the drilled part given in %.

C	Si	Mn	P	S	Cr	Al	N
0.14–0.22	≤0.12	1.00–1.50	≤0.035	0.02–0.035	0.80–1.30	0.015–0.040	≤0.015

The STS 14220 material is a high-grade structural steel for cementation, which is well thermoformable. It is well machinable and weldable after soft and cold annealing. This steel is used for the production of machine parts with the possibility of refining up to diameters of 35 mm for cementation with high core strength, with the use for the production of shafts, gears, camshafts or gear couplings. This semi-finished product is supplied as a form of die forging.

2.3. Pilot-Hole Drilling

When using a gun-drill tool on conventional machines, it is necessary to drill a pilot hole as a guide, the length of which must be between 1 and 1.5 times the gun-drill diameter. For correct guidance, the hole must be within a D8 tolerance. The pilot hole is usually drilled at a greater depth, where the carbide auger has a significantly higher feed rate and, thus, reduces machining time. For a specific part, the gun drill is designed with respect to the given hole profile and the design of the auger is more graded. The pilot gun-drill tool Carbide 10.5/14.5, shown in Figure 5, is made of K40UF carbide from Konrad Friedrichs.

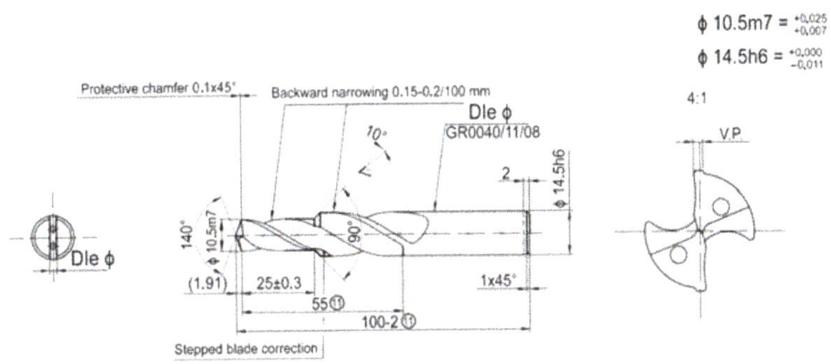

Figure 5. Pilot gun-drill tool Carbide 10.5/14.5 used in the experiment.

3. Statistical Evaluation and Optimization

3.1. Dependence of the Number of Drilled Holes up to the Moment of Tool Damage When Changing the Speed-Constant Feed Rate

The manufacturer-recommended values were used to test the technological parameters of a given part of the transmission shaft, with the cutting speed ranging from 70 to 100 m·min^{-1} for the given workpiece material. Figure 6 shows a plotted representation of the course of the number of drilled holes up to the moment the tool was damaged at different speeds and a constant feed rate of 0.04 mm·min^{-1}.

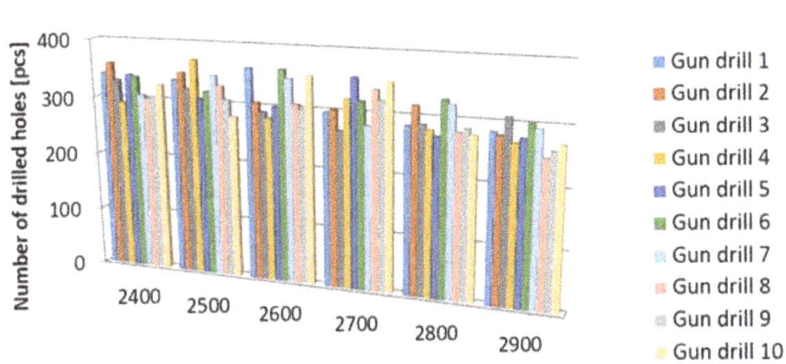

Figure 6. Plotted number of holes drilled up to the moment the tool was damaged when the speed changed and the feed rate remained constant.

The cutting speed of 70–100 m·min^{-1} represents the spindle speed, which ranged from 2100 to 3000 min^{-1}. The selected spindle test range speed was 2400–2900 min^{-1}. The tests were performed on a randomly selected gun-drill tool, while new augers and those that had been re-ground several times were excluded. Figure 7 shows the dependence of the number of drilled holes up to the moment of the tool damage and the length of the drilling time on the tool spindle speed.

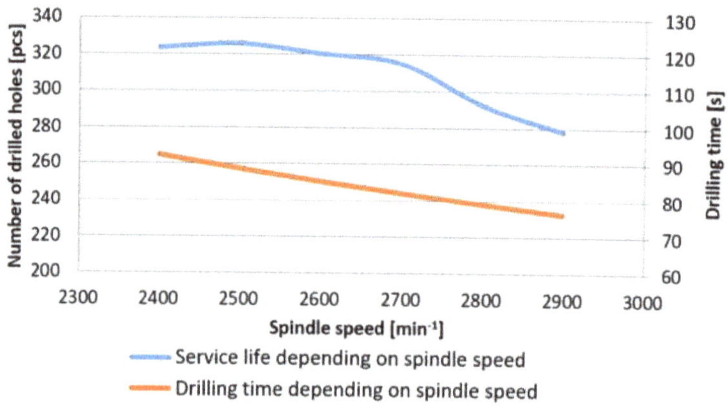

Figure 7. Dependence of the number of holes drilled up to the moment of tool damage on the cutting speed.

The resulting values of the number of holes drilled up to the moment the tool was damaged and, thus, the life of the gun-drill tool under different speeds, show a significantly decreasing service life at speeds above 2700 min^{-1}, as shown in Figure 7. The speed of 2700 min^{-1} corresponded to the cutting speed of 90 m·min^{-1}. This value seems to be the most suitable both in terms of the tool life and in terms of the operation time; at the cutting speed of 90 m·min^{-1}, the length of the drilling operation was 82 s.

3.2. Dependence of the Number of Drilled Holes up to Moment of Tool Damage When Changing the Feed Rate-Constant Speed

For testing the technological parameters on the selected test part, the speed was also based on the values recommended by the manufacturer, while the feed rate for the given workpiece material was given in the range of 0.03–0.06 mm·min^{-1}. This range was significantly lower than the range already used at the thickness planer. Feed rate values of 0.065 mm·min^{-1} were used as a standard. A test interface of 0.05–0.08 mm·min^{-1} was chosen for the testing. A constant value of 2600 min^{-1} was chosen for the speed.

Figures 8 and 9 show plotted dependencies representing the number of holes drilled up to the moment the tool was damaged due to a change in the feed rate, as well as both the dependence of the number of holes drilled up to the moment of tool damage and the duration of the drilling operation on the change in feed rate.

The resulting values of the number of holes drilled up to the moment of damage to the gun-drill tool at different feed rates showed a significantly decreasing tool life at feed rates above 0.07 mm·min^{-1}. The values of the number of holes drilled up to the moment the tool was damaged as a function of the feed are shown in Figure 9. In the tests, the feed rate of 0.55 mm·min^{-1} resulted in the longest gun-drill tool life, but the drilling time was 98.8 s, which is economically inefficient. In light of the economic efficiency of the deep-drilling operation, a value of 0.07 mm·min^{-1} appears to be the most suitable feed-rate value, with the operation lasting 82 s in this case.

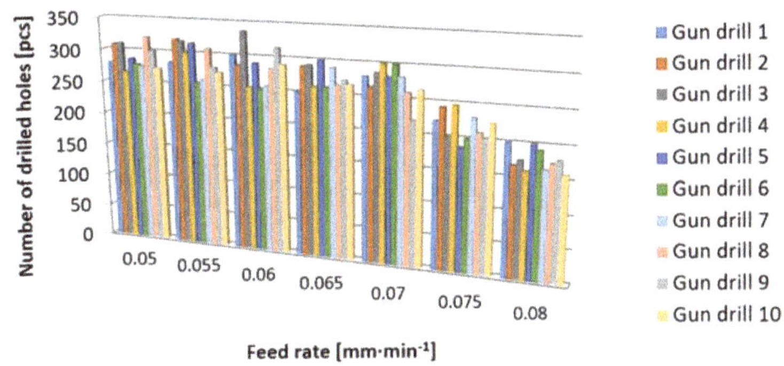

Figure 8. Plotted number of holes drilled until the tool was damaged, when the feed rate changed and the speed remained constant.

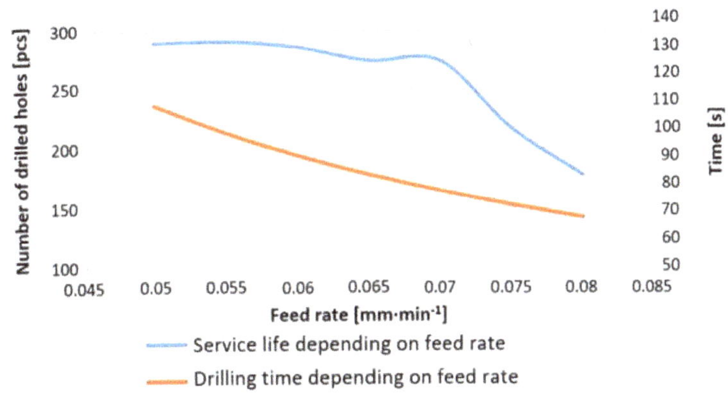

Figure 9. Dependence of the number of drilled holes up to the moment of tool damage and the operation's duration on the change in feed rate.

The reduction in the number of holes drilled by increasing the feed was due to poorer chip removal from the tool and the consequent increase in tool temperature. The heat generation raised the temperature above the melting point of the alloying elements (e.g., Co and W) and the consequent mechanical properties, hardness and chemical wear resistance of the drill bit decreased dramatically due to the peeling of the alloying elements. Heat shock generated vertical or oblique cracks on the blade. These cracks widened and merged and, subsequently, broke the blade.

3.3. Statistical Evaluation and Optimization of the Dependence of the Auger Life at a Constant Feed Rate and Different Spindle Speeds

Figures 10 and 11 show a comparison between the plotted number of holes drilled up to the moment of tool damage and the speed used in the experiment. The results of the comparisons subject to the specified parameters are subsequently presented in Tables 2 and 3, which compare the obtained measurement results based on the number of drilled holes up to the moment of tool damage in terms of the change in experimental speed.

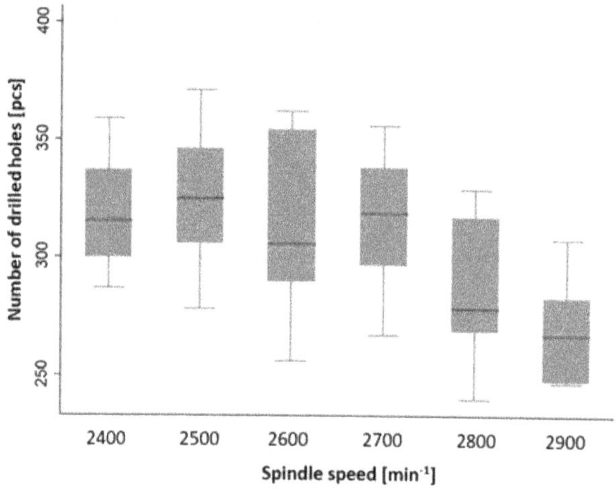

Figure 10. Plotted comparison of the number of drilled holes up to the moment of tool damage and the speed used in the experiment.

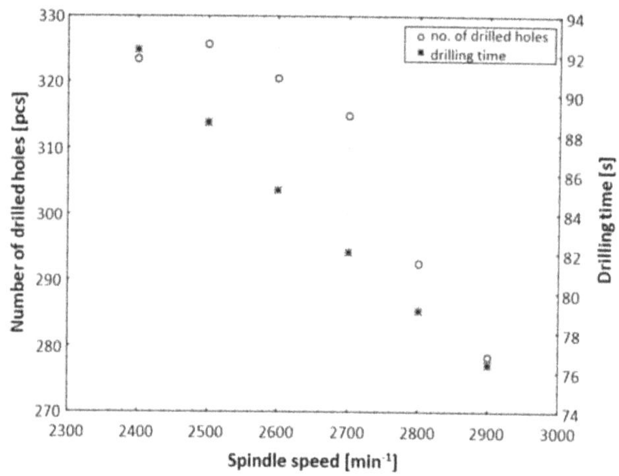

Figure 11. Dependence of the number of holes drilled up to the moment of tool damage and drilling time on speed used in the experiment.

Table 2. Results of a comparison of the number of holes drilled up to the moment of tool damage in terms of speed change.

	Kruskal–Wallis ANOVA by Ranks. Number of Drilled Holes (Analysis). Independent (Grouping) Variable: Spindle Speed. Kruskal–Wallis Test: H (5, N = 60) = 21.13129, p = 0.0008			
	Code	**Valid N**	**Sum of Ranks**	**Mean Rank**
2400 min^{-1}	1	10	376.5000	37.65000
2500 min^{-1}	2	10	410.5000	41.05000
2600 min^{-1}	3	10	360.0000	36.00000
2700 min^{-1}	4	10	354.5000	35.45000
2800 min^{-1}	5	10	203.5000	20.35000
2900 min^{-1}	6	10	125.0000	12.50000

Table 3. Results of multiple comparisons of p values of the number of drilled holes up to the moment of tool damage in terms of speed.

	Multiple Comparisons of p Values (Two-Tailed). Number of Drilled Holes (Analysis). Independent (Grouping) Variable: Spindle Speed. Kruskal–Wallis Test: H (5, N = 60) = 21.13129, p = 0.0008					
	2400 min^{-1} R: 37.650	**2500 min^{-1} R: 41.050**	**2600 min^{-1} R: 36.000**	**2700 min^{-1} R: 35.450**	**2800 min^{-1} R: 20.350**	**2900 min^{-1} R: 12.500**
2400 min^{-1}		1.000000	1.000000	1.000000	0.401362	0.019220
2500 min^{-1}	1.000000		1.000000	1.000000	0.120608	0.003851
2600 min^{-1}	1.000000	1.000000		1.000000	0.676409	0.039334
2700 min^{-1}	1.000000	1.000000	1.000000		0.797884	0.049479
2800 min^{-1}	0.401362	0.120608	0.676409	0.797884		1.000000
2900 min^{-1}	0.019220	0.003851	0.039334	0.049479	1.000000	

The value of the achieved level of significance of the Kruskal–Wallis variance analysis (p = 0.0008) makes it possible to state that there was a significant relationship between the speed and the number of drilled holes up to the moment of tool damage, with a significance level of 5%. In other words, speed has a statistically significant effect on tool life. A closer analysis of the multiple comparisons of the p values shows that there was a statistically significant difference between the number of holes drilled up to the moment of tool damage at: 2900 min^{-1} and 2400 min^{-1}; 2900 min^{-1} and 2500 min^{-1}; 2900 min^{-1} and 2600 min^{-1}; and 2900 min^{-1} and 2700 min^{-1}, respectively. To analyze the dependence of the number of drilled holes up to the moment of tool damage on speed, four different models were considered (Table 4), indicating their basic characteristics.

Table 4. Comparison table of the model for analyzing the dependence of the number of holes drilled up to the moment of tool damage and speed.

Model		AICc	BIC	SSE	RMSE	R-Square
Linear		58.579379	45.954658	304.02571	8.7181666	0.8377591
Exponential 2P		59.133867	46.509145	333.46133	9.1304618	0.822051
Quadratic		75.245678	34.412715	32.944643	3.3138418	0.9824194
Logistic 3P		78.806526	37.973564	59.638621	4.4586478	0.9681743

A plotted representation of the individual models examining the dependence of the number of holes drilled up to the moment of tool damage on speed, according to Table 4, is shown in Figure 12.

The quadratic model was chosen for the resulting analysis, which is shown in Figure 13, since the highest value of the adjusted determination index demonstrated the ability of the model to describe the data. The resulting general expression of the model is as follows:

$$a + b * v + c * v^2 \quad (1)$$

where a is the absolute term, v is the speed, b is the linear term and c is the quadratic term.

$$-1324.217 + 1.3334511v - 0.000269v^2 \quad (2)$$

The numerical values of the individual terms are given in Table 5. A plotted representation of the quadratic model in the ranges of the speeds used is shown in Figure 14.

In general, we obtain stationary points if we set the first derivation of the dependence function to zero:

$$\mathrm{dv}\left(-1324.217 + 1.3334511v - 0.000269v^2\right) = 0 \quad (3)$$

$$1.3334511 - 0.000538v = 0 \quad (4)$$

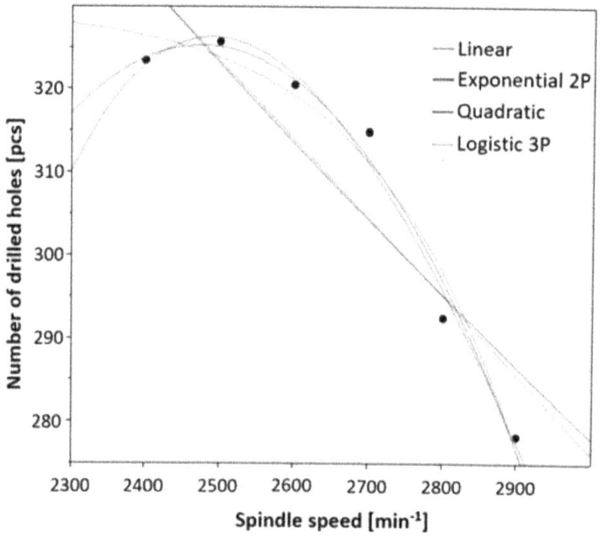

Figure 12. Plotted representation of individual models examining the dependence of the number of the holes drilled up to the moment of tool damage on speed change.

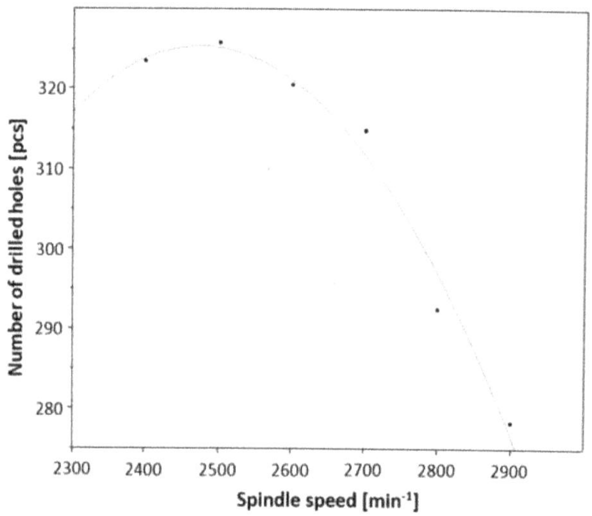

Figure 13. Plotted selected quadratic model of the dependence of the number of holes drilled up to the moment of tool damage on speed change.

Table 5. Basic parameters of the quadratic model of dependence of the number of holes drilled up to the moment of tool damage on speed change.

Parameter	Estimate	Std. Error	Lower 95%	Upper 95%
Intercept	−1324.217	379.86959	−2068.748	−579.6862
Slope	1.3334511	0.2875571	0.7698495	1.8970527
Quadratic	−0.000269	−0.000542	−0.000376	−0.000163

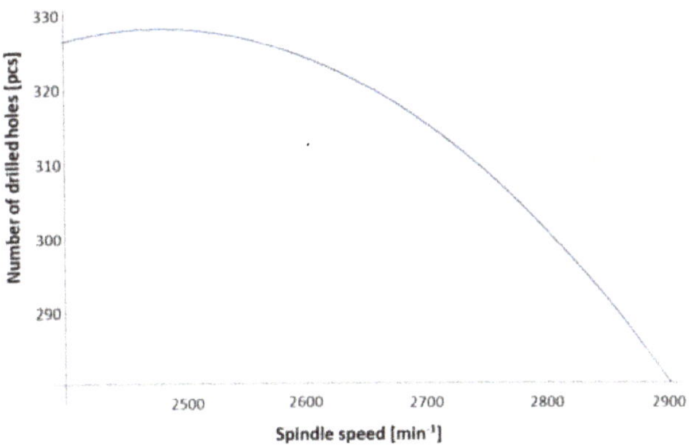

Figure 14. The course of dependence of the number of holes drilled up to the moment of tool damage in the range of the speeds used.

A plotted representation of the first derivation of the investigated function is shown in Figure 15, presenting the course of the first derivation of the number of drilled holes up to the moment of tool damage and speed change.

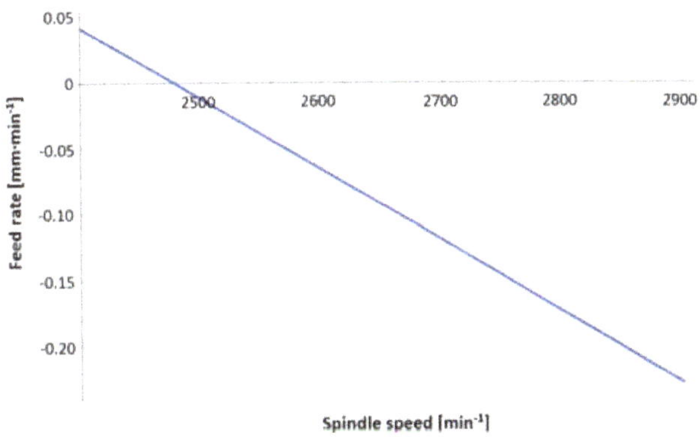

Figure 15. The course of the first derivation of dependence of the number of drilled holes up to the moment of tool damage and speed change.

By solving the following equation (Solve $(1.3334511 - 0.000538v = 0, v)$) we arrive at the stationary point:

$$\{\{v \to 2478.533643122677\}\} \quad (5)$$

By inserting the value of the stationary point into the original dependence of the number of drilled holes up to the moment of tool damage depending on the change in speed, we obtain the value of the local function maximum sought (due to the fact that the second derivation of the function <0), that is to say, the number of drilled holes up to the moment of tool damage at a speed of 2478.53 min^{-1}:

$$\{328.2847064044704, \{v \to 2478.5336431226765\}\} \quad (6)$$

3.4. Statistical Evaluation and Optimization of the Dependence of the Number of Holes Drilled up to the Moment of Tool Damage When Changing the Feed Rate at a Constant Speed

Figures 16 and 17 show a graph illustrating a comparison of the number of holes drilled and the drilling time up to the moment of tool damage under the different feed rates used in the experiment. The results of the comparisons subject to the specified parameters are subsequently presented in Tables 6 and 7, which compare the measurement results obtained based on the number of holes drilled up to the moment of tool damage in terms of the changed experimental feed rate.

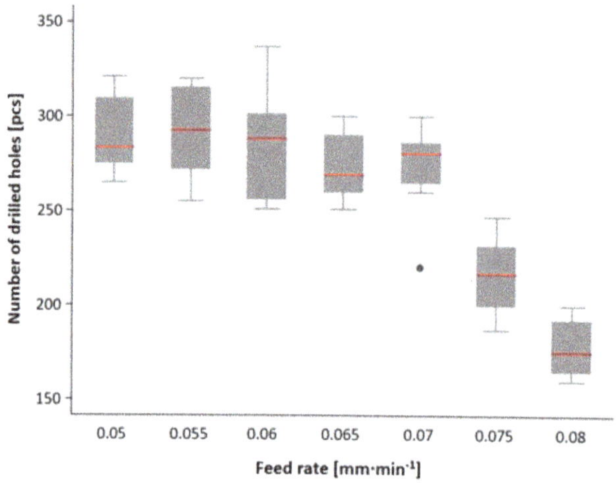

Figure 16. Plotted comparison of the number of holes drilled and the feed rates used in the experiment.

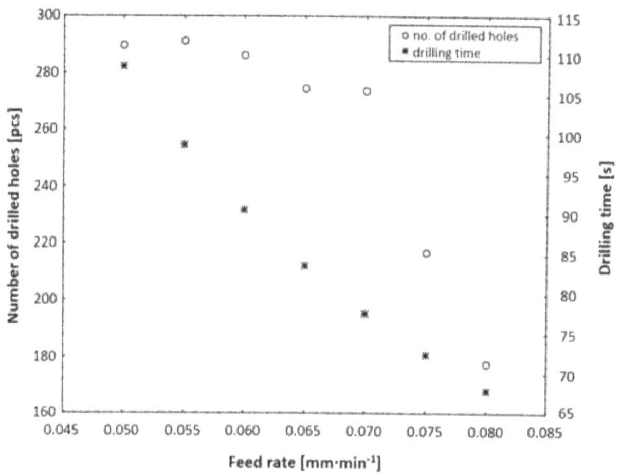

Figure 17. Dependence of the number of drilled holes and drilling time on feed rates used in the experiment.

The value of the achieved level of significance of the Kruskal–Wallis variance analysis ($p = 0.0008$) makes it possible to state that there was a significant relationship between the speed and the number of holes drilled up to the moment of tool damage, with a significance level of 5%. In other words, feed rate has a statistically significant effect on tool life. A closer analysis of the multiple comparisons of p values shows that there was a statistically significant difference between the number of holes drilled at of the following feed rates: 0.075 mm·min^{-1} and 0.05 mm·min^{-1}; 0.075 mm·min^{-1} and 0.055 mm·min^{-1}; and 0.075 mm·min^{-1} and 0.06 mm·min^{-1}, respectively. Furthermore, this also applied to the following feed rates: 0.08 mm·min^{-1} and 0.05 mm·min^{-1}; 0.08 mm·min^{-1} and

0.055 mm·min^{-1}; 0.08 mm·min^{-1} and 0.06 mm·min^{-1}; 0.08 mm·min^{-1} and 0.065 mm·min^{-1}; and 0.08 mm·min^{-1} and 0.07 mm·min^{-1}, respectively.

Table 6. Results of the comparison of the number of holes drilled up to the moment of tool damage in terms of feed rate.

	Kruskal–Wallis ANOVA by Ranks. Number of Drilled Holes (Analysis). Independent (Grouping) Variable: Feed Rate. Kruskal–Wallis Test: H (6, N = 70) = 44.78770 p = 0.0000			
	Code	**Valid N**	**Sum of Ranks**	**Mean Rank**
0.05 (mm·min^{-1})	1	10	496.0000	49.6000
0.055 (mm·min^{-1})	2	10	499.5000	49.9500
0.06 (mm·min^{-1})	3	10	468.5000	46.8500
0.065 (mm·min^{-1})	4	10	400.0000	40.0000
0.07 (mm·min^{-1})	5	10	406.0000	40.6000
0.075 (mm·min^{-1})	6	10	155.0000	15.5000
0.08 (mm·min^{-1})	7	10	60.0000	6.0000

Table 7. Results of multiple comparison of p values of the number of holes drilled up to the moment of tool damage in terms of change in the feed rate.

	Multiple Comparisons of p Values (Two-Tailed). Number of Drilled Holes (Analysis). Independent (Grouping) Variable: Cut Feed Kruskal–Wallis Test: H (6, N = 70) = 44.78770 p = 0.0000						
	0.05 mm·min^{-1} R: 49.60	**0.055 mm·min^{-1} R: 49.950**	**0.06 mm·min^{-1} R: 46.850**	**0.065 mm·min^{-1} R: 40.000**	**0.07 mm·min^{-1} R: 40.600**	**0.075 mm·min^{-1} R: 15.500**	**0.08 mm·min^{-1} R: 6.0000**
0.05 (mm·min^{-1})		1.000000	1.000000	1.000000	1.000000	0.003762	0.000035
0.055 (mm·min^{-1})	1.000000		1.000000	1.000000	1.000000	0.003226	0.000029
0.06 (mm·min^{-1})	1.000000	1.000000		1.000000	1.000000	0.012011	0.000151
0.065 (mm·min^{-1})	1.000000	1.000000	1.000000		1.000000	0.149184	0.003930
0.07 (mm·min^{-1})	1.000000	1.000000	1.000000	1.000000		0.122183	0.003018
0.075 (mm·min^{-1})	0.003762	0.003226	0.012011	0.149184	0.122183		1.000000
0.08 (mm·min^{-1})	0.000035	0.000029	0.000151	0.003930	0.003018	1.000000	

To analyze the dependence of the number of drilled holes on the feed rate, five different models were considered, as presented in Table 8, which provides their basic characteristics.

Table 8. Comparison table of the model for analyzing the number of holes drilled up to the moment of tool damage and feed rate.

Model		**AICc**	**BIC**	**SSE**	**RMSE**	**R-Square**
Quadratic		637.77225	646.15085	32,812.781	22.130138	0.7725013
Cubic		637.74334	648.04832	31,729.031	21.92585	0.7700151
Quartic		639.73713	651.89476	31,547.316	22.030509	0.781275
Linear		672.86817	679.25002	55,944.286	28.682923	0.6121251
Exponential 2P		679.18499	685.56684	61,227.507	30.006741	0.5754953

A plotted representation of the individual models according to Table 8 is shown in Figure 18.

The quadratic model was chosen for the resulting analysis, since the highest value of the adjusted determination index demonstrated the ability of the model to describe the data. The resulting general expression of the model is as follows:

$$a + b * f + c * f^2 \tag{7}$$

where a is the absolute term, f is the feed rate, b is the linear term and c is the quadratic term.

The numerical values of the individual terms are given in Table 9, which lists the basic parameters of the quadratic model of the dependence of the number of holes drilled up to the moment of tool damage on the feed rate.

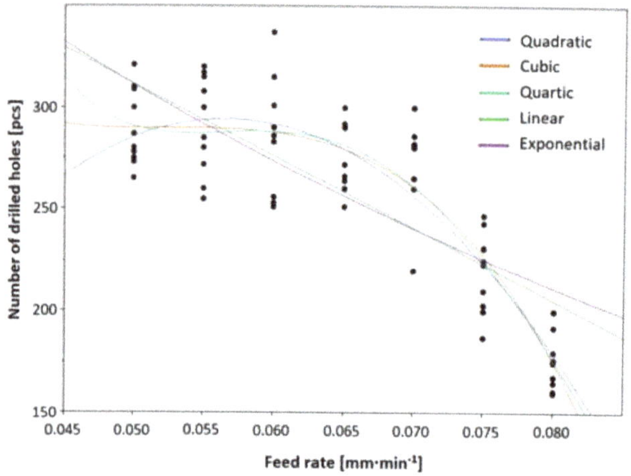

Figure 18. Plotted representation of individual models for examining the dependence of the number of holes drilled up to the moment of tool damage on the feed rate.

Table 9. Basic parameters of the quadratic model of the dependence of the number of drilled holes up to the moment of tool damage on the feed rate.

Parameter	Estimate	Std Error	Lower 95%	Upper 95%
Intercept	−376.4143	127.18298	−625.6884	−127.1402
Slope	23,736.19	3979.3245	15,936.858	31,535.523
Quadratic	−209,904.8	30,542.492	−269,766.9	−150,042.6

A plotted representation of the quadratic model in the range of the feed rates used is shown in Figure 19.

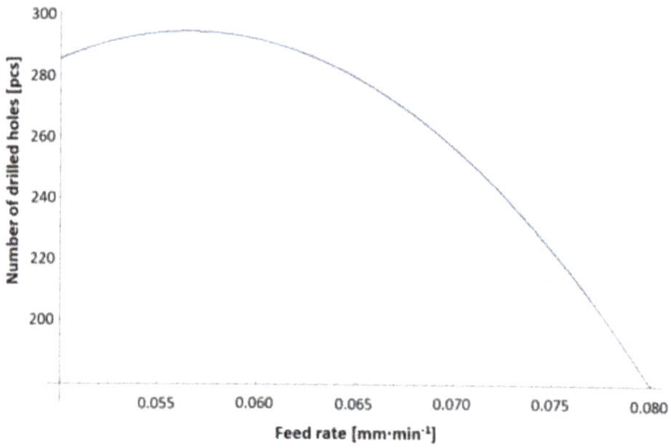

Figure 19. Course of dependence of the number of holes drilled up to the moment of tool damage in the range of the feed rates used.

In general, we obtain stationary points if we set the first derivation of the dependence function to zero:

$$df\left(-376.414 + 23736.19f - 209904.8f^2\right) = 0 \quad (8)$$

$$23736.19 - 419809.6f = 0 \quad (9)$$

A plotted representation of the first derivation of the investigated function is shown in Figure 20.

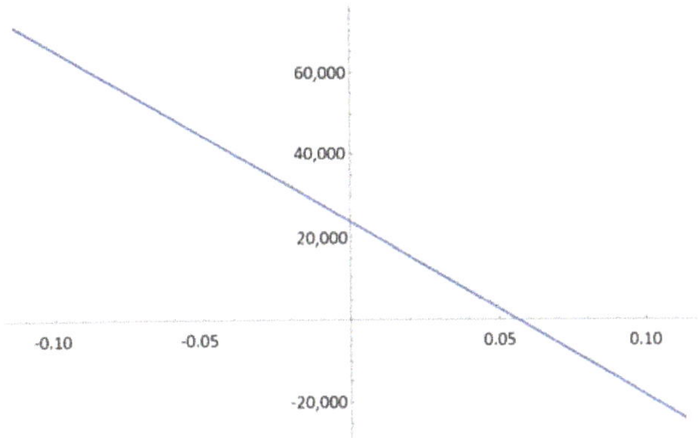

Figure 20. The course of the first derivation of the dependence of the number of holes drilled up to the moment of tool damage and the feed rate.

By solving the following Equation (10), we arrive at the stationary point:

$$\{\{f \to 0.056540369729515476\}\} \quad (10)$$

By inserting the value of the stationary point into the original dependence of the number of holes drilled up to the moment of tool damage on the feed rate, the value of the local function maximum sought is obtained (due to the fact that the second derivation of the function <0), that is to say, the number of the holes drilled up to the moment of tool damage at the feed rate of 0.0565 mm·min^{-1} is:

$$\{294.6121792850139, \{f \to 0.056540369729515476\}\} \quad (11)$$

4. Recommended Conditions for Drilling with Gun-Drill Tool on a CNC Machine Using a Pilot-Hole Guide

Currently, there is a great demand in the engineering industry for a radical reduction in the preparation time and rationalization of the development process for the rapid application of new methods and techniques. Global competition and increasing demands in the field of product life require not only the modernization of production capacities, but also new, innovative methods ensuring the smooth running of technological means. One such method is the deep-drilling process, which is used when it is necessary to create holes with depths of up to ten times their diameters. Based on long-term tests and the statistical processing of technological and physical parameters, it is possible to establish the following recommendations for deep-drilling technology.

Gun drill of the L10.5 × 380 DIN 1835A 20 × 50 CHIP GROOVE type.

Procedure for inserting a gun-drill tool:

- Stop the auger before the hole-rapid traverse.

- Slowly insert the auger at a counter-clockwise speed (or without speed, at a maximum of 50 min^{-1}) with cooling.
- Once the auger has been inserted into the pilot hole, switch on full speed and cooling.
- Start the drilling process.
- When the full depth of the hole is reached, switch off the feed rate, stop the spindle speed with a delay and switch off the high-pressure cooling with a small delay.
- Slide the auger out of the hole at rapid traverse.

Testing conditions and recommendations for maximizing service life in terms of the speed and feed-rate settings:

- A feed rate of 0.07 $mm \cdot min^{-1}$.During the testing of the stable service life at the given feed rate, in the event of a change in input conditions (deviation in the workpiece material), it is recommended to reduce the feed-rate value to 0.06 $mm \cdot min^{-1}$.
- A speed of 2700 min^{-1}. During the testing of the stable service life at the given speed, in the event of a change in input conditions (deviation in the workpiece material), it is recommended to reduce the speed value to 2600 min^{-1}.

Conditions specified by the manufacturer and verified by testing:

- Coolant pressure >30 bar; the manufacturer recommends that it be above 35 bar.
- Emulsion percentage >9% (the manufacturer recommends over 12%).
- The pilot hole must be 1–1.5xD long.
- The pilot-hole tolerance must be in H8.
- The pilot-hole apex angle should be >140°.
- The cutting speed varies, depending on the workpiece material used (the tested cutting speed, which showed the highest service life values, was 90 $m \cdot min^{-1}$).
- The feed rate varies, depending on the workpiece material used (the material tested by us showed the highest tool life at a feed rate of 0.07 $mm \cdot min^{-1}$).
- The pilot-hole drilling deviation should be <0.02 mm in all axes.
- Gun-drill tool deviation <0.02 mm in all axes.
- The minimum length of the drilling part of the gun-drill tool should be 1.5xD.
- Fixed clamping in the hydraulic clamp.

In order to successfully maximize the service life of gun-drill tools, no limit deviations from the prescribed properties and composition of the material being machined may occur. The service life of the augers is also affected by the functional condition of the machine tools–spindle backlash, machine vibration, guide backlash and leaks in the fluid supply conduit, as well as the consequent loss of pressure.

5. Conclusions

Through measurement, the general recommendations of the manufacturer were verified and proven to be informative in real operation conditions. Based on extensive tests and the resulting statistically processed data, the parameters were correlated with the subsequent methodology for deep-drilling technology. Various technological parameters were tested in a series of tests performed during real engineering production, which were subjected to statistical evaluation. As part of the statistical analysis of the dependences of the monitored input factors, such as the feed rate and speed, a nonlinear regression analysis was performed in order to identify the most suitable optimization function. The quadratic function was chosen as the most suitable function based on the criterion of the best possible description of the experimentally obtained results, while the adjusted determination index was chosen as the criterion for the dependence of the tool life on the feed rate. Based on this dependence, it reached the value of 77.25%. Thus, it can be stated that the chosen model can explain 77.25% of the variability in the measured values. The optimization itself was based on the quadratic optimization function, reaching its local maximum at a feed rate of 0.0565 $mm \cdot min^{-1}$, with the predicted value of the number of holes drilled up to the moment of tool damage being 294.61. For the analysis of the dependence of the tool life on the speed, the quadratic function was chosen again as a nonlinear regression analysis based on the criterion of the adjusted determination index as the ability of the model to explain the variability in the measured data. Based on this dependence, the value of the adjusted determination index was 98.24%; therefore, the model could not explain the remaining 1.76% of the measured data. Based on the analysis, the optimization function reached its maximum at a speed of 2478.334 min^{-1}, with the predicted value of the number of holes drilled up to the moment of tool damage being 328.285.

Further research will examine other parameters, such as temperature and pressure of the coolant, with a comparison of the selection of carbide tools for making holes using deep-drilling technology.

Author Contributions: M.P. processed and evaluated the data and wrote the paper; M.K. conceived and designed the experiments; J.P. performed the experiments and measurements; S.D.G. and V.C. performed the data curation. All authors have read and agreed to the published version of the manuscript.

Funding: This research received no external funding.

Data Availability Statement: Not applicable.

Acknowledgments: This paper has been elaborated in the framework of the project KEGA 038TUKE-4/2021, VEGA 1/0051/20 and project APVV-18-0316.

Conflicts of Interest: The authors declare no conflict of interest. The founding sponsors had no role in the design of the study; in the collection, analyses, or interpretation of data; in the writing of the manuscript; or in the decision to publish the results.

References

1. Neo, D.W.K.; Liu, K.; Kumar, A.S. High throughput deep-hole drilling of Inconel 718 using PCBN gun drill. *J. Manuf. Processes* **2020**, *57*, 302–311. [CrossRef]
2. Strodick, S.; Berteld, K.; Schmidt, R.; Biermann, D.; Zabel, A.; Walther, F. Influence of cutting parameters on the formation of white etching layers in BTA deep hole drilling. *Tech. Mess.* **2020**, *87*, 674–682. [CrossRef]
3. Ozerkan, H.B.; Cogun, C. Electrochemical small diameter deep hole drilling of powder metal steel. *Trans. Famena* **2020**, *44*, 47–58. [CrossRef]
4. Vasilko, K.; Murcinkova, Z. Experimental study of chip shapes in grinding by unique quick stop method and the ground subsurface layers micro-hardness. *J. Mech. Sci. Technol.* **2019**, *33*, 1341–1347. [CrossRef]
5. Zhang, Z.; Liu, S.; Zhang, Y.; Wang, C.H.; Zhang, S.; Yang, Z.; Xu, W. Optimization of low-power femtosecond laser trepan drilling by machine learning and a high-throughput multi-objective genetic algorithm. *Opt. Laser Technol.* **2022**, *148*, 107688. [CrossRef]
6. Steininger, A.; Bleicher, F. In-process monitoring and analysis of whirling motions in boring and trepanning association deep drilling. *MM Sci. J.* **2019**, *2019*, 3122–3128. [CrossRef]
7. Arrospide, E.; Bikandi, I.; Larranaga, I.; Cearsolo, X.; Zubia, J.; Durana, G. Harnessing deep-hole drilling to fabricate air-structured polymer optical fibres. *Polymers* **2019**, *11*, 1739. [CrossRef]
8. Garza, C.; Das, R.; Shterenlikht, A.; Pavier, M. Measurement of assembly stress in composite structures using the deep-hole drilling technique. *Compos. Struct.* **2018**, *202*, 119–126. [CrossRef]
9. Li, L.; He, N.; Hao, X.Q.; Yang, Y.F. Deep-hole gun drilling mechanics model of Ti6Al4V alloy based on Johnson and Cook flow stress model. *Int. J. Adv. Manuf. Technol.* **2019**, *104*, 4497–4508. [CrossRef]
10. Haddag, B.; Nouari, M.; Moufki, A. Experimental analysis of the BTA deep drilling and a new analytical thermomechanical model for assessment of cutting forces and BTA drill design. *Int. J. Adv. Manuf. Technol.* **2020**, *106*, 455–469. [CrossRef]
11. Panda, A.; Duplak, J. Comparison of theory and practice in analytical expression of cutting tools durability for potential use at manufacturing of bearings. *Appl. Mech. Mater.* **2014**, *616*, 300–307. [CrossRef]
12. Dobransky, J.; Mikus, R.; Ruzbarsky, J. Comparison of cooling variants by simulation software. *Mater. Technol. Qual. Assur.* **2013**, *801*, 75–80. [CrossRef]
13. Gao, T.; Li, C.H.; Wang, Y.; Liu, X.; An, Q.; Li, H.N.; Zhang, Y.; Cao, H.; Liu, B.; Wang, D.; et al. Carbon fiber reinforced polymer in drilling: From damage mechanisms to suppression. *Compos. Struct.* **2022**, *286*, 115232. [CrossRef]
14. Lu, H.; Zhao, X.; Tao, B.; Ding, H. A state-classification approach for light-weight robotic drilling using model-based data augmentation and multi-level deep learning. *Mech. Syst. Signal Processing* **2022**, *167*, 108480. [CrossRef]
15. Bornschlegel, B.; Haasler, D.; Finger, J.; Grizmann, D.; Trautz, M.; Gillner, A. Deep drilling of structural timber with high power ultrashort pulsed laser radiation. *J. Laser Micro Nanoeng.* **2020**, *15*, 195–203. [CrossRef]
16. Hong, J.; Zhou, J.H.; Chan, H.L.; Zhang, C.; Xu, H.; Hong, G.S. Tool condition monitoring in deep hole gun drilling: A data-driven approach. In Proceedings of the 2017 IEEE International Conference on Industrial Engineering and Engineering Management (IEEM), Singapore, 10–13 December 2017; pp. 2148–2152. [CrossRef]

17. Wang, Y.; Jia, W.; Zhang, J. The force system and performance of the welding carbide gun drill to cut AISI 1045 steel. *Int. J. Adv. Manuf. Technol.* **2014**, *74*, 1431–1443. [CrossRef]
18. Zhang, X.; Leong, T.G.; Liu, K.; Anantharajan, S.K. Effect of apex offset inconsistency on hole straightness deviation in deep hole gun drilling of Inconel 718. *Int. J. Mach. Tools Manuf.* **2018**, *125*, 123–132. [CrossRef]
19. Kirsanov, S.V.; Babaev, A.S. The study of deep holes accuracy and surface roughness after gun drilling. In Proceedings of the 2014 International Conference on Mechanical Engineering, Automation and Control Systems (MEACS), Tomsk, Russia, 16–18 October 2014; pp. 1–4. [CrossRef]
20. Mann, J.B.; Saldana, C.; Guo, Y.; Yeung, H. Effects of controlled modulation on surface textures in deep-hole drilling. *SAE Int. J. Mater. Manuf.* **2013**, *6*, 24–32. [CrossRef]
21. Zhang, W.; He, F.; Xiong, D. Gundrill life improvement for deep-hole drilling on manganese steel. *Int. J. Mach. Tools Manuf.* **2004**, *44*, 327–331. [CrossRef]
22. Schnabel, D.; Oezkaya, E.; Biermann, D.; Eberhard, P. Modeling the motion of the cooling lubricant in drilling processes using the finite volume and the smoothed particle hydrodynamics methods. *Comput. Methods Appl. Mech. Eng.* **2018**, *329*, 369–395. [CrossRef]
23. Wang, Y.; Yan, X.; Li, B.; Tu, G. The study on the chip formation and wear behavior for drilling forged steel S48CS1V with TiAlN-coated gun drill. *Int. J. Refract. Met. Hard Mater.* **2012**, *30*, 200–207. [CrossRef]
24. Wegert, R.; Guski, V.; Mohring, H.C.; Schmauder, S. Temperature monitoring in the subsurface during single lip deep hole drilling. *Tech. Mess.* **2020**, *87*, 757–767. [CrossRef]
25. Aized, T.; Amjad, M. Quality improvement of deep-hole drilling process of AISI D2. *Int. J. Adv. Manuf. Technol.* **2019**, *69*, 2493–2503. [CrossRef]
26. Arunkumar, N.; Thanikasalam, A.; Sankaranarayanan, V.; Senthilkumar, E. Parametric optimization of deep-hole drilling on AISI 1045 steel and online tool condition monitoring using an accelerometer. *Mater. Manuf. Processes* **2018**, *33*, 1751–1764. [CrossRef]
27. Chu, N.H.; Nguyen, D.B.; Ngo, N.K.; Nguyen, V.D.; Tran, M.D.; Vu, N.P.; Ngo, Q.H.; Tran, T.H. A new approach to modelling the drilling torque in conventional and ultrasonic assisted deep-hole drilling processes. *Appl. Sci.* **2018**, *8*. [CrossRef]
28. Han, C.; Luo, M.; Zhang, D.H.; Wu, B.H. Iterative learning method for drilling depth optimization in peck deep-hole drilling. *J. Manuf. Sci. Eng. Trans. ASME* **2018**, *140*, 121009. [CrossRef]
29. Valicek, J.; Harnicarova, M.; Hlavaty, I.; Grznarik, R.; Kusnerova, M.; Hutyrova, Z.; Panda, A. A new approach for the determination of technological parameters for hydroabrasive cutting of materials. *Mater. Werkst.* **2016**, *47*, 462–471. [CrossRef]
30. Hoekstra, B.; Shekarian, A.; Kolasangiani, K.; Oguamanam, D.C.D.; Zitoune, R.; Bougherara, H. Effect of machining processes on the damage response and surface quality of open hole hybrid carbon/flax composites: An experimental study. *Compos. Struct.* **2022**, *285*, 115244. [CrossRef]
31. Panda, A.; Duplak, J.; Jurko, J.; Pandova, I. Roller bearings and analytical expression of selected cutting tools durability in machining process of steel 80MoCrV4016. *Appl. Mech. Mater.* **2013**, *415*, 610–613. [CrossRef]

machines

MDPI

Article

Computer-Aided System for Parametric Design of Ship Hull Structures—CADS-Hull

Sergey Ryumin [1] and Vladimir Tryaskin [2,*]

1 Department of Ship Design, Faculty of Shipbuiding and Ocean Engineering, Saint Petersburg State Marine Technical University, Lotsmanskaya ul., 3, 190121 Saint Petersburg, Russia; rumin@smtu.ru
2 Department of Ship Structure and Technical Operation, Faculty of Shipbuiding and Ocean Engineering, Saint Petersburg State Marine Technical University, 190121 Saint Petersburg, Russia
* Correspondence: vladimir.tryaskin@smtu.ru

Abstract: The problem of the parametric design of ship hull structures is considered here. The parametric design of ship hull structures is intended to determine the dimensions of structural elements that meet the requirements of regulatory documents (Rules of Classification Societies/Strength Standards). It is shown that, in all existing shipbuilding CAD-CAM systems, the problems of the parametric design of ship hull structures are essentially not affected. A modern methodology for the parametric design of hull structures is proposed, which defines a set of principles for the structural and logical organization of the design process, methods, and technical means for implementing design procedures. The most characteristic methodological principles of the parametric design of ship hull structures and other offshore engineering structures are identified: A systematic approach; modeling as one of the main methods for solving design problems; and a rational design strategy based on iterative search procedures that implement the methodological principle of successive complication of models and procedures for designing structures. The article considers the architecture and methodological foundations of the system for automated parametric design of ship hull structures (CADS-Hull), developed at the Saint Petersburg State Marine Technical University, Russia. The system is intended for use in design bureaus when designing civil ship hull structures on the basis of regulatory requirements of the Russian Maritime Register of Shipping and IACS General Rules in terms of large-capacity tankers, bulk carriers, and container ships.

Keywords: ship hull structure; computer-aided design of structure; database; function soft block

Citation: Ryumin, S.; Tryaskin, V. Computer-Aided System for Parametric Design of Ship Hull Structures—CADS-Hull. *Machines* **2022**, *10*, 262. https://doi.org/10.3390/machines10040262

Academic Editors: Marek Kočiško and Martin Pollák

Received: 16 March 2022
Accepted: 2 April 2022
Published: 7 April 2022

1. Introduction

Ship hull structure design is a process of forming a geometric, structural, parametric, and structural-and-technological description of the structure and developing the technical documentation (on this basis) necessary for structure manufacturing in the conditions of ship hull production. Often, structural design is understood as the development of a hull structure using the universal software of modern computer graphic systems such as AutoCAD [1] and COMPAS-3D [2] or the corresponding blocks of "heavy" CAD/CAM systems (FORAN [3], AVEVA [4], etc.). Actually, the mentioned systems mainly provide the implementation of computer-aided construction processes. At the same time, the computer screen performs the function of an "electronic drawing board", since mainly routine graphic procedures are automated in such graphic systems. Computer-aided construction can be implemented only if there is information about structure shape parameters and its elements' dimensions. In this regard, ship hull structure design can be represented as a combination of two interlinked processes—computer-aided construction and parametric design. Computer-aided construction is intended for the formation of visual parameterized information regarding the structural composition of hull structures, as well as its geometric and construction concepts. One of the computer-aided design tasks is the development of structural drawings using modern office equipment. Parametric computer-aided design is intended to determine the dimensions of structural elements that meet the regulatory requirements (Rules of Classification Societies/Structures strength standards).

The problem with the parametric design of ship hull structures is hardly affected in existing shipbuilding CAD-CAM systems. To an extent, for the parametric design of ship hull structures, one can use (during verification calculations) Classification Societies' computer-aided systems: MARS 2000 [5], VeriStar [6]—BV; NauticusHull [7], Poseidon [8]—DnV-GL; SafeHull, CSR Prescriptive Analysis software [9]—ABS; ATLAS [10]—Russian Maritime Register of Shipping (RMRS), and others, intended to check the compliance of the ship's hull with the requirements of the Rules of Classification Societies. Parametric design automation requires specialized software. It is a very common notion that structural design comprises the structure's calculations using software that implements the finite element method (FEM). It is known that, in order to perform finite element analysis, it is also necessary to obtain complete (even detailed) information about the structure, which does not exist or is not sufficient at the initial stages of designing.

Some basic information on the parametric settings of the hull shape and internal structures is given in the works of A. Papanikolau [11,12]. The basic principles of modeling ship structures based on an object-oriented approach, which is very relevant in terms of software development, are given in [13]. Ship structures are physical objects with a complex hierarchical structure, which is very well-suited to creating an information structure and methods for its processing in computer systems, which are programmed using an object-oriented approach.

A large amount of information on the modeling of ship structures is contained in works that reveal the problem of automating data preparation for FEM analysis (for example, in [14,15]). The preparation of such data is usually carried out separately from the parametric design of structures, but the system described in this article has special data translators for transferring the model to FEM analysis systems. The approach proposed in [14,15] was partially used to develop these translators in terms of describing plate and beam elements. The creation of structural models at the early stages of design, which are later used for FEM analysis and production preparation, are given in [16].

Historically, the computer modeling of ship structures originates from the preparation of documentation and drawings for the construction of a ship and the planning of the production process. A modern approach to the formation of this kind of information on the ship's hull is shown in [17]. An example of the implementation of search optimization methods for the design of ship structures to minimize their mass can be found in a number of works [18–20].

A large set of studies, related to the development of the methodology, algorithms, and software for the computer-aided design of hull structures of merchant ships and floating docks, was carried out in the Ship Design Department of the Saint Petersburg State Marine Technical University (SPbSMTU, Russia) [21]. The research results are based on the analysis of the experience in ship design and construction, taking into account the evolution of design methods. Conclusions and practical proposals are confirmed by department specialists' longstanding works on the problem of computer-aided design of structures. A modern methodology was proposed for hull structure design, which defines a set of structural and logical organization principles of the design process, as well as methods for design procedure implementation. It also identified the most characteristic methodological principles of parametric design of ship hull structures and other engineering structures: A system approach, modeling as one of the main methods for design task solving, and a rational design strategy. The suggested strategy is based on iterative search procedures that implement the methodological principle of successive complications of models and structure design procedures, from the design of structural elements (plates, stiffeners), to the design of entire structures as a set of stiffeners and primary support members or a set of primary support members in accordance with the requirements of local or transverse strength, and finally, to the design of hull girder longitudinal members in accordance with the requirements to the ship's hull general strength. The values of design parameters, obtained as a result of lower-level task-solving, are used in the mathematical model of the next-level design task as the "lower" bound. A possible change of these parameters is a (search) value increase in the case of design condition dominance at the considered design level. The proposed approach's effectiveness is substantiated by many years of experience in solving tasks of ship hull structure designing.

The problem of structures designing is presented as a general mathematical programming problem. The objective function is usually a characteristic of the structure mass. Inequality constraints are formed on the basis of the requirements of RMRS Rules or IACS General Rules for double-hull oil tankers, bulk carriers, and containerships. On the back of the proposed methodology, a specialized system for parametric computer-aided design of ship hull structures was developed—CADS-Hull—in which the initial information preparation for design procedures is implemented using simplified methods, algorithms, and software created on this basis. The regulatory framework of the system is the Rules and Regulations for the Construction and Classification of Sea-Going Ships [22] and Common Structural Rules for Bulk Carriers and Oil Tankers [23].

2. General Structure of the Computer-Aided System for Parametric Design of Ship Hull Structures

The structure of the computer-aided system for the parametric design of ship hull structures is determined by a set of software blocks and their interrelation, the functional purpose of each block (the totality of problems to be solved), and the sequence of program procedure implementation. The general system structure is shown in Figure 1.

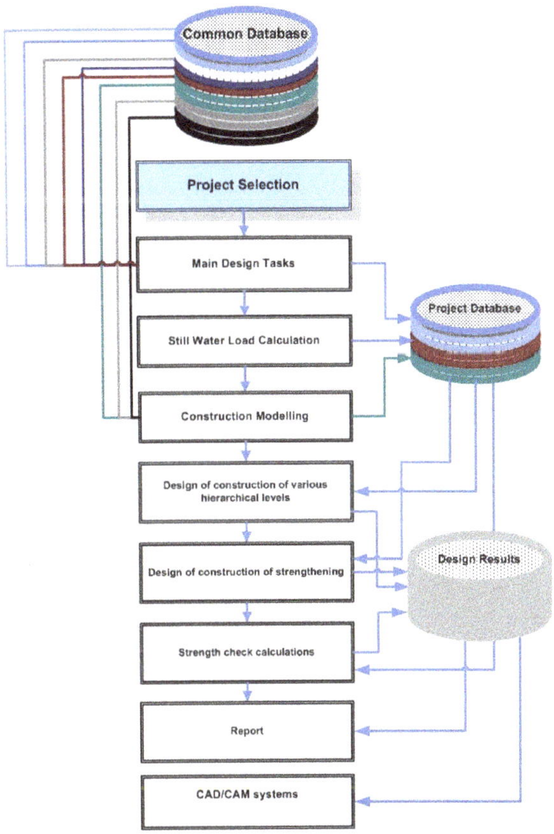

Figure 1. General structure of parametric computer-aided design system.

Experience in the field of computer-aided systems for parametric structure design made it possible to determine the necessary blocks of tasks that should ensure the effective use of such systems in the design bureau practice:

- Block of general design tasks.
- Block of constructional modeling.
- Block of procedures for designing various hierarchical levels structures.
- Block of procedures for designing reinforcement structures.
- Block of verification calculations of structural strength.
- Block of creation of the reporting documentation.
- Block of transferring design results to the CAD-CAM system of the design bureau.

Effective automation of the parametric design of hull structures is not possible if a number of general design tasks have not been solved previously. The task list is presented in Figure 2. Before starting work, it is necessary to select a project. It can be one of the projects that have already been worked on or a new one.

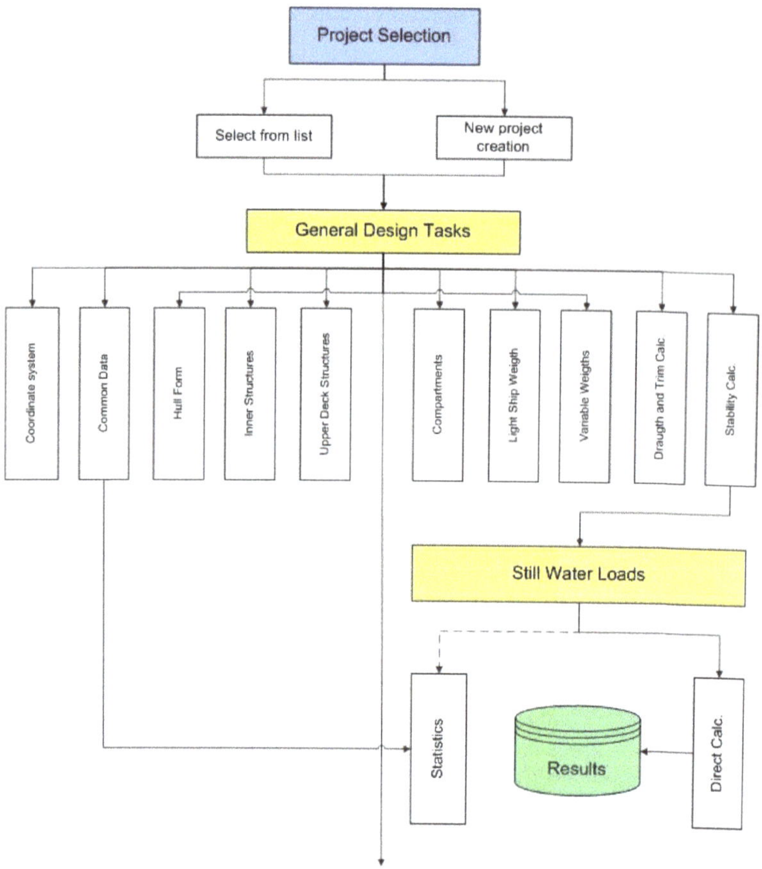

Figure 2. General design block of the parametric computer-aided design system.

Determination of the coordinate system is one of the fundamental problems that must be solved. *The block of general data determining* is intended to generate information such as the ship type, operating limits, ice-strength class, and life cycle. The main ship dimensions, computational length, block coefficient, and displacement constitute the next group of general data. Furthermore, the parameters of superstructures (deckhouses) are set, and their construction type is determined.

The block of the ship hull geometry description is one of the key factors in the computer-aided design system. The hull geometric modeling problem can be solved by various methods: Importing a geometric model created by other systems (for example, AVEVA, FORAN); data entry point by point, which defines the geometry of each frame section; and geometric model creation by transforming the shape of the prototype ship hull. The block operation result is a ship hull wireframe model (Figure 3).

Figure 3. Ship hull wireframe model.

Concerning the *block of internal structures' geometry description* (inner bottom, inboard, intermediate decks and platforms, longitudinal and transverse bulkhead, etc.), the need for such an information block in the system is obvious. The structure's shape is one of the most important characteristics that directly affects its structural arrangement and design calculations results. In the present computer-aided design system, a method of simplified description of the internal structures' geometry was implemented, which is specified by the values of parameters in the so-called control sections. The control section is considered to be the one in which the characteristic shape parameters of the present structure can be easily determined, or even set by specific dimensions borrowed from the general arrangement drawings. The shape parameters between control sections are determined by the linear interpolation method. Special software makes it possible to obtain a parametric description of the internal structures' geometry in the form of planes or triangulated surfaces (Figure 4). Structures that spill over the body surface are automatically "cut off" along the intersection line.

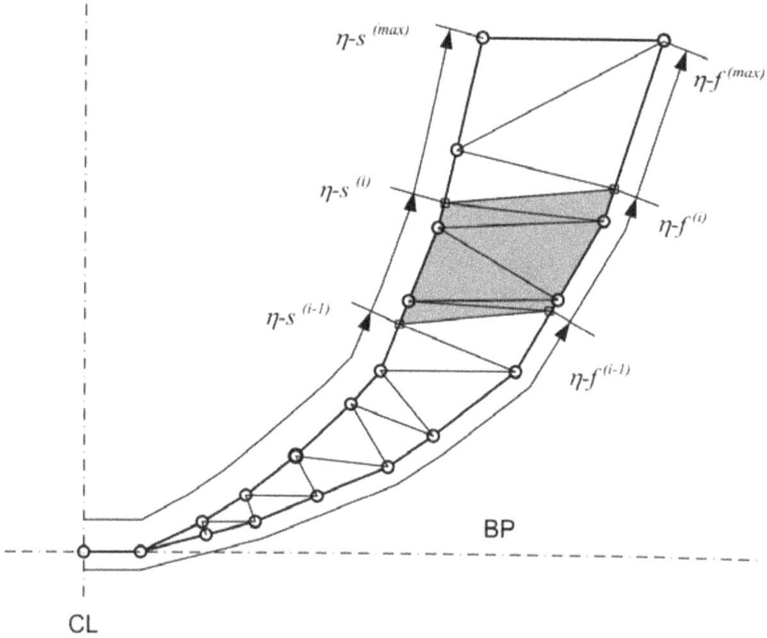

Figure 4. Wireframe triangulation of ship surfaces.

The block of superstructures' geometry description. The superstructures' geometry can be set in the block of the ship hull geometry description. In the proposed computer-aided system, it is possible to represent each superstructure geometry as a separate geometric object, with its description in the same format as the main hull. Such a methodical approach expands the possibilities of geometric modeling, including in relation to the complex stroke of the main hull (bulbous bow, skegs, stern-post elements, etc.). The solution to the problem of construction modeling of upper deck sections under deckhouses and the visualization of its results is simplified. The possibility of construction modeling and parametric design of superstructures (deckhouses) is provided, regardless of the main hull.

The block of ship compartments/tanks description. The ship *compartments'* geometric characteristics are used in the problem of the ship variable load layout for the purpose of following calculations of the ship draft and trim and the loads on the hull in still water. The list and purpose of ship spaces/compartments are generated in the process of creating a current project database. In addition to geometric parameters, the block of ship compartments/tanks description includes the following information: The load type in the compartment, stacking height of general or timber cargo, the level of the free surface of liquid or bulk cargo, the value of the stowage factor, or the average cargo density. For bulk cargo, the angle of internal friction is additionally set. Compartment loading information can be obtained from the data on a permissible double bottom or deck load. For tanks, there is also the introduction of parameters, which make it possible to determine the test load value.

The block of light ship load layout. The light ship load layout is one of the problems that is associated with significant difficulties, especially in the early stages of ship design. These difficulties are due to the lack of initial information. In the proposed computer-aided system, the light ship load formation can be carried out in two ways:

- Manual data entry in the presence of relevant initial information—load tables.
- Calculation of load components according to statistical dependencies—in the absence of load tables.

The block of variable load layout. The layout of the *variable* load is implemented in an interactive mode. A special interface allows one to select a concrete compartment from the corresponding database of the current project to indicate the cargo type, its characteristics (stowage factor or density, angle of internal friction, etc.), the amount of cargo in it, or the filling percentage.

The block of equilibrium draft and trim calculation allows one to evaluate the permissibility of the accepted ship loading variant by the value of the average draft and trim.

The block of ship stability calculating allows one to conduct additional checks of the permissibility of the accepted ship loading variant by the parameters of initial stability, stability at a large heeling angle, or by the value of the weather criterion. The presence of geometric hull and superstructure models makes it possible to automate the calculation of the lateral projection area, the parameters of which must be known for stability calculations.

The information prepared by the previous blocks is used in *the block of still-water load calculations.* The software created during the development of the computer-aided system makes it possible to perform the still-water load calculations with general load distribution and the actual ship position on the water. The ship weight distribution can be represented as a stepped configuration. There are software-implemented procedures for still-water load calculations with the actual weight load distribution, which is obtained as a result of the operation of the following blocks: *The block for the light ship load layout and the block for the variable ship load layout* (Figure 5). The results for still-water load calculations for all variants of ship loading are stored in the database.

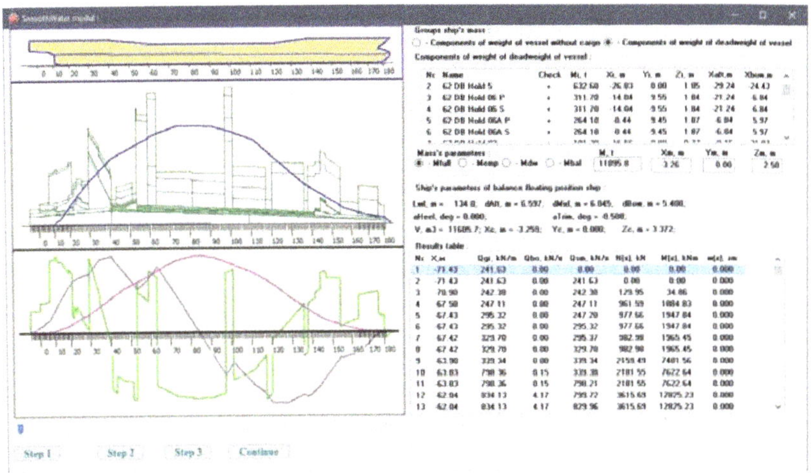

Figure 5. Calculations of still water loads based on the actual distribution of the light ship and variable ship loads.

The block of construction modeling. The computer-aided system provides the possibility of implementing three modes of hull construction modeling:

- Structure modeling in a given cross-section.
- Hull modeling by describing the design of each of its structural components (substructures).
- Hull modeling by describing the "structural compartment" (a set of substructures bounded by transverse planes passing through given practical frames).

The first modeling mode is convenient in cases where it is necessary to obtain information about the structural element dimensions only in a given cross-section over the length or in a set area with a characteristic (unchanged) topological (structural) scheme (Figure 6). The modeling of hull structures in a given cross-section is provided by a specialized editor that allows one to interactively divide plate structures into separate plates, arrange the longitudinal stiffeners and primary support members, and assign the dimensions based on the design calculation results.

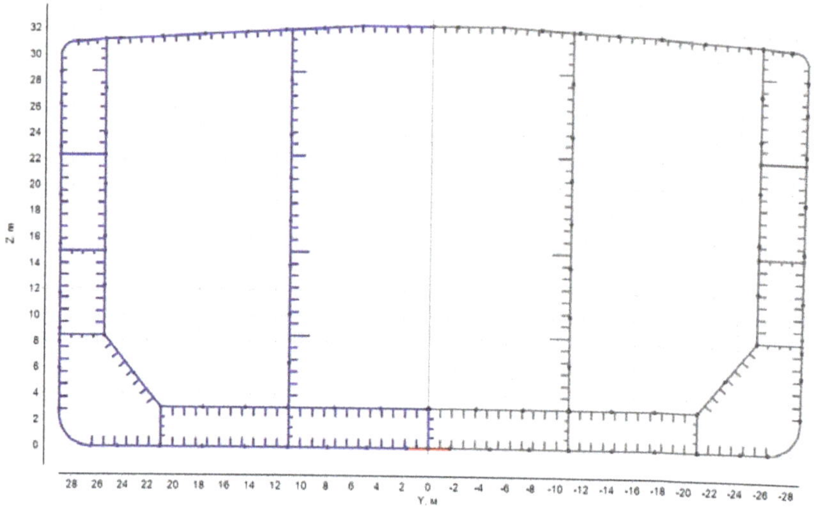

Figure 6. Structural modeling of the ship hull cross-section (first mode).

The second modeling mode allows one to describe the structural scheme of the substructure as a whole (the upper deck, intermediate deck, platform, inner bottom, shell expansion along their entire length, etc.) or its separate parts in a given area (Figure 7). Structural modeling of the outer shell and internal hull structures is implemented by various software blocks due to the modeling algorithms' features. The presence of structural modeling created by the second-mode software makes it possible to obtain a structural model of the hull cross-section in the plane of a given frame or in a plane with a given X-coordinate by processing the database information on the hull design.

The third design mode allows one to obtain a description of the hull at the level of the structural compartment. In practice, it is implemented as an assembly of the results of structural modeling of the outer shell and internal hull substructures (separate parts of the outer shell and internal substructures). The structural modeling results can be used to perform structural verification calculations via the method of finite element analysis of the hull strength at the level of a compartment and up to three compartments in total.

Such an analysis has to be performed in some design cases, for example, with an alternating pattern of heavy cargo loading of a bulk carrier hold, when analyzing the deck displacement in its plane for an RO-RO ship, when assessing the hull tension in local areas, etc.

The block of various hierarchical levels' structure designing implements the parametric design concept proposed in the work, based on the successive complication of models and design algorithms. At this stage, the design procedures based on the requirements of the Rules common to all ships are implemented: The design of outer shell plating, bottom and side framing for sea water pressure, design of deck structures, double bottom for cargo pressure, etc.

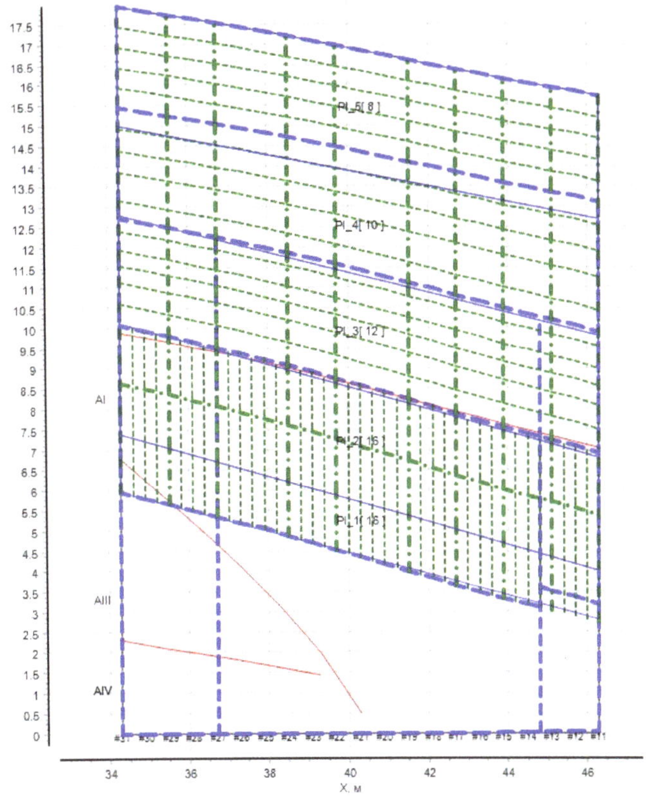

Figure 7. Structural modeling of the ship hull (second mode).

In accordance with the parametric design concept based on the ship hull decomposition and the successive complication of models and design algorithms, the block of structural design can be divided into three levels:

1. Level of substructure elements design.
2. Level of design of the structure as a whole.
3. Level of a set of structures design.

Each level has its own models and structural design algorithms.

At the first (lower) level, procedures of plates and beam structural elements' (stiffeners) parametric design are implemented according to the requirements for local strength and buckling using the simplest models. The design begins with plate elements. The results of this design stage are then used in beam element design since the plate element design parameter (plate thickness) is the thickness of the attached plate of the beam element.

At the first design level, the requirements for local strength and buckling of plates and beams can be only implemented as a first approximation, since the allowable stresses from local bending, as well as the calculated compressive forces and shear loads that can lead to collapse, are dependent on the structural element dimensions defined in the higher design levels.

In most cases, when solving design problems of the first level with a given layout scheme and selecting structural elements material, calculated dependences can be directly found relative to the desired structural element parameters. More complex computer-aided design procedures for plate and beam elements can be used to justify the layout scheme of the structure (choosing a framing system or scheme of arrangement of stiffening elements) and the choice of material strength. The need to solve such problems may arise at the early (research) stages of ship hull structure design. The structure of the requirements for sheet and beam structural elements is such that it is expedient to formulate these problems in the form of nonlinear programming problems with constraints.

At the second level, procedures of plate and beam structural elements' parametric design are implemented according to the requirements for local strength and buckling using models of a higher hierarchical level: Plane beam systems, structurally orthotropic plates, and the complex plane web frame system (transverse frame and other frames). In most cases, at this stage, there are primary supporting members of structures designed (web frames, girders etc.). Ordinary beams (stiffeners) can either be ignored or included as a component of their (web beams) attached plates.

Such design models usually do not provide an unambiguous solution. Regulatory requirements for strength and/or buckling can be provided, with various combinations of structural elements dimensions. Therefore, it is advisable to use search procedures at this design stage. Structure behavior modeling using the design of experiments method can improve the efficiency and simplify the solution of such a problem.

At the third (highest) level, procedures for parametric structural design are implemented according to the requirements for buckling, strength, and stiffness in the cause of ship hull general deformation (during general bending). The hull model used at this stage is a hull girder. The model is formed on the basis of the construction modeling results contained in project database tables. The results of structural design, obtained in the first two levels, are used as initial data. The design of the hull structure is carried out in two stages:

1. Verification calculation of the hull girder; analysis of compliance with the requirements for strength, stiffness, and buckling during general bending.
2. Correction dimensions of certain longitudinal elements of the hull if the conditions of general strength and/or stiffness and/or buckling are not met.

In the second case, the structural design is implemented in the form of a search procedure.

The block of structural reinforcements design (ice strengthening, slamming strengthening, etc.). Hull reinforcements, caused by additional external efforts or architecture and structural features (the presence of cutouts, deckhouse endings), apply to the corresponding hull areas. Reinforcements can extend to a small area—these are considered local reinforcements (cutouts areas, deckhouse endings, etc.). In some cases, the extent of the reinforcement area is significant (ice reinforcements, reinforcements for load accommodation during slamming, etc.). The requirements for hull reinforcement structures are singularized into separate sections in the regulatory documents (Rules of Russian Maritime Register of Shipping). Accordingly, additional software is also provided in the computer-aided design system.

The block of verification calculations of strength is intended for solving design problems in the verification calculation mode. This approach is appropriate for tasks that require complex structural modeling (for example, based on FEM), which can be solved only with known (given) values of structure elements' dimensions. This block provides the possibility of solving a problem using search procedures and modeling on the basis of the general approach based on the methodology of numerical experiment planning.

3. The General Structure of the Database and the Methodological Foundations of Its Formation

The organization of the process of parametric computer-aided design is greatly simplified with the presence of a database. As a part of the system, it is advisable to have a common database and a database of the project on which the user is working. In the present computer-aided system, *a database* is understood as a set of information blocks organized in the form of files of various structures and formats. The file structure is ordered in a certain way: A directory, folder, or workbook.

In a common database, it is advisable to store information (by definition) of a general nature, which can be used to form a database for a specific project for ships of various architectural and structural types. The experience of system development has shown that, in order to solve the problems of ship hull structure design, the common database should contain the following information blocks:

- General hierarchical directory of structures.
- Directory of ship compartments/tanks.
- Directory of ship structures material grades.
- Assortment of plates, profiles, and standardized compound profiles (T-profiles).

The structure of the common database should be able to grow as the system improves and its functionality capabilities expand.

The general hierarchical directory of structures (Figure 8) is necessary for the implementation of the initial stage of database formation according to the specific project hull design. The directory structure corresponds to the accepted general structural decomposition of the ship hull, which does not contradict the ship hull design fundamentals. A rigidly fixed hierarchical system of structural element coding is associated with the structural decomposition of the ship hull. Formalization of the requirements of the applicable rules to the present hull structures is carried out on the basis of the structural elements code. The top-level directory (Figure 9a) can be supplemented with new groups of structural components. Catalogs of lower levels (Figure 9b) are formed, taking into account the following fundamental principle—the directory structure should be as unified as possible for different groups of structures, even if this leads to code redundancy (i.e., some code levels may never be used).

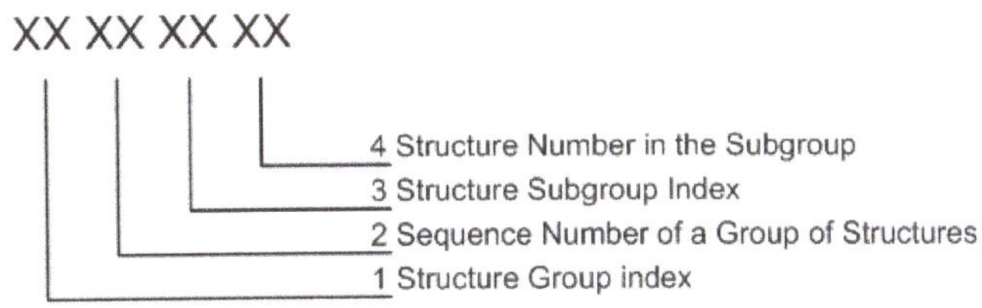

Figure 8. Ship structure classification.

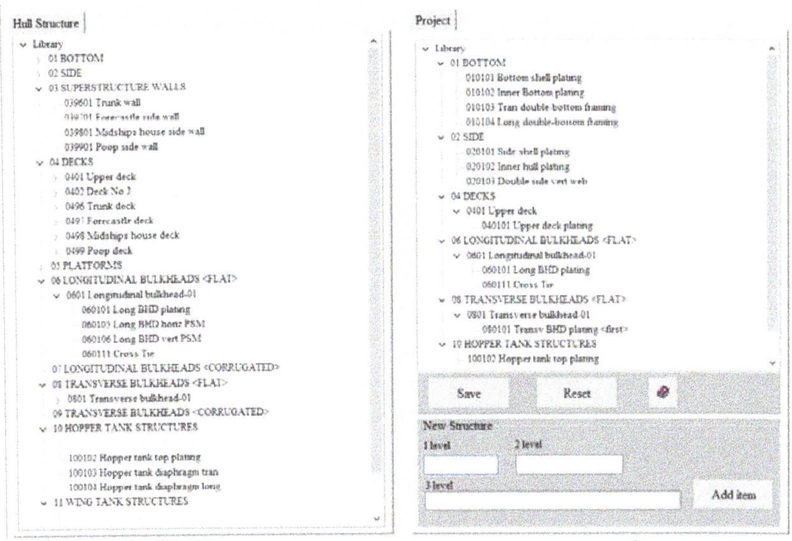

Figure 9. General hierarchical structure directory: (**a**) Directory of top-level structural groups; (**b**) directory of structural groups of second and third levels of the present project.

The directory of the current (specific) project is created in accordance with the following algorithm:

- The formation of the structure list showing the actual construction decomposition of the project hull structure.
- The coding of selected structures.

The formation of the structures list is carried out in the process of selecting the desired structure from the general hierarchical structure directory. At the same time, the user works with the text variables that reflect the generally accepted names of ship structure elements: "Bottom plate", "Shearstrake", "Deck stringer", "Frame", etc. The coding of selected structures is carried out automatically. A portion of code is generated that uniquely determines the selected structure (Figure 8).

Parametric provisioning of the database on the ship hull design is performed at the stage of ship hull structural modeling, using a specialized interface and software. To solve this problem, it is first necessary to perform the database formation according to the geometry of the hull and its structural components.

To create a database on ship compartments/tanks, materials, and assortments, the following general approach is used:

- When developing a common system database, there is a general directory created for compartments/tanks of the ship hull, ship structure materials, and an assortment of plates and profile materials.
- According to the general design documentation, a hierarchical directory of the compartments/tanks of the ship under consideration is created.
- Based on the hull design documentation, lists of the materials used, types and dimensions of profiles used for the frame beam manufacturing, and standard sizes of plate products are created; when designing a new ship, the restrictions associated with the ordered material for ship construction must be taken into account.
- The formation of a database of compartments/tanks, materials, and assortments for a specific project is reduced to selecting the corresponding records (rows) from the common database tables in an interactive mode. As a result, project directories are formed, with which the system then provides the opportunity to work. Usually, these directories are much smaller than the original ones. This fact simplifies the working conditions.

In this article, as an illustration of the proposed methodological approach, the creation of a Directory on ship compartments/tanks is considered. Classification of the ship compartments/tanks is proposed, mainly focused on the tasks of ship structure design, as well as the structure of a general directory of ship compartments/tanks (Table 1). A "truncated" directory is created, the structure of which reflects the needs of civil ship structure design (Figure 10).

Table 1. Classification of the ship compartments/tanks.

Group Code	Compartment's Group Name
00	External environment
01	Special compartments
02	Service spaces
03	Ship storage tanks
04	Water ballast tanks
07	Chain bins
08	Cofferdams
09	Dry compartments
10	Void compartments

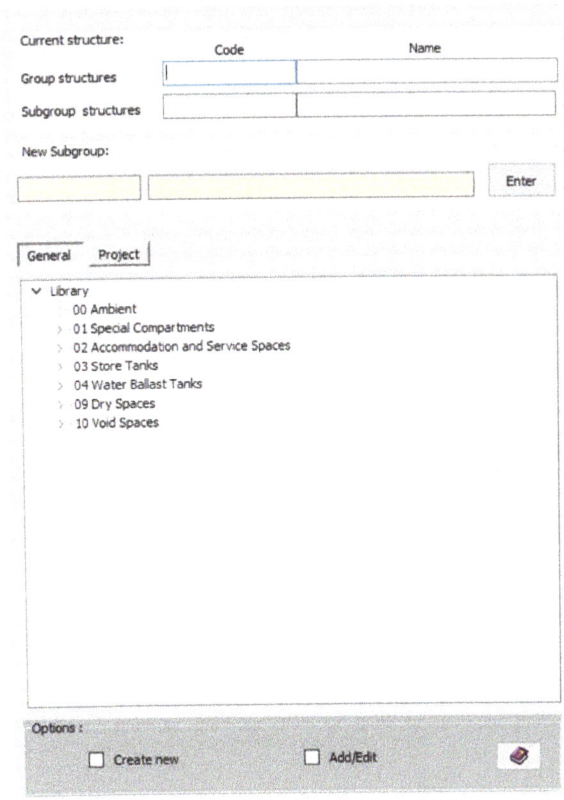

Figure 10. Top-level directory (compartments groups).

Each compartment/tank of the ship hull has a unique code, which generally consists of eight digits (Figure 11).

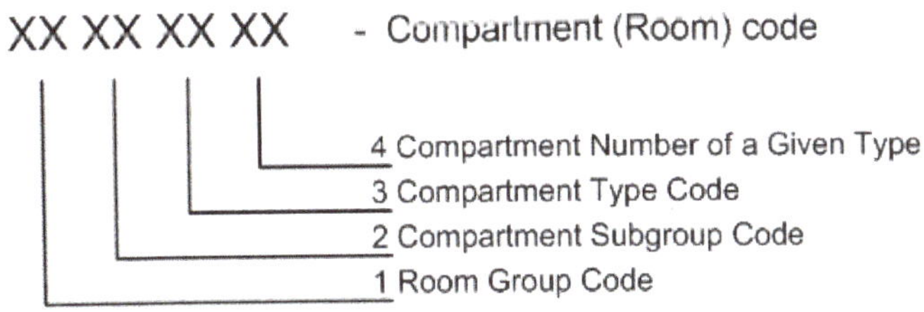

Figure 11. Ship structure classification.

The directory includes a group with code 00, which defines the conditional compartment "Ambient <environment>". The presence of such a code makes it possible to automate the calculation of the corrosion wear rates of the elements of the outer shell plating and open decks.

Group 01, "Special spaces", is for cargo ships, and combines various subgroups and groups of cargo compartments (see Table 2).

Group 02, "Service spaces", in the system of computer-aided design of civil ship structures, includes only engine and boiler spaces since the register rules impose a number of additional requirements on hull structures in these ship areas.

Table 2. Subgroups and groups of ship compartments/tanks.

Code		Name	Supporting Code
01		Special compartments (group)	
0101		Cargo compartments (subgroup)	
0101	01	Upper cargo deck	LUD
0101	02	Dry-cargo tween deck	LTW
0101	03	Dry-cargo hold	LTR
...			
0101	20	Cargo tanks	LLT
0101	21	Segregated ballast tank	SBT
...			
0101	30	Bulk cargo hold	BTR
0101	31	Bulk cargo and water ballast hold	B_WTR
...			
02		Service spaces (group)	
0201		Main machinery spaces (subgroup)	
0201	01	Engine room	ER
...			
03		Ship storage tanks (group)	
0301		Fuel oil tanks	DO
0302		Lubricating oil tanks	LO
...			
0307		Freshwater tanks	FW
...			
04		Water ballast tanks (group)	WB
0401		Forepeak	FP_WB
0402		Afterpeak	AP_WB
0403		Water ballast tank	WB
...			
09		Dry compartment (group)	DC
...			

Subgroups of spaces related to groups 03–04 and 07–09 can be divided into types (for example, 030101—reserve bunker oil tank, 030102—daily bunker oil tank, etc.), but this information will be redundant in the case of computer-aided structural design. Therefore, such a code hierarchy is not provided for these compartment groups. In addition to the unique hierarchical (numerical) code for each ship hull compartment/tank, a supporting letter code (automatically generated) is provided, which allows the user to freely navigate the compartment type, since it has a mnemonic meaning.

The general directory of the ship hull spaces/compartments can be supplemented or edited.

Figure 12 shows the results of the creation of a directory of a particular ship's compartments/tanks. For this purpose, the groups, subgroups, and compartments/tank types that are in the present project are selected from the general directory of the ship compartments/tanks.

structures and methodological foundation for its development. The generalization of the obtained representation is as follows:

1. The structure of the computer-aided design system is determined by a set of soft blocks and their interrelation, the functional type of each block (the totality of problems to be solved), and the sequence of program procedure implementation.
2. It is expedient to have a common database as part of the parametric computer-aided system for ship hull structure design. The common database stores information that is (by definition) common to ships of all projects.
3. The availability of a common database and specialized software makes it possible to form a project database and ensure the convenience and simplification of the functional soft blocks' development and the organization of the parametric design process.
4. Catalogs (or classifiers) of ship structures, compartments/tanks of the ship hull, data on materials, and profiles are of great practical importance.
5. It is fundamentally important to develop a special system for coding the ship hull structure elements and ship hull spaces. Such information is absolutely necessary for the development of the program code of the computer-aided system modules.

Author Contributions: Conceptualization, methodology, V.T.; project administration, V.T.; system structure, software, analysis, S.R.; writing the draft article, S.R.; editing the article, V.T. All authors have read and agreed to the published version of the manuscript.

Funding: This research was carried out as a part of the strategic academic leadership program "Prioritet-2030" for the period 2021–2022 and beyond until 2030, funded by the Ministry of Education and Science of the Russian Federation and as part of the World-class Research Center program: Advanced Digital Technologies (contract No. 075-15-2020-903 dated 16 November 2020).

Data Availability Statement: Not applicable.

Conflicts of Interest: The authors declare no conflict of interest. The funders had no role in the design of the study; in the collection, analyses, or interpretation of data; in the writing of the manuscript, or in the decision to publish the results.

References

1. Available online: https://www.autodesk.com/ (accessed on 1 April 2022).
2. Available online: https://ascon.ru/ (accessed on 1 April 2022).
3. Available online: https://www.plm.automation.siemens.com/global/en/products/foran/ (accessed on 1 April 2022).
4. Available online: https://www.aveva.com/ (accessed on 1 April 2022).
5. 2D Ship Structural Assessment Software. Available online: https://marine-offshore.bureauveritas.com/mars-2000-2d-ship-structural-assessment-software (accessed on 1 April 2022).
6. Available online: https://www.veristar.com/portal/veristarinfo (accessed on 1 April 2022).
7. Ship Structural Analysis and Design—Nauticus Hull. Available online: https://www.dnv.com/services/ship-structural-analysis-and-design-nauticus-hull-1061 (accessed on 1 April 2022).
8. Strength Assessment of Hull Structures—POSEIDON. Available online: https://www.dnv.com/services/strength-assessment-of-hull-structures-poseidon-18518 (accessed on 1 April 2022).
9. Engineering Software. Available online: https://ww2.eagle.org/en/rules-and-resources/engineering-software.html (accessed on 1 April 2022).
10. Aleksashin, D.; Boyko, M.; Kuteynikov, M.; Sotskov, S. IACS Common Structural Rules development and RS design verification during ship plan approval. *Trans. Krylov State Res. Cent.* **2018**, *2*, 59–63. [CrossRef]
11. Papanikolaou, A. *A Holistic Approach to Ship Design. Volume 1: Optimisation of Ship Design and Operation for Life Cycle*; Springer Nature Switzerland AG: Athens, Greece, 2019; pp. 24–26, 231–233. [CrossRef]
12. Papanikolaou, A. *A Holistic Approach to Ship Design. Volume 2: Application Case Studies*; Springer Nature Switzerland AG: Athens, Greece, 2021; pp. 340–345. [CrossRef]
13. Suh, S.-W.; Kang, W.-S.; Lee, K.-Y.; Lee, K.-O. A study on the data structure and implementation techniques for the object oriented ship structure modeling. *J. Soc. Nav. Archit. Korea* **1994**, *31*, 1–11.
14. Roh, M.-I.; Lee, K.-Y.; Yoo, S.-J. An algorithm for generating the hull structural analysis model using the seam information of the hull structure at the initial design stage. *J. Ship Ocean Technol.* **2006**, *10*, 24–33.
15. Shigehiro, M.; Masahiro, M.; Song, X.; Takao, Y. Structural optimization of the mid-ship section by applying genetic algorithm and response surface method. *J. Jpn. Soc. Nav. Arch. Ocean Eng.* **2014**, *20*, 109–117. [CrossRef]

16. Lee, K.-Y.; Lee, W.-J.; Roh, M.-I. Development of a semantic product modeling system for initial hull structure in shipbuilding. *Robot. Comput. Manuf.* **2004**, *20*, 211–223. [CrossRef]
17. Roh, M.-I.; Lee, K.-Y. An initial hull structural modeling system for computer-aided process planning in shipbuilding. *Adv. Eng. Softw.* **2005**, *37*, 457–476. [CrossRef]
18. Na, S.-S.; Karr, D.G. Development of Pareto strategy multi-objective function method for the optimum design of ship structures. *Int. J. Nav. Arch. Ocean Eng.* **2016**, *8*, 602–614. [CrossRef]
19. Jang, C.-D.; Na, S.-S. Development of optimum structural design system for double hull oil tankers. *J. Soc. Nav. Archit. Korea* **2000**, *37*, 118–126.
20. Nobukawa, H.; Zhou, G. Discrete optimization of ship structures with genetic algorithms. *J. Soc. Nav. Archit. Jpn.* **1996**, *179*, 293–301. [CrossRef]
21. Tryaskin, V.N.; Ryumin, S.N.; Kuteynikov, M.A. *The ATLAS Software. Objective and Architecture. Methodological Basis. Test Results*; Reseatch Bulletin by Russian Maritime Register of Shipping, №54/55; Russian Maritime Register of Shipping: Saint-Petersburg, Russia, 2019; pp. 106–127.
22. Russian Maritime Register of Shipping. *Rules for the Classification and Construction of Sea-Going Ships*; Russian Maritime Register of Shipping: Saint Petersburg, Russia, 2021.
23. IACS. *Common Structural Rules for Bulk Carriers and Oil Tankers*; IACS: London, UK, 2021.

machines

MDPI

Article

Feasibility of Predictive Models for the Quality of Additive Manufactured Components Based on Artificial Neural Networks

Sorin D. Grozav [1], Alexandru D. Sterca [1,*], Marek Kočiško [2], Martin Pollák [2] and Vasile Ceclan [1]

1 Faculty of Industrial Engineering, Robotics and Production Management, Technical University of Cluj-Napoca, Bulevardul Muncii nr. 103-105, 400641 Cluj-Napoca, Romania; sorin.grozav@tcm.utcluj.ro (S.D.G.); vasile.ceclan@tcm.utcluj.ro (V.C.)

2 Faculty of Manufacturing Technologies with a Seat in Presov, Technical University of Kosice, Bayerova 1, 080 01 Presov, Slovakia; marek.kocisko@tuke.sk (M.K.); martin.pollak@tuke.sk (M.P.)

* Correspondence: alex.d.sterca@gmail.com; Tel.: +40-752141875

Abstract: Additive manufacturing technologies present a series of advantages such as high flexibility, direct CAD to final product fabrication, and compact production techniques which make them an attractive option for fields ranging from medicine and aeronautics to rapid prototyping and Industry 4.0 concepts. However, additive manufacturing also presents a series of disadvantages, the most notable being low dimensional accuracy, low surface quality, and orthotropic mechanical behaviour. These characteristics are influenced by material properties and the process parameters used during manufacturing. Therefore, a predictive model for the characteristics of additive manufactured components is conceivable. This paper proposes a study on the feasibility of implementing Deep Neural Networks for predicting the dimensional accuracy and the mechanical characteristics of components obtained through the Fused Deposition Modelling method using empirical data acquired by high precision metrology. The study is performed on parts manufactured using PETG and PLA materials with known process parameters. Different Deep Neural Network architectures are trained using datasets acquired by high precision metrology, and their performance is tested by comparing the mean absolute error of predictions on training and validation data. Results show good model generalisation and convergence at high accuracy, indicating that a predictive model is feasible.

Keywords: machine learning; artificial neural network; additive manufacturing; high precision metrology; CAD; predictive model

Citation: Grozav, S.D.; Sterca, A.D.; Kočiško, M.; Pollák, M.; Ceclan, V. Feasibility of Predictive Models for the Quality of Additive Manufactured Components Based on Artificial Neural Networks. *Machines* **2022**, *10*, 128. https://doi.org/10.3390/machines10020128

Academic Editor: Panagiotis Kyratsis

Received: 30 December 2021
Accepted: 7 February 2022
Published: 11 February 2022

1. Introduction

Additive manufacturing technologies are becoming an essential tool in a wide variety of fields ranging from medicine and aeronautics to industry 4.0 concepts, rapid prototyping, reverse engineering, and hobby use. This rapid increase in the exploitation of additive manufacturing techniques is a direct result of the many advantages presented by these technologies, which include high flexibility, low cost, direct computer-aided design (CAD) to finished product manufacturing, reduced tooling requirements, the capability of realising high complexity geometries in a short time and at low cost, etc. At the same time, additive manufacturing techniques also suffer from a series of disadvantages, most notably low dimensional accuracy, low surface quality, and orthotropic material behaviour. The Fused Deposition Modelling (FDM) in particular has seen a great increase in utilisation, mostly due to the ease of use and availability of low-cost machines and raw material for this technique. Due to the low cost of the machines and materials, the FDM technique is becoming important for rapid prototyping, hobby use, manufacturing of replacement parts, and even as a supplement for the more productive technologies such as Injection Molding, being employed in the form of "3D printing farms" composed of many FDM machines to compensate for the low productivity, thus enhancing the flexibility of the production lines where they are implemented. However, of all the additive manufacturing techniques, FDM suffers the most from the disadvantages mentioned above; therefore, extensive postprocessing of the manufactured parts is employed, which is costly and time consuming.

2. State of the Art

Research has been done [1] with good results on increasing the accuracy of FDM manufactured parts by fabricating components followed by parametric compensation of the 3D model using data obtained by high precision metrology.

Studies by Yadav D. et al. [2] have been performed on improving the tensile strength of additive manufactured parts by employing neural networks trained on parameters such as infill density, extrusion temperature, and material density. The study was performed on parts obtained from materials such as Acrylonitrile Butadiene Styrene (ABS) and Polyethylene Terephthalate Glycol (PETG) and showed good results, improving the tensile strength by 4.54% in subsequent samples.

Research on predicting optimal manufacturing parameters by Deshwal S. et al. [3] to increase the tensile strength of parts obtained from Polylactic Acid (PLA) material has been performed by training neural networks on process parameters such as nozzle temperature, printing Speed, and infill density. The study reported an increase in tensile strength of ~2 MPa using predictions from a neural network trained at 99.89% accuracy.

Research studies by O.A.Mohamed et al. [4] have been performed regarding the ability of artificial neural networks to predict dimensional characteristics of parts obtained through the FDM process by analysing parameters such as slice thickness, the number of perimeters, deposition angle, part print direction, and raster to raster air-gap. The study compared the precision of neural networks to classic regression techniques and showed that neural networks can account for dimensional deviations.

Tura A.D. et al. [5] performed studies on predicting the effects on surface roughness of FDM process parameters such as layer thickness, orientation angle, and infill angle with good results showing a high impact of layer thickness on surface roughness and the ability of neural networks to predict the effects.

Lyu et al. [6] researched the ability of neural networks to predict dimensional characteristics of FDM manufactured parts by analysing the complex relationship between nozzle temperature, layer thickness, and infill density and their effects on dimensional characteristics.

Previous research by A. D. Sterca et al. [7] was conducted to determine if the main FDM process parameters, e.g., temperature, speed, and part/layer orientation, influence the dimensional accuracy of 3D printed parts. The study was performed using regression analysis on data obtained by high precision metrology on a set of test parts. The study showed a strong correlation between the main process parameters and dimensional accuracy. The study also showed that 40% of the dimensional deviation from nominal is caused by a combination of factors that are not accounted for and by complex interactions between parameters which can not be modelled by a simple linear regression function.

Based on previous research [7], a correlation exists between geometric feature size and dimensional deviation, accounted for by the material shrinkage rate.

The research performed in this study characterises the materials used in terms of their shrinkage rate with the aim of increasing the accuracy of dimensional deviation predictions. In addition, the study takes printing speed into account, which has been shown to affect dimensional deviations through a combination of effects [7]. Part warping has been shown to be the most important factor in dimensional accuracy and is strongly influenced by part and layer cooling times and material shrinkage rates. Cooling times for both parts and layers are determined by a number of parameters such as printing speed, layer cooling fan speed, nozzle temperature, and heated bed temperature. These parameters all affect the temperature gradient throughout the part and thus the degree to which the part warps. Therefore, this study proposes including all of these parameters as inputs for training the neural network, filling gaps in the field by accounting for the fundamental parameters such as printing speed and material shrinkage rate, which have been shown to influence part quality but have not been studied in a correlative manner. The study also proposes the development of separate neural networks for each predicted characteristic (mechanical behaviour, dimensional deviations) in a cross-platform environment, allowing for modularity and the ability to develop more complex networks that can make design decisions using predictions produced by the different modules.

The study presented in this paper aims to determine the ability of artificial neural networks to learn and approximate the complex functions needed to predict dimensional deviations and mechanical behaviour of parts obtained by the FDM process using the main process parameters as well as material properties as inputs and empirical data obtained from high precision metrology and tensile strength measurements as outputs. A predictive mathematical model can provide information in the form of deviation from nominal, which can be used to improve dimensional accuracy in the design stage by applying corrections based on predictions to a parametric 3D model of the part, reducing or eliminating the need to manufacture and measure a series of parts to compensate for dimensional accuracy. Predictions of the mechanical characteristics of 3D printed parts can be used in the design process for guiding decisions regarding layer orientation and feature dimensions to ensure proper mechanical behaviour of the manufactured part. The predictions for dimensional and mechanical characteristics can be used to train an additional artificial neural network to provide a good compromise between dimensional accuracy and mechanical properties.

The ability to predict the characteristics of additive manufactured parts can increase the quality and interchangeability of these components and can even lead to a tolerance standard for components obtained through these techniques.

3. Materials and Methods

A test part geometry is designed and manufactured from PETG and PLA materials using different printing speeds, nozzle temperatures, and orientations. The parts obtained by the FDM process are measured using Coordinate measuring techniques and 3D scanning techniques. Tensile strength measurements are performed on the parts to determine if and how the mechanical characteristics are affected by different process parameters.

Different Artificial Neural Network (ANN) architectures are developed and trained using empirical data acquired through measurements and material properties data obtained from the manufacturer and public databases. The performance of Artificial neural networks is determined by comparing predictions with real-world data. The predictions can be tested either by manufacturing a new set of parts to be used as test data or by splitting the existing dataset into Training and Validation Data. In this study, the second method will be used, splitting the dataset into two subsets, 80% being used as training data and 20% for validation. The data splitting is done at random, providing an unbiased test of the performance. The performance of the ANN is determined by the ability of the model to generalise [8]. The experimental process is described in the flowchart presented in Figure 1, where the entire experimental process from part design through measurement, data management, and neural network training and performance analysis processes are set in order. Neural network performance analysis through comparisons between training accuracy and validation accuracy can be done at the same time after each training iteration, or a network can be validated on new data after the training has been performed to a satisfactory accuracy. This study performs validation after each training iteration, providing a plot of the performance at every step. This allows for close monitoring of the performance and early identification of unsatisfactory models.

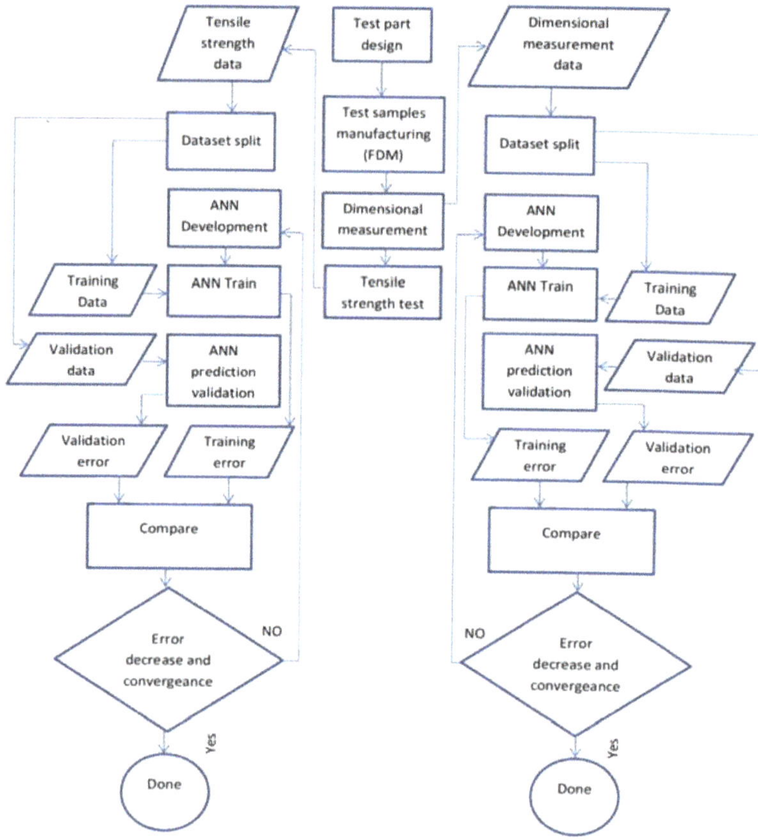

Figure 1. Experimental process flowchart.

A test part presented in Figure 2 is designed using the SolidWorks [9] software suite. The geometry of the test part is chosen to amplify the effects of process parameters on the dimensional and mechanical characteristics of the manufactured parts. A high length to thickness ratio increases the effect of part warping during cooling. A low surface area zone (marked as Width3 in Figure 2) is used as a breaking point for tensile strength measurements as well as for amplifying the effects of different parameters on surface quality. A transition from a low to a high surface area defined by Angle1 and Angle2 is used to determine the variation of surface quality while at the same time providing a possible clamping feature for tensile strength measurements. The test part geometry includes different linear and circular features, which can provide measurement data for different layer orientations. The part is designed and annotated according to ISO 1101 [10].

A total of 36 parts are manufactured, 12 using DevilDesign PETG material [11] and 24 using DevilDesign PLA [12]. The parts are manufactured on two FDM 3D printers, an Anycubic ProMax2 [13] presented in Figure 3a for the PLA parts and a LeapfrogXS [14] as seen in Figure 3b for the PET-G parts. Manufacturing is carried out in sets of two parts simultaneously, one in horizontal and one in a vertical orientation, as seen in Figure 3b, to ensure identical conditions for both orientations. The parameters of nozzle temperature and print speed is varied for each set of parts according to Table 1. and the values are chosen to represent the minimum, medium, and maximum temperature and speed for the material and machine used. A fractional factorial design is preferred to force the neural network to learn correlative functions. The parameter variation for the fractional factorial design is chosen in a range to be statistically significant, providing two extremes (minimum and maximum) and a median point. A minimum of three sample points for the values is needed for the neural network to identify them as variable instead of categorical data. If a neural network shows good performance in terms of generalisation and convergence, a full factorial design can be implemented to increase the prediction accuracy. All parts were manufactured using a 0.5 mm nozzle, 0.25 mm layer height, 5 mm wall thickness, 1 mm top/bottom thickness, and 30% infill using a linear infill pattern. The 5 mm wall thickness ensures a 100% infill for the 5 × 5 mm area designated as the breaking point for tensile strength tests. A sample of the manufactured parts can be seen in Figure 3c.

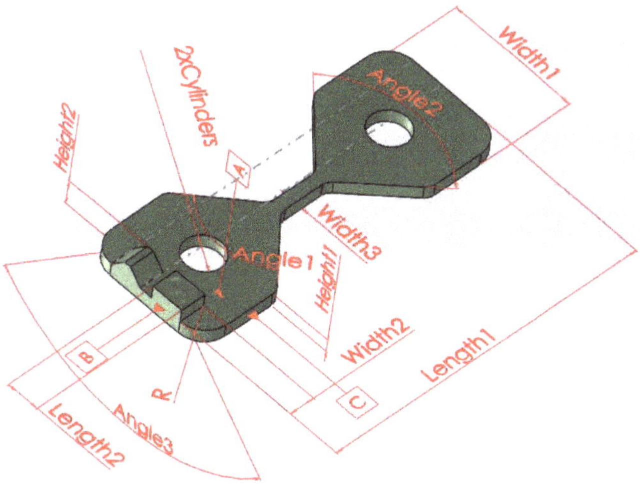

Figure 2. Test part CAD model.

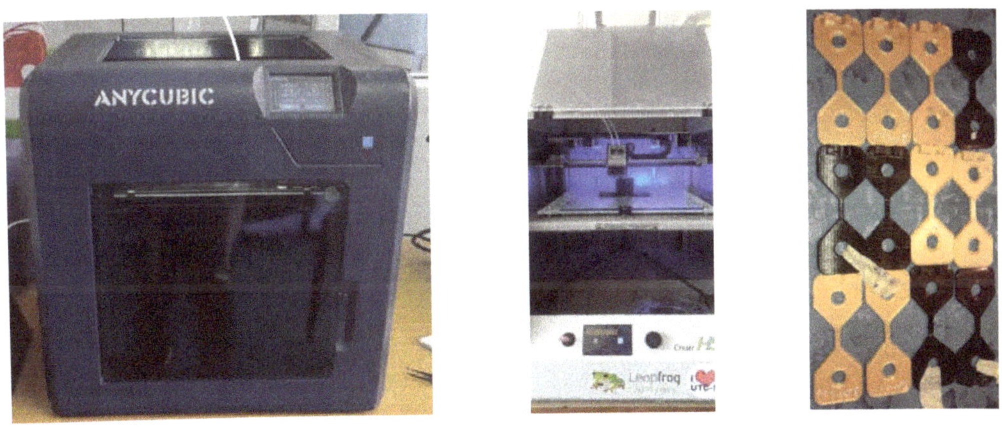

(a) (b) (c)

Figure 3. Machines used for manufacturing test parts, (**a**) Anycubic 4Max Pro 2.0, (**b**) Leapfrog Creatr HS and (**c**) sample of manufactured parts.

Table 1. Process parameters used for manufacturing test part pairs.

Pair No.	Material	Temperature (°C)	Speed (mm/s)
1	PET-G	220	45
2	PET-G	230	60
3	PET-G	230	70
4	PET-G	230	90
5	PET-G	240	45
6	PET-G	250	45
7	PLA	190	45
8	PLA	190	60
9	PLA	190	70
10	PLA	190	90
11	PLA	200	45
12	PLA	200	60
13	PLA	200	70
14	PLA	200	90
15	PLA	210	45
16	PLA	210	60
17	PLA	210	70
18	PLA	210	90

The naming of the test parts within the context of this paper is done according to Table 1 and takes the form of Temp_Speed_Orientaion_Material; thus, a part with the designation 210_90_V_PLA is identified as the test part manufactured with a nozzle temperature of 210 °C at a printing speed of 90 mm/s in the vertical orientation using PLA material.

Dimensional and geometric measurements are performed using two main techniques, coordinate measuring and 3D scanning. High precision measurements are obtained using a Cimcore Stinger 2 [15] Coordinate Measuring Arm (Figure 4a) and Autodesk PowerInspect [16] software. Three-dimensional surface scan data is acquired using a Gom Atos II (Compact Scan) [17] structured light 3D scanner (Figure 4c). Computed Tomography data is acquired using a Werth XS Tomoscope [18] (Figure 4d), providing in-volume and surface 3D scans of the parts. Data obtained from optical 3D scans and computed tomography is used for CAD for actual mesh comparison using the Gom Inspect suite [19] software, providing a detailed analysis of geometric and dimensional deviations. These methods of measurement are chosen to provide as much detail as possible concerning dimensional and geometric characteristics of the manufactured parts, which are needed to guide decisions on how the input and output data should be chosen for ANN training as well as to provide information regarding the architecture of the ANN. Multiple points are sampled for each measured feature, and the results are averaged to reduce outliers and avoid sampling manufacturing artefacts such as holes or material overextrusion. Three-dimensional optical scans are used to corroborate results from CMM measurements and to provide details on surface topology and part warping. The ability to sample a large number of points for dimensional measurements using 3D scan data allows for better averaging of deviations and the eliminations or reduction of outliers caused over and under extrusion artefacts. Scans of the parts in the form of STL files can be used in the future for developing a more advanced neural network trained using printing parameters, CAD nominal model, and actual part 3D model.

Tensile strength measurements are performed on an Instron 3366 [20] tensile tester (Figure 4b). The data provided by these measurements can be used to model the mechanical behaviour of the manufactured parts, which can then be used to create an orthotropic material profile for CAD systems that can inform decisions of layer orientation and feature dimensions in the design stage, as well as provide data for the training of an ANN that can predict the mechanical behaviour of a part. The tests are performed on all test parts, and the results are plotted for pairs representing the horizontal and the vertical orientations for each combination of parameters, providing a comparative study of tensile strength for the different orientations used.

(**a**) (**b**) (**c**) (**d**)

Figure 4. Machines used for performing measurements. (**a**) Cimcore Stinger 2, (**b**) Instron 3366, (**c**) GOM Atos II (CompactScan), (**d**) Werth Tomoscope XS.

Artificial neural network development is performed using the Python [21] programming language through the PyCharm [22] Integrated Development Environment. To facilitate universality and cross-platform availability of the code, as well as to use various functions and tools optimised for machine learning, the neural network architectures are developed using the TensorFlow [23] machine learning library with the Keras [24] frontend. The architecture and performance of the tested neural networks are analysed using the TensorBoard [25] visualisation kit for TensorFlow. The performance analysis is displayed as a plot of 'Training accuracy' Vs. 'Predictions on validation data' accuracy in the form of the Mean Absolute Error (MAE) between predicted and actual values. A tendency of the MAE values to decrease indicates the ability of the model to learn the complex correlative function and the ability to generalise. The Mean Absolute Error metric is defined by Equation (1),

$$\mathrm{MAE} = \frac{\sum_{i=1}^{n} |y_i - x_i|}{n} \tag{1}$$

where MAE = Mean Absolute Error, y_i is the prediction, x_i is the true value, and "n" is the number of samples.

A base program is developed to create every neural network analysed in this study. The source code is written in the Python Programming language and is available for reference in Appendix A. The neural network is implemented as a Sequential Model inside the base program to facilitate the implementation of different model architectures. A neural network architecture in this context is defined by the number of layers and the number of nodes in every layer. A deep neural network (DNN) is defined as a neural network that contains multiple layers (also known as hidden layers) between the input and output layers. All of the neural networks analysed in this study are of the Deep Neural Network Type. Different architectures are developed by modifying the number of hidden layers and nodes in the sequential model implemented in the base program.

In the context of this paper, the neural network architecture will be defined as the number of hidden layers times the number of nodes per layer because all architecture will employ hidden layers with equal numbers of nodes; therefore, 3×32 indicates an architecture containing 3 hidden layers and 32 nodes per layer.

This study is performed on a number of neural networks composed of 2, 3, and 4 fully interconnected hidden layers containing 8, 32, and 64 nodes each, for a total of 9 architectures; thus, the number of weights (trainable parameters) is dependent on the number of nodes and the number of hidden layers used. The optimisers [26] used are Stochastic Gradient Descent and ADAdelta. The selection of the optimiser and the parameters for the optimiser (learning rate) is determined by observing the performance of the network under test (speed of convergence, accuracy of predictions). The weights and bias values of the neural networks are initiated as zeroes and adjusted by the optimiser during the training process. The number of hidden layers and nodes-per-layer are selected to represent networks with increasing parameter spaces (trainable parameters) in order to determine the optimal combination for best accuracy without overfitting the data. As the number of hidden layers and nodes-per-layer is highly dependent on the number of samples in the training dataset as well as the complexity of the problem to be solved, these parameters are subject to fine-tuning based on the observed performance of the network under test, as there is currently no clear mathematical formula for determining the exact architecture for a specific problem. The smallest network contains 2 hidden layers and 8 nodes per layer and is selected to provide a number of trainable parameters smaller than the number of samples in the present dataset and serves as a good starting point. The developed architectures are presented in Table 2 as well as the activation function for the hidden layers and the number of trainable parameters for each architecture.

Table 2. Neural network architectures.

Network architecture	2×8	2×32	2×64	3×8	3×32	3×64	4×8	4×32	4×64
Hidden layers	2	2	2	3	3	3	4	4	4
Number of nodes per layer	8	32	64	8	32	64	8	32	64
Activation function	ReLU	ReLU	ReLU	ReLU	ReLU	ReLU	ReLU	ReLU	ReLU
Number of trainable parameters	113	1217	4481	185	2273	8641	257	3329	12,801

The number of epochs [27] for training the neural networks is not predetermined. The networks are allowed to train until they diverge or until the decrease in prediction error becomes insignificant.

Input and output layer shapes are determined by the number of input and output parameters, and the activation function for the output layer is Linear.

Strategies to prevent overfitting can be employed in the form of l2 regularisers [28] and Dropout layers [29] which can be implemented after each hidden layer. Dropout layers do not affect the number of trainable parameters and are employed only in architectures that show a tendency to overfit. The parameters for l2 regularisers and dropout layers are chosen by fine-tuning for best performance.

Separate neural networks are employed for predicting dimensional characteristics and mechanical characteristics. This modular approach allows for separate training of the neural networks according to the availability of data, while at the same time allowing for interconnectivity with other networks, i.e., the predictions for dimensional deviations and mechanical behaviour can be used as inputs for another network that can inform decisions on the best compromise between dimensional accuracy and mechanical properties.

The data used for training is provided in the form of comma separated values (CSV) tables containing both input and output data. The input data used for training the neural network employed for predicting dimensional deviations are in the form of the values for feature length, nozzle temperature, printing speed, orientation, material shrinkage rate, print cooling fan speed, and heated bed temperature. The material shrinkage rate is provided as a minimum, medium, and maximum shrinkage. The minimum and maximum shrinkage for PLA are 0.3% and 0.5%, and for PET-G 0.2% and 1% [30]. The medium shrinkage value is calculated as the average between the minimum and maximum values. The output data is in the form of deviations from the nominal dimension of the feature, the value of which is obtained through high precision metrology measurements. An example of the table head and a sample of values is presented in Table 3.

Table 3. Sample of input data table head and values used for training dimensional characteristics neural network.

Length (mm)	Temp (°C)	Speed (mm/s)	Orientation	Shr_min (%)	Shr_med (%)	Shr_max (%)	Cooling (%)	Bed_temp (°C)	Deviation (mm)
12.5	220	45	0	0.2	0.6	1	0	80	−0.082

The neural networks employed in predicting mechanical properties have input data in the form of values for nozzle temperature, printing speed, material and orientation. The value for "material" represents categorical data; therefore, each material is encoded in the form of a number "1" for PETG and "2" for PLA to avoid not a number (NaN) errors during training. Considering that for this dataset, the orientation also takes on only two values, horizontal and vertical, this feature is also encoded as a number "1" for horizontal and "2" for vertical. For datasets where the orientation takes on many different values, these can be provided as angles. The output data used for training the neural network are provided in the form of tensile strength values in [Mpa] obtained from tensile strength measurements. An example of a table head and a sample of values is presented in Table 4.

Table 4. Sample of input data table head and values for training mechanical behaviour neural network.

Temp (°C)	Speed (mm/sec)	Material	Orientation	Tensile Strength (Mpa)
190	45	2	2	16.54

A graphical representation of a neural network analysed in this study showing the path of the data and the operations performed throughout the network during training is presented in Figure 5, showing the path of the data through the neural network. Data enters through the dense layers (dense, dense_1, dense_2), where an inference (prediction) is made. The result of the inference is used in combination with the actual value to compute the Mean Absolute Error (MAE). The error is used to adjust the value of the weights and biases to a degree determined by the optimiser employed, i.e., for Figure 5, the optimiser used is Stochastic Gradient Descent (SGD). The network design also employs functions to prevent NaN scenarios. The data is normalised (values scaled between 0 and 1) externally to allow for better portability of the model. Normalisation is performed to increase the speed of convergence by reducing the difference in scale between the various parameter values. Data normalisation is optional; however, it can increase model performance and decrease convergence times. The normalised data takes values from 0 to 1 but preserves the relative ratios between the original values.

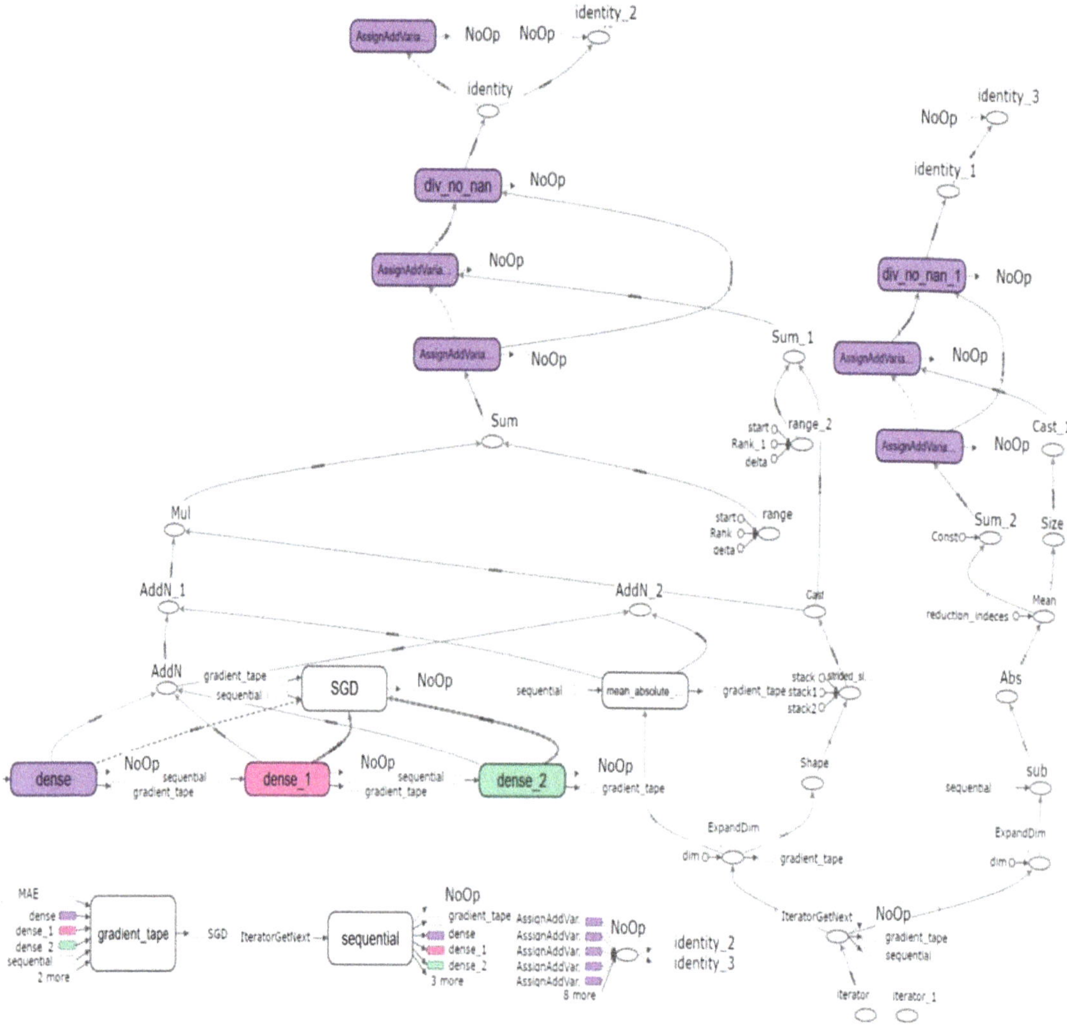

Figure 5. Graphical representation of a neural network with 3 hidden layers.

4. Results

4.1. Part Measurement Results

Dimensional measurements provide the empirical data in the form of deviations from nominal feature length used to train the neural networks. The measurements are performed using two independent techniques, Coordinate measuring and 3D scanning, enabling cross-validation of data which can be used to eliminate outliers and reduce the noise inherent in any measurement process. Besides dimensional measurements, geometric analysis of the manufactured parts is very important for informing decisions regarding the selection of the training dataset. An in-depth analysis of the geometric and dimensional deviations of the manufactured parts indicates that material properties, in particular, material shrinkage rate, has a significant influence on the quality of the 3D printed components. Measurement results show that most of the deviations fall within the values predictable by the shrinkage rate of the material used, while the extent to which the parts deform due to this effect is influenced by nozzle temperature, print speed, and part cooling fan speed, parameters which influence layer and part cooling time directly. Analysis shows that parameter combinations that lead to short cooling times for the layers and the part, in general, tend to produce parts with higher dimensional and geometric deviations. This can be explained in part due to high internal tensions produced during rapid cooling that lead to more severe part warping and contraction. These results informed the decision of including material shrinkage rates and part cooling fan speed as training parameters for the artificial neural networks analysed in this study. Part warping can be analysed by performing a Surface comparison on CAD using data obtained from 3D scanning. Examples of severe part warping and mild part warping are presented in Figure 6, where (a) is the 230_60_H_PETG test part and (b) is the 220_45_H_PETG test part. The severity of part warping can be determined by the relative difference between the deviations of the part height along its full length, denoting the curvature of the part. The deviations in part height can be seen in the form of deviation labels in Figure 6. Measurement results for all manufactured parts show a high value for the deviation of the total length (130 mm) that can not be accounted for by the effects of contraction alone. Analysis of the effects of warping indicates that the high value for the deviation in total length is due to the curvature of the part. Therefore, a decision can be made regarding the inclusion of this specific feature length in the training data. The complete dimensional measurement results are found in Supplementary Materials Table S1.

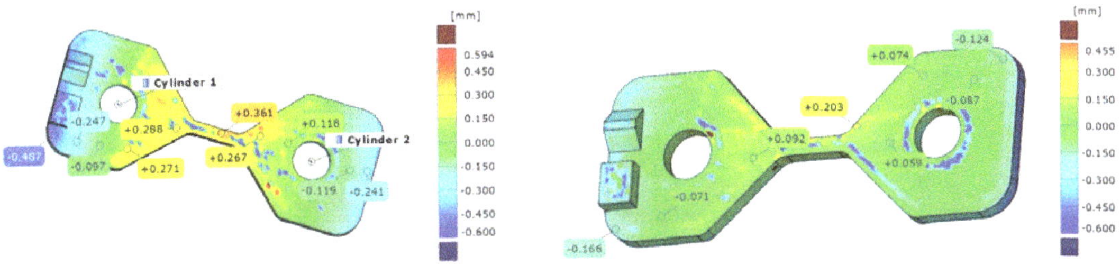

(a) (b)

Figure 6. Surface comparison on CAD. (**a**) severe part warping, (**b**) mild part warping.

4.2. Tensile Strength Measurement Results

The results of tensile strength measurements are presented in Supplementary Materials Table S2 in the form of tensile stress at maximum load (at breaking point) expressed in [Mpa]. The plots of tensile strain vs. tensile stress for two pairs of test parts are presented in Figure 7, where (a) represents the test results for the pair composed of the 200_70_H_PETG and the 200_70_V_PETG part, while (b) represents the results for the pair composed of the 190_90_H_PLA part and the 190_90_V_PLA test part. Sample 1 represents horizontal part orientation and Sample 2 vertical part orientation. The results show that for the vertical orientation, the tensile strength is determined by layer adhesion, which is influenced by the temperature used during the manufacturing process; therefore, the results indicate that the tensile strength of parts manufactured in the vertical orientation is directly proportional to the nozzle temperature.

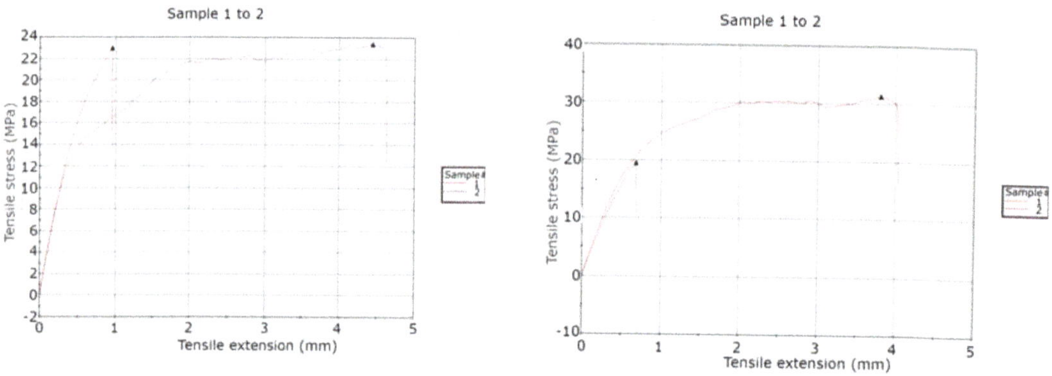

(a) (b)

Figure 7. Sample of tensile stress vs. tensile strain results. (**a**) 200_70_PLA pair, (**b**) 190_90_PLA pair.

Tensile strength measurement results for test parts manufactured in the horizontal orientation show that, in this case, tensile strength is inversely proportional to temperature. This result can be explained by the increased porosity of parts manufactured at high temperatures and the thermal degradation of the material. An example of differences in porosity can be seen in Figure 8, where (a) is the 250_45_H_PETG part and (b) is the 220_45_H_PETG part. A higher porosity can be observed on the high-temperature part in the form of small holes in the volume of the material; however, although the low-temperature part presents a lower porosity, the effects of increased layer decohesion can be observed. These results must be taken into account when choosing manufacturing process parameters.

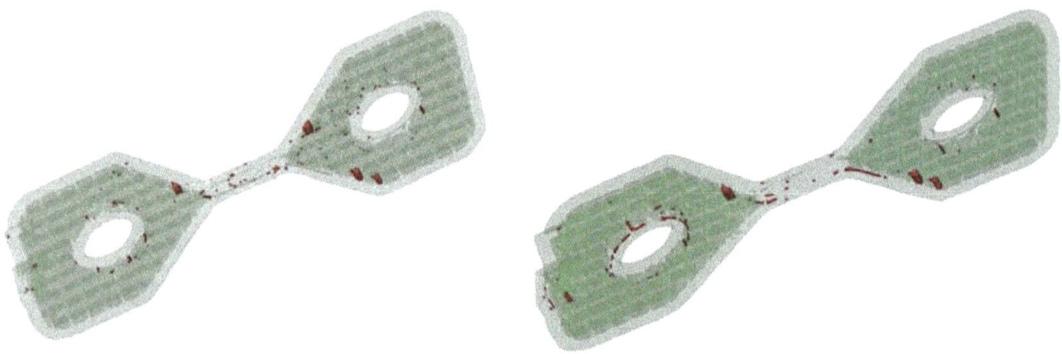

(a) (b)

Figure 8. The difference in porosity due to printing temperature. (**a**) high-temperature part (**b**) low-temperature part.

One reason for using a complex geometry test part is to determine the mechanical behaviour in scenarios closer to real-world cases. One such behaviour is the part failing in other areas than the designated failure zone before the part breaking. A comparison between a part breaking in the designated failure zone and a part that fails in other areas before breaking is shown in Figure 9. The part in Figure 9a represents expected behaviour, breaking in the designated failure zone. The part in Figure 9b presents the separation at the interface between the top and bottom layers and the part perimeter. This behaviour is accounted for by a low overlap percentage between the top/bottom layers and the part perimeter and is present for all materials tested (PETG and PLA). These results must be taken into account when choosing a top and bottom thickness and the overlap percentage. Another factor that influences this behaviour is infill pattern and density.

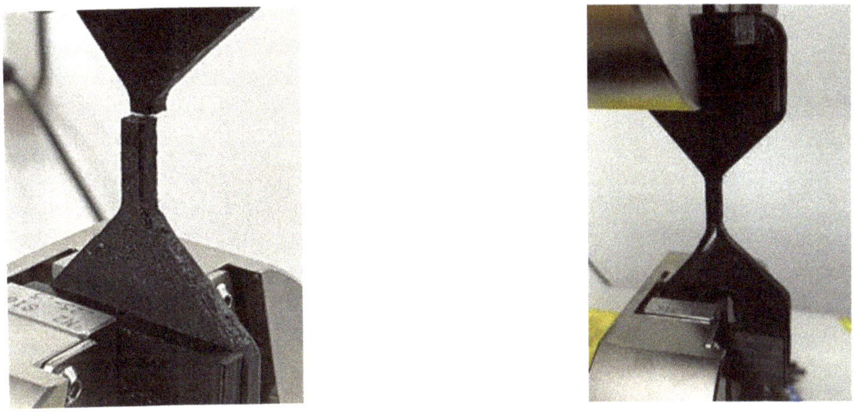

(a) (b)

Figure 9. Comparison of test part failure modes. (**a**) part breaking at designated failure point (**b**) part failing before breaking.

For training the neural network, only the data from parts breaking in the designated failure zone are used.

The results of dimensional and tensile strength measurements are presented in Table 5 in the form of deviations from nominal for dimensional measurements and maximum stress in MPa for tensile strength measurements. The maximum stress in MPa for parts that failed before breaking, as exemplified earlier and shown in Figure 9b, is also presented.

Table 5. Dimensional and tensile strength measurement results.

Measured Characteristic	Length 1 (mm)	Length 2 (mm)	Width 1 (mm)	Width 2 (mm)	Width 3 (mm)	Tensile Strength (Mpa)	Part Failure before Breaking Maximum Stress (Mpa)
Nominal	130	12.5	50	10	5		
220_45_H_PETG	−0.4236	−0.1002	−0.1388	−0.0851	0.258	21.41	
220_45_V_PETG	−0.0842	0.1069	−0.0615	0.1694	0.079	15.56	
230_60_H_PETG	−0.4156	−0.1758	−0.1403	−0.1758	0.383	34.53	
230_60_V_PETG	−0.0872	−0.1191	−0.0364	0.2913	0.117	36.69	
230_70_H_PETG	−0.5384	−0.2518	−0.3019	−0.0553	0.269		45.86
230_70_V_PETG	−0.1714	−0.162	−0.0766	0.2713	0.159	36.76	
230_90_H_PETG	−0.767	−0.2647	−0.527	0.0068	0.3		32.35
230_90_V_PETG	0.0639	−0.1415	−0.1882	0.3302	0.361	37.24	
240_45_H_PETG	−0.2899	−0.0184	−0.0498	0.1928	0.365		28.27
240_45_V_PETG	−0.0793	−0.0029	−0.0755	0.1137	0.051		25.4
250_45_H_PETG	−0.2731	−0.0632	−0.0301	0.1881	0.413	38.61	
250_45_V_PETG	−0.0976	0.0298	−0.048	0.2461	0.071	33.23	
190_45_H_PLA	−0.623	−0.384	−0.319	−0.52	−0.209	49.57	
190_45_V_PLA	−0.727	0.336	−0.19	−0.327	−0.133	16.54	
190_60_H_PLA	−0.587	−0.422	−0.052	−0.36	−0.327	38.48	
190_60_V_PLA	−0.971	0.1204	−0.2238	−0.354	−0.4883	13.56	
190_70_H_PLA	−0.915	−0.213	−0.602	−0.316	−0.415		35.54

Table 5. *Cont.*

Measured Characteristic	Length 1 (mm)	Length 2 (mm)	Width 1 (mm)	Width 2 (mm)	Width 3 (mm)	Tensile Strength (Mpa)	Part Failure before Breaking Maximum Stress (Mpa)
190_70_V_PLA	−0.924	0.047	−0.019	−0.359	0.052	18.37	
190_90_H_PLA	−0.634	−0.214	−0.065	−0.068	−0.151	31.39	
190_90_V_PLA	−0.772	0.292	−0.023	−0.114	−0.471	19.52	
200_45_H_PLA	−1.25	−0.453	−0.494	−0.503	−0.703		18.65
200_45_V_PLA	−1.311	−0.072	−0.487	−0.539	0.12		15.82
200_60_H_PLA	−1.326	−1.326	−0.573	−0.006	−0.229		22.44
200_60_V_PLA	−1.344	−0.122	−0.585	−0.097	0.081	14.62	
200_70_H_PLA	−0.229	−0.229	−0.115	−0.038	−0.306	23.4	
200_70_V_PLA	−0.762	0.402	0.057	−0.144	−0.189	23	
200_90_H_PLA	−1.373	−0.473	−0.473	−0.568	−0.568		31.71
200_90_V_PLA	−1.205	0.035	−0.295	−0.412	0.159	11.97	
210_45_H_PLA	−1.459	0.082	−0.695	−0.209	0.126		30.42
210_45_V_PLA	−1.195	−0.038	−0.409	−0.398	0.082	12.83	
210_60_H_PLA	−1.309	−0.553	−0.405	−0.635	−0.377		37.85
210_60_V_PLA	−0.22	0.13	−0.326	−0.388	0.179	15.6	
210_70_H_PLA	−1.204	−0.416	−0.308	−0.578	−0.21		35.67
210_70_V_PLA	−1.514	0.124	−0.315	−0.516	−0.605	12.48	
210_90_H_PLA	−1.005	−0.467	−0.11	−0.605	−0.284		39
210_90_V_PLA	−1.276	0.084	−0.362	−0.408	−0.625	12.28	

4.3. *Artificial Neural Network Performance Analysis Results*

The performance of the tested ANN architectures is evaluated by plotting the Mean Absolute Error (MAE) of the training predictions versus the Mean Absolute Error of the predictions performed on the validation data. The validation data is not seen by the neural network during training and, at the same time, represents real-world data obtained by splitting the dataset between training and validation data, thus allowing an unbiased performance analysis applicable to real scenarios.

The interpretation of the plot is as follows: Two lines are plotted, one for training predictions and one for validation data predictions, as mentioned above. If the two lines show a tendency, however small of their respective mean absolute error, it indicates that the model can approximate a correlative predictive function for the input and output parameters and is also capable of generalising. If the lines show a tendency to diverge, i.e., the MAE of the training predictions decreases while the MAE of the validation data predictions increases, the model is considered to overfit the data, i.e., it starts to memorise data rather than approximate a predictive function. If the lines show a random distribution on the plot, this indicates that the model can not find a predictive function and a correlation between the input and output parameters.

A wide range of architectures was tested with 2 and 3 hidden layers and 8, 32, and 64 nodes per layer. The performance of each tested architecture is used to inform decisions regarding optimisation and fine-tuning of subsequent networks leading to an architecture that best fits the available dataset.

The performance plots for two architectures used for predicting tensile strength are presented in Figure 10. The architectures tested in the two plots are 3 × 64 and 3 × 32. From the plot, it can be seen that the model with the 3 × 64 tends to overfit the data. The best performance on the available dataset is reached by the model with 3 × 32 architecture. The performance plot for the 3 × 32 architecture shows good model generalisation and a tendency for the MAE to decrease; thus, the model shows the ability to learn a correlative and predictive function using the input and output parameters provided. Due to the low number of parameters and strong correlation between them, the neural network for predicting tensile strength is able to approximate a predictive function in a short time on a lower number of nodes, while at the same time, the low number of parameters leads to overfitting on large models.

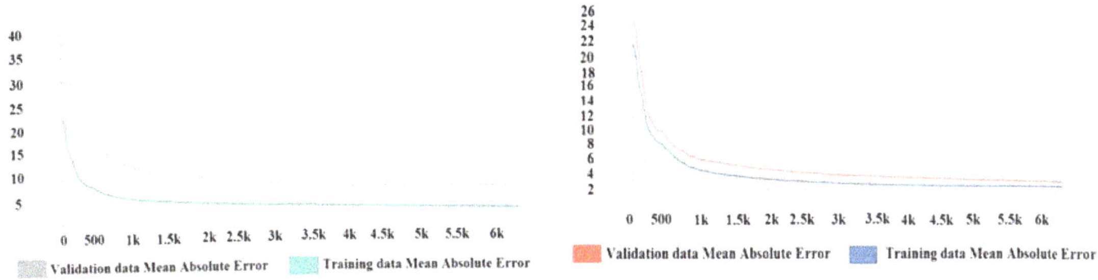

(a) (b)

Figure 10. Performance plots for tensile strength predictive models. (**a**) 3 × 64 architecture, (**b**) 3 × 32 architecture.

Performance plots for dimensional deviation predictive models are presented in Figure 11. Due to the increased complexity of the problem that needs to be solved for this model and the high number of input parameters, a larger model is required to approximate a correlative and predictive function. Because of the low performance of 3 × 8 and 3 × 32 architectures, a 4 × 64 architecture was also analysed for performance to provide a comparison point for the 3 × 64 architecture. An analysis of the plots shows that the model with 4 × 64 architecture tends to overfit the data. The best performance was achieved by the model with 3 × 64 architecture, showing a tendency to reduce the Mean Absolute Error for both the training and validation data predictions. Due to the high complexity of the problem to be solved by the model and the high number of input parameters, the predictive model for dimensional deviations takes longer times to train than the simpler model for predicting tensile strength. However, the model reached a value of 0.05 for the mean absolute error of the predictions for validation data after 120,000 training epochs [23] while still showing a tendency to improve. A value of 0.05 for MAE is insignificant; therefore, the model can produce usable predictions.

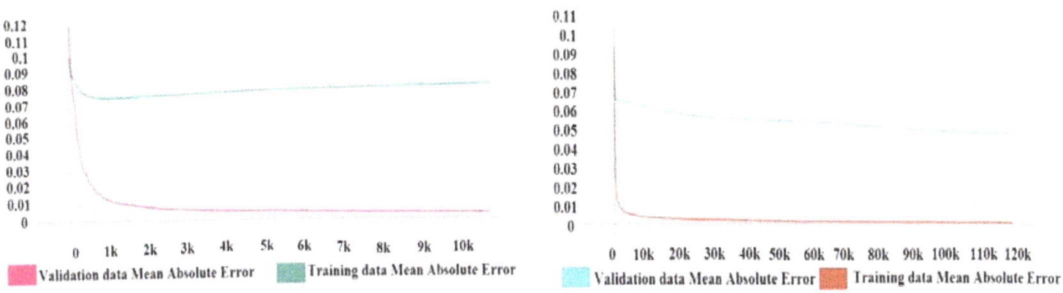

(a) (b)

Figure 11. Performance plots for dimensional deviation predictive models. (**a**) 4 × 64 model architecture (**b**) 3 × 64 model architecture.

The neural networks can be optimised for better generalisation and performance by choosing adequate optimisers [24], increasing the number of layers and nodes, and implementing techniques such as layer weight regularisation [25] and dropout layers [26]. The input data needs to be curated as well, outliers need to be eliminated, and input parameters with low variety should also be eliminated; otherwise, the neural network will consider them as categorical data. An example of this behaviour can be observed in the data for shrinkage, cooling fan speed, and bed temperature. These parameters have only two values in the dataset. For large datasets that are comprised of parts manufactured from more materials at a variety of printed bed temperatures and cooling fan speed, these parameters can be accounted for and included. A performance plot for an optimised neural network is presented in Figure 12. The neural network employs the Adadelta optimiser and techniques to prevent overfitting. The parameters with a low variety were removed, and training was performed strictly on the main process parameters, temperature, speed and feature length. The plot indicates that the model can generalise, with both MAE values tending to decrease.

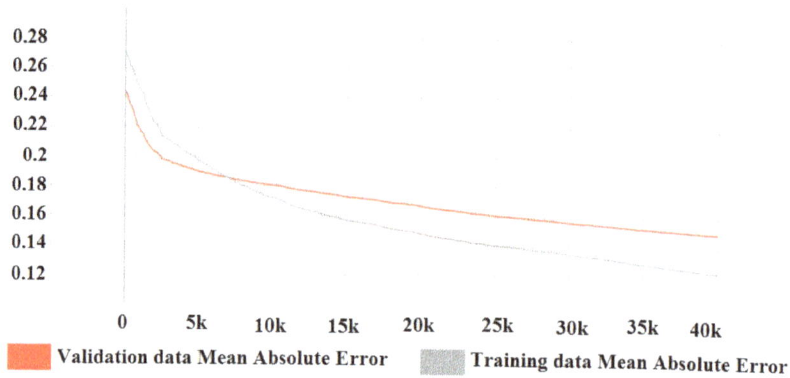

Figure 12. Performance plot for an optimised neural network with 3 × 64 architecture.

5. Discussion

The study shows that predicting dimensional deviations and mechanical behaviour in the form of tensile strength is feasible. The ability of the model to generalise is dependent on the quantity, quality, and variety of the input data. The dataset used in this study is limited, however, it is sufficient to prove that predictive models can be created, encouraging future research on large datasets. The models present predictions with higher accuracy than the predictions obtained through linear regression in previous studies. The predictions provided by the models can be used for guiding decisions regarding the choice of parameters for manufacturing and for compensating dimensional deviations in the design phase, leading to a lower number of failed parts and the reduction or elimination of postprocessing operations.

The findings are consistent with the results of previous studies in the field.

The inclusion of additional parameters such as shrinkage rate, printing speed, part fan cooling speed, and heated bed temperature as input training data does not diminish the neural network's ability to learn and approximate the predictive function, indicating that a correlation exists between these parameters and dimensional deviations. However, the dataset needs to be improved in variety and quantity for these parameters to be statistically significant and to increase prediction accuracy. An absolute minimum of three different values for each of the parameters must be used, i.e., three different materials, three different heated bed temperatures, etc.

The study shows a strong correlation between parameters that affect part and layer cooling times and dimensional deviations. Therefore, ambient temperature plays an important role; however, the study was performed under controlled laboratory conditions of constant temperature and humidity; thus, future datasets must include environmental variables such as ambient temperature and relative humidity.

The neural networks developed in this study can be used to provide predictions in the form of dimensional deviation or tensile strength of a part for a given parameter combination or for determining the best parameter combination for a given dimensional deviation or tensile strength value by reversing the input and output configuration of the model. The predicted values for dimensional deviations can be used to adjust a parametric CAD model of the part to compensate for the errors. The predicted tensile strength values can guide decisions for optimal part orientation and manufacturing parameter selection. However, while the networks developed for this study provide simple predictions of deviations and tensile strength that can be used in real-world applications for improving part quality and mechanical behaviour, the main utility of these networks is for informing the development of more complex networks or as modules in interconnected network designs.

6. Conclusions

The results show that artificial neural networks are capable of learning complex predictive functions for the quality of FDM manufactured parts, encouraging and informing future work in the field.

The limiting factors were determined to be the size and variety of the available dataset on which the performance of the neural networks depends; therefore, the authors propose the creation of an open public database where individuals and institutions can submit and access measurement data on additive manufactured parts in the form of 3D scans, high precision metrology, and tensile strength measurements.

Guided by the findings of this study and the ability of artificial neural networks to approximate and learn complex functions and to work on large datasets and input data, a future study is proposed to develop a neural network that takes the point cloud, or STL file of an FDM manufactured part and the process parameters as input and provides, as output, a compensated STL file, using the predictions to modify the coordinates of the points which define the 3D model. Such a network will most probably be in the form of a generative model, i.e., a Generative Adversarial Neural network or a Variational Autoencoder.

Supplementary Materials: The following supporting information can be downloaded at: https://www.mdpi.com/article/10.3390/machines10020128/s1, The data used for training the neural networks is published alongside the manuscript in the form of tables, Table S1: Dimensional training data, Table S2: Tensile strength training data.

Author Contributions: Conceptualization, S.D.G., A.D.S. and V.C.; methodology, S.D.G., A.D.S. and V.C.; software, A.D.S.; validation, S.D.G. and V.C.; formal analysis, S.D.G., A.D.S. and V.C.; investigation, S.D.G., A.D.S. and V.C.; resources, S.D.G. and V.C.; data curation, A.D.S.; writing—original draft preparation, A.D.S.; writing—review and editing, M.K., M.P., V.C. and A.D.S.; visualization, A.D.S.; supervision, S.D.G.; project administration, V.C.; funding acquisition, S.D.G. All authors have read and agreed to the published version of the manuscript.

Funding: This research was funded by the Technical University of Cluj-Napoca, and The APC was funded by the Technical University of Cluj-Napoca.

Acknowledgments: Special thanks to our colleagues at the Technical University of Cluj-Napoca, Eugen GUTIU for manufacturing the test parts, Assoc. Prof. Pop GRIGORE for providing high precision metrology measurements, Vlad BOCANET for providing 3D scans. High precision metrology measurements were also provided through a collaboration with the Technical University of Vienna facilitated by the CEEPUS program and the OeAD. Special thanks go to the team from TU-WIEN: Numan M. Durakbasa, Osman BODUR and Res. Assist. Eva Maria WALCHER. Computer Tomography measurements were made possible by collaborating with the Werth Messtechnik Österreich GmbH company.

Conflicts of Interest: The authors declare no conflict of interest.

Appendix A

```
import pandas as pd
from time import time
from keras import backend as K
from sklearn import preprocessing
from keras.layers import Dropout
from keras import regularizers
from tensorflow.python.keras.callbacks import TensorBoard
from keras.callbacks import ModelCheckpoint
import tensorflow as tf
from tensorflow.keras import layers
from sklearn.model_selection import train_test_split
printed_train = pd.read_csv("force train.csv",names=["temp", "speed", "material", "orientation", "mpa"], skiprows=1)
datasetf = printed_train.values
x = datasetf[:, 0:4]
y = datasetf[:, 4]
min_max_scaler = preprocessing.MinMaxScaler()
xs = min_max_scaler.fit_transform(x)
xt, xv, yt, yv = train_test_split(xs, y, test_size=0.1, shuffle=True)
printed_model = tf.keras.Sequential([
layers.Dense(64,kernel_regularizer=regularizers.l2(0.001),activation='relu',input_shape=(4,)),
layers.Dropout(0.2),
layers.Dense(64, kernel_regularizer=regularizers.l2(0.001),activation='relu'),
layers.Dropout(0.1),
layers.Dense(64,kernel_regularizer=regularizers.l2(0.001), activation='relu'),
layers.Dropout(0.1),
layers.Dense(1, activation='linear')
])
tensorboard = TensorBoard(log_dir="logs/{}".format(time()))
printed_model.compile(loss=tf.keras.losses.MeanAbsoluteError(),optimizer=tf.optimizers.SGD(0.01),metrics=[tf.keras.metrics.MeanAbsoluteError()])
checkpoint = ModelCheckpoint("best_model_valforce.hdf5", monitor='val_mean_absolute_error', verbose=1,save_best_only=True, mode='min')
earlystop=tf.keras.callbacks.EarlyStopping(monitor='val_mean_absolute_error',mode='min', min_delta=1, patience=2000),
printed_model.fit(xt, yt, epochs=30000, shuffle=True, validation_data=(xv, yv),
                callbacks=[checkpoint,earlystop, tensorboard]).
```

References

1. Bodur, O.; Stepanek, V.; Walcher, E.M.; Durakbasa, N. Precision in additive manufacturing optimization and evaluation of the accuracy of 3D printer based on GPS system. In Proceedings of the 31st DAAAM International Symposium, Mostar, Bosnia and Herzegovina, 21–24 October 2020; Katalinic, B., Ed.; DAAAM International: Vienna, Austria, 2020; pp. 0963–0972, ISBN 978-3-902734-29-7. [CrossRef]
2. Yadav, D.; Chhabra, D.; Garg, R.K.; Ahlawat, A.; Phogat, A. Optimization of FDM 3D printing process parameters for multi-material using artificial neural network. *Mater. Today Proc.* **2020**, *21*, 1583–1591. [CrossRef]
3. Deshwal, S.; Kumar, A.; Chhabra, D. Exercising hybrid statistical tools GA-RSM, GA-ANN and GA-ANFIS to optimize FDM process parameters for tensile strength improvement. *CIRP J. Manuf. Sci. Technol.* **2020**, *31*, 189–199. [CrossRef]
4. Mohamed, O.A.; Masood, S.H.; Bhowmik, J.L. Modeling, analysis, and optimization of dimensional accuracy of FDM-fabricated parts using definitive screening design and deep learning feedforward artificial neural network. *Adv. Manuf.* **2021**, *9*, 115–129. [CrossRef]
5. Tura, A.D.; Mamo, H.B.; Jelila, Y.D.; Lemu, H.G. Experimental investigation and ANN prediction for part quality improvement of fused deposition modeling parts. *IOP Conf. Ser. Mater. Sci. Eng.* **2021**, *1201*, 012031. [CrossRef]
6. Lyu, J.; Manoochehri, S. Dimensional Prediction for FDM Machines Using Artificial Neural Network and Support Vector Regression. In Proceedings of the International Design Engineering Technical Conferences and Computers and Information in Engineering Conference, Anaheim, CA, USA, 18–21 August 2019. [CrossRef]
7. Sterca, A.D.; Calin, R.A.; Cristian, L.; Walcher, E.M.; Bodur, O.; Ceclan, V.; Grozav, S.D.; Durakbasa, N.M. Evaluation of Fused Deposition Modeling Process Parameters Influence on 3D Printed Components by High Precision Metrology. In *Digitizing Production Systems*; Lecture Notes in Mechanical Engineering; Durakbasa, N.M., Gençyılmaz, M.G., Eds.; Springer: Cham, Switzerland, 2022. [CrossRef]
8. Understand Neural Networks & Model Generalization, Towardsdatascience.com. Available online: https://towardsdatascience.com/understand-neural-networks-model-generalization-7baddf1c48ca (accessed on 27 December 2021).
9. Dassault Systems SolidWorks. Available online: https://www.solidworks.com/domain/design-engineering (accessed on 27 December 2021).
10. ISO 1101. Geometrical product specifications (GPS)—Geometrical tolerancing—Tolerances of form, orientation, location, and run-out. 2017. Available online: https://www.iso.org/obp/ui/#iso:std:iso:1101:en (accessed on 10 January 2022).
11. DevilDesign PET-G Product Description. Available online: https://devildesign.com/download/PET-G_-_product_card.pdf (accessed on 27 December 2021).
12. DevilDesign PLA Product Description. Available online: https://devildesign.com/download/PLA_-_product_card.pdf (accessed on 27 December 2021).
13. AnyCubic 4Max Pro 2.0. Available online: https://www.anycubic.com/products/4max-pro-2-0 (accessed on 27 December 2021).

14. Leapfrog Creatr HS User Manual. Available online: http://support.lpfrg.com/support/solutions/articles/11000083231-user-manual (accessed on 27 December 2021).
15. Hexagon Manufacturing Intelligence Coordinate Measuring Arms. Available online: https://www.hexagonmi.com/?utm_source=DomainRedirect&utm_medium=cimcore (accessed on 27 December 2021).
16. Autodesk PowerInspect. Available online: https://www.autodesk.com/products/powerinspect/overview (accessed on 27 December 2021).
17. Gom Atos II (Compact Scan). Available online: https://www.gom.com/en/products/3d-scanning/atos-compact-scan (accessed on 27 December 2021).
18. TomoScope®XS, Werth, Inc. 2018. Available online: http://werthinc.com/products/tomoscope-xs/ (accessed on 27 December 2021).
19. GOM Inspect Suite. Available online: https://www.gom.com/en/products/gom-suite/gom-inspect-pro (accessed on 27 December 2021).
20. Instron 3360 Series Mechanical Testing Systems. Available online: https://www.instron.com/~{}/media/literature-library/products/2011/06/3300-series-table-model.pdf (accessed on 27 December 2021).
21. Python Programming Language. Available online: https://www.python.org/ (accessed on 27 December 2021).
22. PyCharm Integrated Development Environment. Available online: https://www.jetbrains.com/pycharm/ (accessed on 27 December 2021).
23. TensorFlow Machine Learning Platform. Available online: https://www.tensorflow.org/ (accessed on 27 December 2021).
24. Keras Frontend for TensorFlow. Available online: https://keras.io/ (accessed on 27 December 2021).
25. TensorBoard Visualisation Kit for TensorFlow. Available online: https://www.tensorflow.org/tensorboard (accessed on 27 December 2021).
26. An Overview of Gradient Descent Optimization Algorithms. Available online: https://ruder.io/optimizing-gradient-descent/ (accessed on 29 December 2021).
27. Epoch Definition. Available online: https://deepai.org/machine-learning-glossary-and-terms/epoch (accessed on 27 December 2021).
28. Layer Weight Regularizers. Available online: https://keras.io/api/layers/regularizers/ (accessed on 29 December 2021).
29. Dropout Layer. Available online: https://keras.io/api/layers/regularization_layers/dropout/ (accessed on 29 December 2021).
30. Shrinkage Values for Different Plastic Materials. Available online: https://omnexus.specialchem.com/polymer-properties/properties/shrinkage (accessed on 27 December 2021).

machines

Article

Effect of the Computational Model and Mesh Strategy on the Springback Prediction of the Sandwich Material

Pavel Solfronk *, Jiří Sobotka and David Koreček

Department of Engineering Technology, Faculty of Mechanical Engineering, Technical University of Liberec, Studentska 1402/2, 46117 Liberec, Czech Republic; jiri.sobotka@tul.cz (J.S.); david.korecek@tul.cz (D.K.)
* Correspondence: pavel.solfronk@tul.cz; Tel.: +420-485-353-374

Abstract: The effect of the computational model and mesh strategy on the springback prediction of the thin sandwich material made of micro-alloyed steel was investigated in this paper. To verify the chosen computational strategy, a comparison of the experimentally obtained specimen (U-bending) with the FEA result was performed. The Vegter yield criterion combined both with the isotropic and kinematic hardening law was used for the calculation. In addition, the effect of the deformation mesh element (surface and volume) on the accuracy of the springback prediction was investigated. It was concluded that the choice of the volume deformation mesh does not significantly improve the accuracy of the results. Moreover, it is quite a time-consuming approach. The much greater influence was monitored by concerning the selection of hardening law, where the anisotropic one was more suitable to be used on the springback prediction of a given sandwich material.

Keywords: sandwich; springback; Vegter yield criterion; numerical simulation; PAM-STAMP 2G; isotropic hardening law; kinematic hardening law; bending; Bauschinger effect

Citation: Solfronk, P.; Sobotka, J.; Koreček, D. Effect of the Computational Model and Mesh Strategy on the Springback Prediction of the Sandwich Material. *Machines* **2022**, *10*, 114. https://doi.org/10.3390/machines10020114

Academic Editors: Marek Kočiško and Martin Pollák

Received: 27 December 2021
Accepted: 2 February 2022
Published: 3 February 2022

1. Introduction

In recent years, there has been a steady development in the range of materials ready to be used in various areas of industry. Such materials include, for example, so-called sandwich materials, which derive benefits from the advantageous utility properties of different material groups (e.g., combination metal-plastic). Nowadays, sandwich materials are mainly used either to reduce the overall weight of given products or due to their specific properties to reduce noise or vibrations.

The increasingly widespread utilisation of various types of sandwich materials, not only in the automotive industry, naturally also brings with it the necessity to know their specific deformation behaviour (generally material characteristics) under various loading conditions. For this reason, sandwich materials have been subjected to continuous testing and determination of their deformation behaviour, e.g., Fischer [1] performed static and dynamic tests (compression, shear, bending and impact tests) of aluminium foldcore specimens and compared the results with the finite element analysis (FEA). The effect of variable mechanical properties of the sandwich beam core in light of elastic loading was investigated by Grygorowicz et al. [2]. Moreover, other authors focused, for example, on the hardness measurement and thermal analysis of aluminium-based sandwich composites [3], mechanical, fatigue and thermal properties of the honeycomb sandwich structures [4–12] or properties of sandwich materials when joining technologies are taken into account. For example, Gladkovsky [13] researched a feature of interface structure at explosive welding and Nilsson [14] tested the fatigue properties of laser stake-welds. Generally, the joining of sandwich materials represents very important research and was investigated by many authors such as, e.g., [15–17]. Nevertheless, before the application of sandwich material, it is very important not only to verify these results but also to incorporate the chosen material and its material behaviour into the whole production process. For this reason, numerical simulations are very often used during the pre-production phase. A general overview of the numerical simulation used in the sheet metal forming is quite clearly described, e.g., by Ablat [18].

Numerical simulations are nowadays commonly used to verify the material properties as well as the deformation behaviour of materials chosen for the production of a given component. Mostly a finite element method is used. These analyses (FEA) then provide very important knowledge about the behaviour of the material during the actual manufacturing process, without the need for real tools, machines and materials.

One of the most important phenomena that affects the dimensional accuracy of given products within the metal forming technologies is so-called springback, which is elastically driven and thus represents the change of the final shape after unloading. There are different methods of how to eliminate the springback (e.g., modification of tool or trial and error method), but it is always preferable to predict the springback magnitude.

In light of the springback prediction, two basic approaches can be generally used. The first is an analytical model and the other one rests in the application of a finite element analysis (FEA). Analytical (or semi-analytical) modelling of springback can be found, e.g., in [19–28]. The greatest disadvantage of such an approach rests in the utilisation of some simplifications, so their accuracy is generally not so high compared to the results of FEA.

On the other hand, the proper use of FEA is largely dependent on the accuracy of its own input data (generally input parameters), which is especially important from the springback prediction point of view. Han et al. [29] considered the input parameters in springback simulation and determined the best combination of such input parameters. Mertin et al. [30] determined the elastoplastic material parameters of HSS by using the inverse modelling approach. Trzepieciński used a genetic algorithm to optimise the number of input parameters of the multilayer perceptron. It was found that the most important variable affecting the springback coefficient is the punch bend depth. Gu [31] increased the accuracy of springback prediction by the use of cyclic sheet metal test methods such as uniaxial tensile-compression and cyclic shear tests. Li [32] investigated the influence of process parameters (e.g., friction and clearance) on the springback at the free bending process. Seo [33] evaluated the effect of constitutive equations on the springback prediction accuracy. It was concluded that it is always very important to use appropriate constitutive equations. Liu et al. [34] found that springback prediction in cold roll forming is significantly improved by incorporating the Young's modulus variation into the FEA simulations.

In addition to these two basic approaches, the use of so-called artificial neutral networks (ANN) in the material science has become more widespread in recent years, e.g., Pouraliakbar [35] developed an ANN model to predict the toughness of HSLA steel. Nowadays, in light of springback prediction, utilisation of ANN represents an alternative tool for springback prediction, especially regarding that in light of nonlinear recovery, FEM has become quite complicated to achieve reliable results as it is concluded, e.g., by [36]. However, it is still a developing procedure, so, e.g., Angsuseranee [37] compared the efficiency of springback and sidewall curl prediction of AHSS in the U-bending process by the FEM and ANN and it was found that FEM was more efficient than the ANN approach. On the other hand, Miranda [38] combined the use of ANN and learning toll with a simulation and data generation tool to conclude that such a methodology can model the complex nonlinear behaviour with sufficient reliability.

Utilisation and development of the different material constitutive models is quite digestedly presented in the work of Zhu et al. [39], where the effect of variable elastic modulus, nonlinear recovery, the isotropic hardening model and hardening model taking into account the Bauschinger effect are summarised. Quite a lot of different yield criterions are mentioned in this paper (e.g., Hill48, Barlat89, Barlat2003, BBC 2005, etc.) [39]. However, the Vegter yield criterion is missing in this case. The effect of the variations in the elastic modulus and its influence on the springback prediction is, e.g., studied by Chatti et al. [40], where a new model to take into account variation in the elastic modulus was proposed. Jung [41] developed an elastoplastic material constitutive model to take into account already the effect of anisotropy change at loading and springback recovery for AHSS materials. A combined isotropic–kinematic hardening model was modified in this case having a good matching with the real samples. Baara et al. [42] improved the accuracy of the finite element software (Abaqus) by introducing the new constitutive hardening model that can predict the final shape of a part that changes its dimensions after forming. The modification of the original Chord model and its combination with nonlinear isotropic–kinematic revealed a higher accuracy of this new model.

However, in light of the springback prediction, it is always important to consider the mutual effect between all major input parameters (elastic modulus, geometry, yield criterion and hardening laws). That is why quite a lot of papers deal with the effect of these parameters on the springback prediction. Mulidrán et al. [43,44] combined different yield criterions and hardening laws to determine the accuracy of springback prediction, in these cases for DP steel and aluminium alloy AA6451, respectively. Trzepiecinski and Lemu [45] studied the springback behaviour of steel DC04 under the different input parameters (geometry, conditions and material) for FEA (here, Abaqus). The work of Naofal et al. [46] presented calibrated plasticity models arising from the appropriate tests to predict springback during the roll forming process. In this case, the Yoshida-Uemori hardening law revealed a higher accuracy to predict springback than the isotropic hardening law. Sumikawa [47] developed and implemented an FEM software material model, which takes into account not only average Young´s modulus and Bauschinger effect, but also elastic and plastic anisotropy. It was concluded that besides the Bauschinger effect and Young´s modulus, there is also a great effect arising from elastic and plastic anisotropy on the results of springback predictions. The effects of the different yield criterions (Hill48 and Hill90), bending radii and directions on the aluminium alloys springback are described by Slota et al. [48].

In this paper, the springback prediction of the sandwich material was performed by using the FEA. Due to the previous results (see above), only the Vegter yield criterion was used and was combined with the two hardening laws (isotropic and kinematic). In addition to that, the effect of mesh strategy (selection of surface or volume elements) on the springback prediction accuracy was investigated as well. This quite complex combination (sandwich material–two hardening laws–type of mesh elements) of input parameters can be considered as the greatest novelty of this paper.

2. Materials and Methods

For the tests, the sandwich material CR270Y340T-LA -EG47/47-E-R-O (designation according to the standard VW 50065) was chosen. It is a micro-alloyed, electro-galvanized sheet designed for the production of car-body outer panels. Concerning the increased requirements for acoustic vibration absorption, this tested material was supplied as a sandwich with a layer of thickness as follows: 0.4–0.1–0.4 mm. The core material was identified by using the Fourier transform infrared spectroscopy (FTIR) on a device Nicolet iS10 manufactured by the company Thermo Scientific (Waltham, MA, USA). The method is based on the absorption of infrared radiation as it passes through the specimen, during which changes in the rotational and vibrational energy levels of the molecule occur as a function of the molecular dipole moment changes. The resulting infrared spectrum is a functional dependence of the energy, usually expressed in percent of transmittance or absorbance units, on the wavelength of incident radiation. For the present analysis, the attenuated total reflection (ATR) technique was used and the measured spectra of the individual materials (32 scans were used) were compared with the spectra in the instrument's database. From the results of the IR spectrometry, it could be concluded that it was an acrylate copolymer. The identification of core material was carried out for further material testing of the sandwich core mechanical properties.

The chemical composition analysis was performed on a Bruker-Quantron spark-optical emission spectrometer Q4 Tasman. Such an analysis was determined based on 5 measurements. The results of the chemical composition analysis are given in Table 1.

Table 1. Chemical composition of the tested sandwich material.

Chemical Element	C	Si	Mn	Al	P	S	Ti	Nb
Composition [wt%]	0.097	0.406	0.746	0.211	0.016	0.013	0.122	0.074

The subsequent structural analysis was carried out on a TESCAN MIRA 3 electron scanning microscope. From the results it is clear that the structure was a ferritic one with the dispersed particles containing an increased content of Mn, Si and Al micro-alloys. The analysed structure can be seen in Figure 1 (left) and detail of the particle is shown in Figure 2 (right). Finally, the EDX bulk analysis arising from the analysed structure is given in Figure 2.

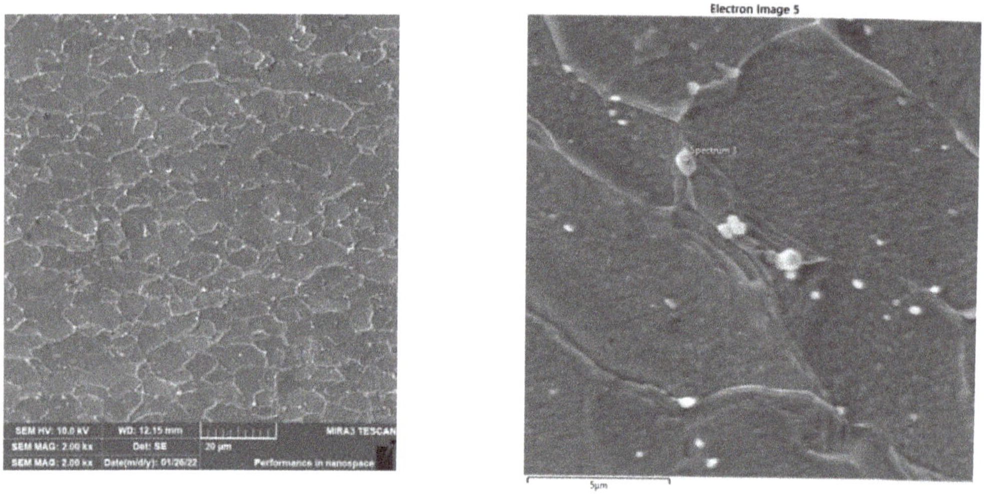

Figure 1. Structure of the tested sandwich material (**left**) and detail of the particle (**right**).

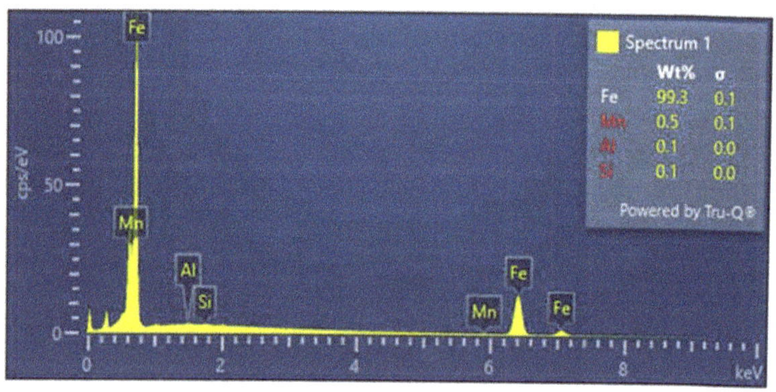

Figure 2. EDX bulk analysis.

2.1. Static Tensile Test

A static tensile test was used to determine the basic material characteristics of the tested material. With a view to the future use of the experimentally obtained data for the definition of the material model in numerical simulations, these tests were already performed for specimens taken in the directions 0°, 45° and 90° concerning the rolling direction. In order to eliminate any material anomalies, material characteristic values were determined each time from a set of 5 measured specimens. The shape and dimensions of the test specimens were chosen in accordance with the standard EN ISO 6892-1. In addition, the methodology of test execution and evaluation was in accordance with the standard above-mentioned EN ISO 6892-1. The static tensile test was carried out on a modernized device TIRA test 2300 using the software Labtest v.4 to evaluate all basic mechanical properties of the tested material (proof yield strength $R_{p0.2}$, ultimate strength R_m, total ductility A_{80mm}, uniform ductility A_g and Young´s modulus E). The plastic strain ratio (Lankford parameter) r was determined in accordance with the standard ISO 10113 in the interval of true plastic strain values from 10% to 20%. The results of the individual material quantities are summarised in Table 2. Engineering stress R (MPa) vs. engineering strain ε (1), thus so-called engineering stress–strain curves (regarding the individual rolling directions) are shown in Figure 3. The evaluation of the mechanical values given in Table 2 was carried out by considering the measured force values to the whole cross-section, i.e., the measured values represent the average values of mechanical properties for the three layers of the tested sandwich material (steel-plastic-steel).

Table 2. Basic material properties of the tested sandwich material.

Direction	$R_{p0.2}$ (MPa)	R_m (MPa)	A_g (%)	A_{80mm} (%)	$r_{10–20}$ (1)	E (MPa)
0°	291.7	426.5	17.8	23.1	1.335	174,283
45°	299.6	418.3	18.9	27.1	1.515	183,718
90°	307.7	428.5	18.1	26.6	1.622	179,928

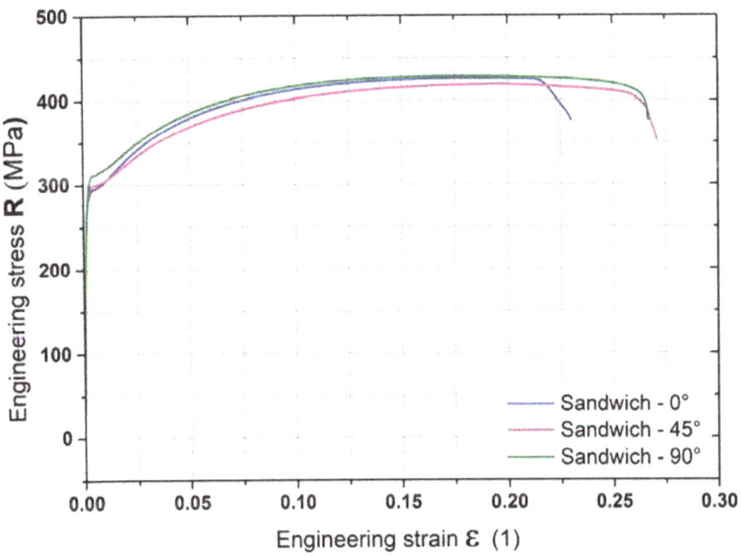

Figure 3. Engineering stress–strain curves (static tensile test) of tested sandwich material.

A numerical simulation has to be a stress–strain curve (hardening curve) entered as a dependence of true stress (effective stress) vs. true plastic strain. Measured dependences of the true stress vs. true plastic strain for the individual rolling directions 0°, 45° and 90° were subsequently approximated (fitted) using the Krupkowski law (power-law function) according to Equation (1).

$$\sigma = K\,(\varepsilon_{pl} + \varepsilon_0)^n \tag{1}$$

where:

σ—true stress (effective stress) (MPa)
K—strength coefficient (MPa)
ε_{pl}—true plastic strain (1)
ε_0—offset true strain (pre-strain) (1)
n—strain hardening exponent (1)

An approximation of the stress–strain curve (hardening curve) was performed within the interval from 1% up to A_g in the software OriginPro 9.0. The advantage of defining the hardening curve by using the Krupkowski hardening law is the simple possibility of integrating this function and thus it is easy to express the equivalent plastic work parametrically in dependence on the true strain and the approximation constants arising from such a function. An example of the approximation result can be seen in Figure 4. The results of the obtained approximation constants are presented in Table 3.

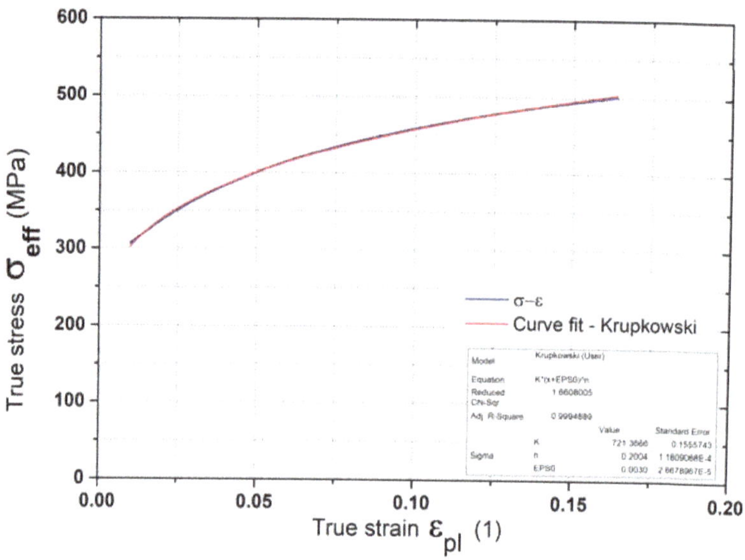

Figure 4. True stress–strain curve (STT) and application of the Krupkowski law.

Table 3. Stress–strain curve approximation constants from the static tensile test.

Direction	K (MPa)	n (1)	ε_0 (1)
0°	720.6	0.1995	0.0030
45°	707.6	0.2025	0.0065
90°	726.9	0.2039	0.0067

2.2. Hydraulic Bulge Test

When testing material by static tensile testing, the test specimen was deformed only by the uniaxial tension (uniaxial tensile stress state). To obtain information about the material deformation behaviour under the multi-axial loading, a hydraulic bulge test (HBT) was mostly performed. The principle of this test rests in stretching the specimen into the circular die by applying the pressure of fluid until a crack occurs. Similar to the static tensile test, the dependence of the true stress vs. true plastic strain (again true stress–strain curve) is evaluated. Figure 5 shows an element of the test specimen together with the parameters used to derive the stress in the test specimen during the HBT.

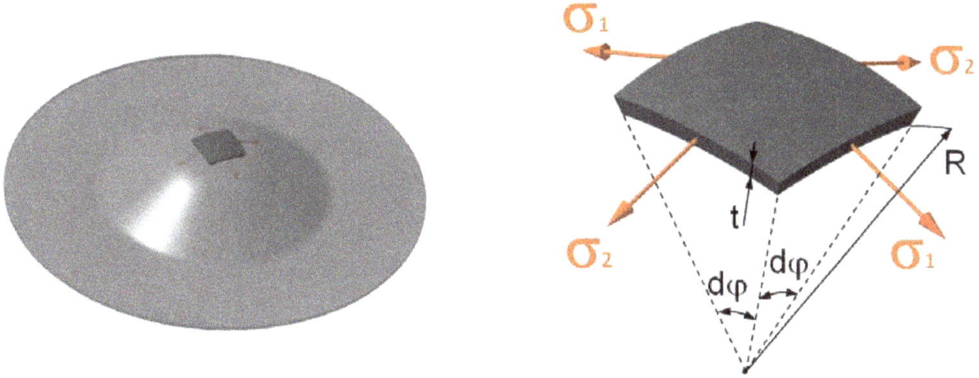

Figure 5. Principle and stress state of the hydraulic bulge test (HBT).

Based on the right part of Figure 5, where the effects of fluid pressure on the specimen element are shown (equilibrium equation), can be acc. to membrane theory derived Equation (2) for the stress in the test specimen wall and so the effective stress σ_{eff} (arising from the Tresca's plastic flow criterion) can be finally expressed by relation (3).

$$\sigma_{HBT} = \frac{p \cdot R}{2 \cdot t} \tag{2}$$

$$\sigma_{eff} = \frac{p \cdot R}{2 \cdot t} + \frac{p}{2} \tag{3}$$

where:

p—hydraulic pressure (MPa)
R—radius of curvature (mm)
t—actual thickness of specimen (mm)

Finally, effective strain ε_{eff} can be calculated using principal strains $\varepsilon_{1,2,3}$ according to Equation (4)

$$\varepsilon_{eff} = \frac{\sqrt{2}}{3} \cdot \sqrt{(\varepsilon_1 - \varepsilon_2)^2 + (\varepsilon_2 - \varepsilon_3)^2 + (\varepsilon_3 - \varepsilon_1)^2} \tag{4}$$

where:

$\varepsilon_{1,2,3}$—principal strains (1)

Actual thickness of the specimen can be subsequently determined through the principal strain ε_3 (sometimes termed also as a thickness reduction) firstly via the constant volume law (5) and subsequently using Equation (6).

$$\varepsilon_1 + \varepsilon_2 + \varepsilon_3 = 0 \tag{5}$$

$$\varepsilon_3 = \ln\left(\frac{t}{t_0}\right) \tag{6}$$

where:

t—actual thickness (mm)
t_0—initial thickness (mm)

To determine the hardening curve in the HBT, it is therefore necessary to measure the dependence of the test specimen deformation on the hydraulic pressure. That is why a contactless photogrammetry method using the software Mercury RT from the company Sobriety, Ltd. (Kuřim, Czech Republic) was used to measure the true strain values. A scheme of the measurement layout is shown in Figure 6.

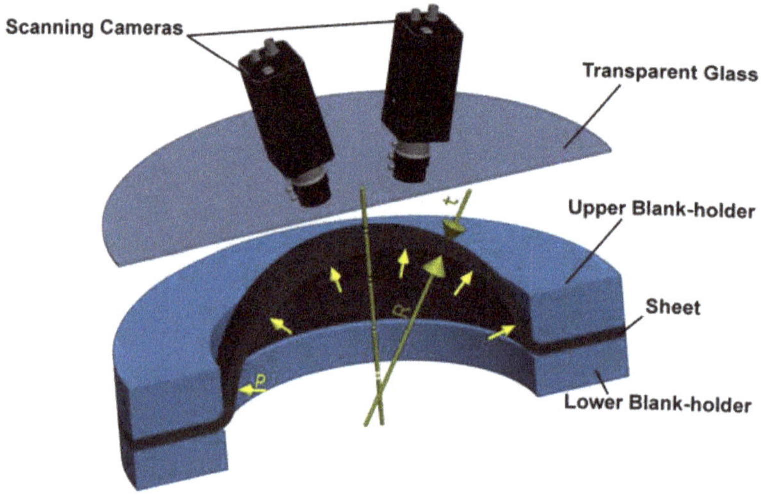

Figure 6. Scheme of the measurement of true stress–strain curve at HBT.

Because stress relaxation occurs when the specimen is loaded, the pressure of the fluid was not increased continuously, but gradually at intervals of 5 s. The value of the pressure increase (pressure step) was chosen to be 0.25 MPa and the own magnitude of the required pressure was controlled by a precision hydraulic servo valve with PID control.

The achievable accuracy of the required pressure was 0.002 MPa. The system of hydraulic pressure activation and its control was solved by a separate hydraulic unit working independently of the hydraulic circuit of the given press.

From the contactless deformation analysis using the system Mercury RT, the magnitudes of major and minor strain (ε_1, ε_2), as well as the radius of curvature R must be detected to subsequently determine the stress–strain curve. For this purpose, two surfaces have been defined on the measured area, in which the system calculates the desired quantities. An example of the analysis area is shown in Figure 7. The larger analysis area (area A) is used to find the radius of curvature R and the smaller area (area B) is used to calculate the major and minor strain, from which the actual thickness of the material can be calculated using Equations (5) and (6). An example of true strain distribution in the relevant area and evolution of the radius of curvature R is shown in Figure 8.

Similar to the static tensile test, a hardening curve was obtained from the measured data, which was then approximated by the Krupkowski law (1). An example of this approximation of the hardening curve from HBT, so under the equi-biaxial stress state, can be seen in Figure 9. The average values of approximation constants obtained under equi-biaxial loading can be seen in Table 4.

Table 4. Stress–strain curve approximation constants from the hydraulic bulge test.

Material	K (MPa)	n (1)	ε_0 (1)
Sandwich	865.1	0.267	0.0076

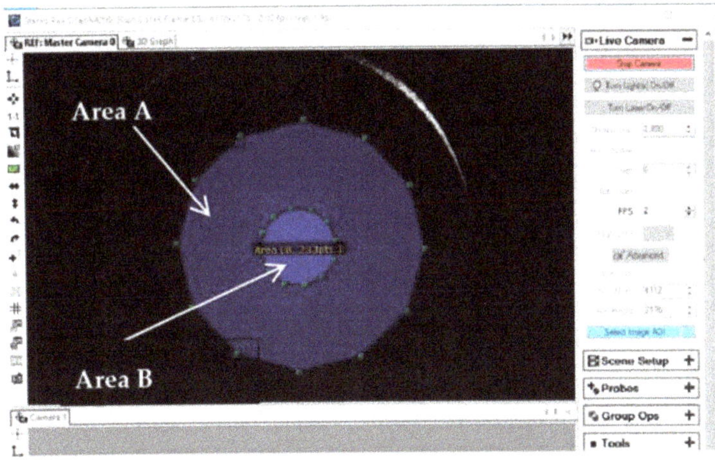

Figure 7. Definition of the analysis areas A and B (software Mercury RT).

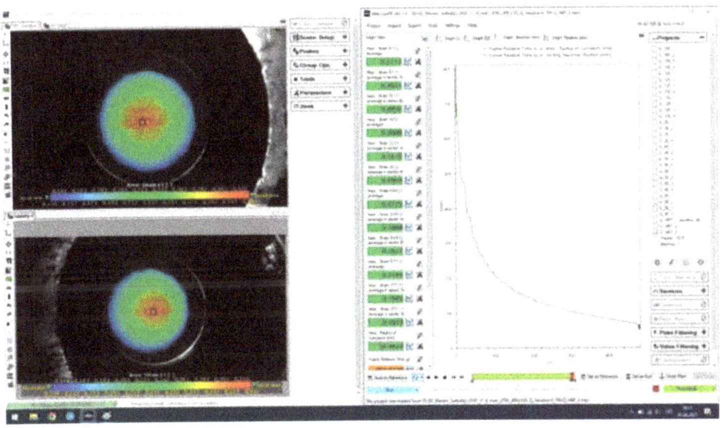

Figure 8. True strain distribution (here ε_1) and evolution of the radius of curvature R.

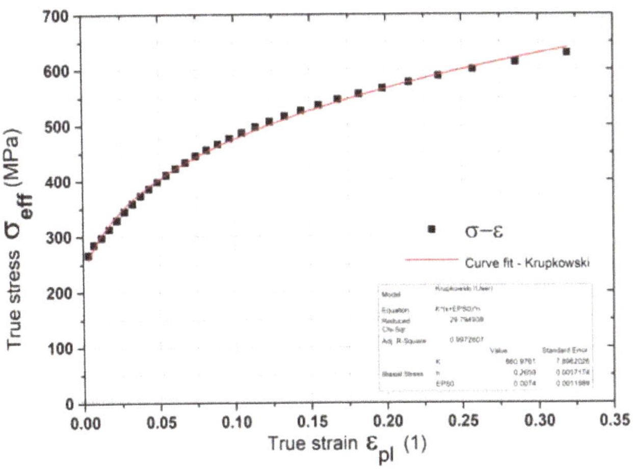

Figure 9. True stress–strain curve from HBT and application of the Krupkowski law.

2.3. Cyclic Test

In addition to the previous tests, we also carried out a cyclic test, which takes the effect of the change of loading direction (+, −), to define the material model termed as Yoshida-Uemori. Due to the compressive stress states, a performance of this test for sheet samples was quite demanding and there was a loss of stability resulting in sample buckling. Thus, a testing jig was proposed as an additional jig for the clamping grips. The testing jig consists of four subdivided supporting grips, which are hydraulically controlled and can prevent the specimen from buckling during the compression. The offset of measured curves from monitored cycles rests in the magnitude of the Bauschinger effect for the tested material. A total of 5 measurements were made under the data scanning frequency of 50 Hz, from which an average curve was determined. The layout of the cyclic test is shown in Figure 10 (left) and results are given in Figure 10 (right).

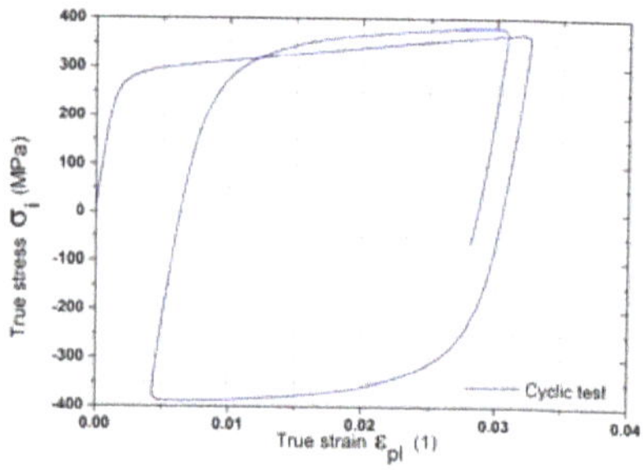

Figure 10. Layout of the cyclic test (**left**) and results of the cyclic test for tested sandwich material (**right**).

2.4. U-Bending of the Specimens

For the experimental determination of the springback, it was suitable to choose such a test when a change of the stress state occurred in the bending area because this case can be fully developed the Bauschinger effect. Concerning the equipment of our laboratory, the U-bending of the workpiece (blank) in the tool with a fixed blank-holder was chosen for the springback analysis. This experiment simulates the process that occurs in a tool at the deep-drawing (stamping) process. During the test, the workpiece is two times bent in the opposite bending direction. The principle of the test is evident from Figure 11, where the scheme of this U-bending test is shown. The test conditions were as follows: distance between fixed blank-holder and die was 0.9 mm (i.e., draw clearance between tool and workpiece was 0 mm). The specimen was in contact with the blank-holder, but there was not force acting at the beginning of the test. The tool is equipped with a lower blank-holder that applies force against the bending punch and prevents the deflection of the sheet. The blank-holder force was developed by a coil spring with an initial force of 4 kN at the start of bending and 6 kN at the end of bending in the bottom dead centre of the bending punch. The die drawing radius was R5, punch drawing radius was R2.5, workpiece dimensions (rectangular shape) 30 mm × 140 mm, bending direction was perpendicular to the rolling direction and own bending was performed on the eccentric press LENP 40.

After finishing the bending test, a specimen was measured by the dimensional and shape analysis on the 3D coordinate measuring device SOMET XZY 464 by the relevant software TANGO1! 3D for own evaluation. As a result of the experimental measurement, there were coordinates of 100 points in *.stp format, which represented the actual shape of the specimen. The principle of measuring the resulting shape is shown in Figure 12.

Software CATIA V5 was used for further processing of the measured points. Through the measured points a curve (spline) was fitted, from which a 30 mm wide area was subsequently extruded. This area had to be further offset by 1.45 mm. This offset value (1.45 mm) was determined by the diameter of the measuring tip (2 mm) and the material thickness

(0.9 mm). The offset plane then represented the midplane of the experimentally measured shape of the specimen. The procedure of obtaining the final shape of the experimental specimen in the Catia environment can be seen in Figure 13. The measured shape of the bent blank was subsequently used as a comparative criterion to verify the match between results of the real experiment and numerical simulation (in this case, software PAM-STAMP 2G was used).

Figure 11. Scheme of the used U-bending experiment.

Figure 12. Determination of the specimen actual shape (device SOMET XZY 464).

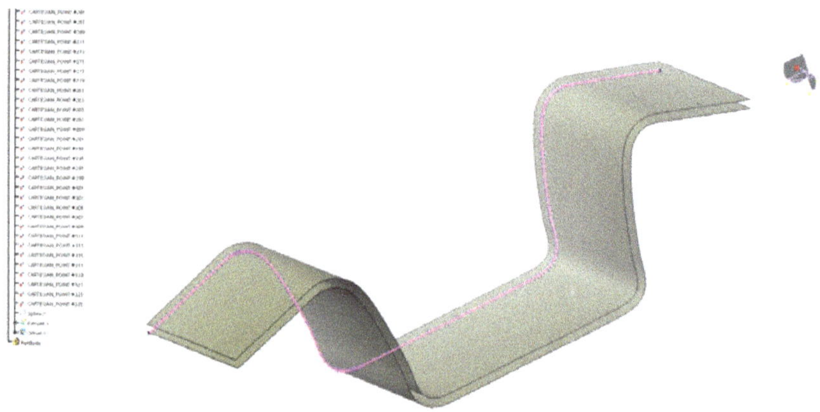

Figure 13. Detection of the experimentally prepared specimen shape (CATIA V5).

3. Numerical Simulation

3.1. Vegter Yield Criterion

The formulation of the Vegter yield criterion (function) is based on the principle of incorporating and using more experimentally measurable data in the definition of this yield criterion. This criterion is established for the general directions of the principal stresses, which do not need to be identical with the direction of sheet anisotropy. Using the suitable orientation of the coordinate system (principal axes), the shear stress components vanish and the stress state in the given zone is defined by using only the normal stress components (thus principal stresses) σ_1 and σ_2. The angle θ denotes the relative rotation between the principal stresses and chosen coordinate system. The principle of the element planar loading and the individual stress components can be subsequently expressed by Equations (7) and (8).

$$\sigma_1 = \frac{\sigma_{xx} + \sigma_{yy}}{2} + \sqrt{\left(\frac{\sigma_{xx} - \sigma_{yy}}{2}\right)^2 + \sigma_{xy}^2} \tag{7}$$

$$\sigma_2 = \frac{\sigma_{xx} + \sigma_{yy}}{2} - \sqrt{\left(\frac{\sigma_{xx} - \sigma_{yy}}{2}\right)^2 + \sigma_{xy}^2} \tag{8}$$

The quadratic Bezier curve is used to mathematically describe the Vegter yield criterion passing through the experimentally determined reference points, as shown in Figure 14. This parametrically determined curve allows the construction of a smooth yield surface with smooth transitions at the reference points (via tangents intersecting in the hinge-points).

Points A and C are experimentally determined reference points defining the yield surface. The tangents to the Bezier curve passing through these reference points are defined by the strain vectors. The intersection of these tangents determines the final shape of the Bezier curve, and it is called a hinge-point [49]. Although the curve does not pass through this hinge-point, its shape strongly depends on this point. Coordinates of any point on the Bezier curve are given by the so-called Bezier parameter μ and position of the hinge-point B (strain vectors). Figure 15 shows part of the curve defining the yield criterion, where coordinates of the point BI denote the yield strength at the equi-biaxial stress state. The position of the point UN denotes the transition between the elastic and plastic state (yield strength) under the uniaxial loading (uniaxial tensile test), and finally point PS defines the yield strength under the plane strain test (so-called plain strain point). Points a and b are the intersections of the tangent line passing through point PS with the tangent line passing through UN and BI, respectively. By introducing an extra input parameter α, which should have a value between 0 and 1, the position of the point PS can be determined (see Figure 15).

However, it is quite difficult to determine the position of the point PS experimentally and for this reason the value $\alpha_{PS} = 0.5$ is very often assumed when defining the yield criterion. A similar problem is arising during the experimental determination of the point SH position characterising the pure shear stress. In this case, the presumption about the symmetry of yield criterion and parameter α_{SH} having the value equal to 0.5 is used, as in the case of plane strain point. In the case of experimental determination of these points, no estimation of such extra input parameter α is necessary and the Vegter yield surface directly passes through these experimentally detected points.

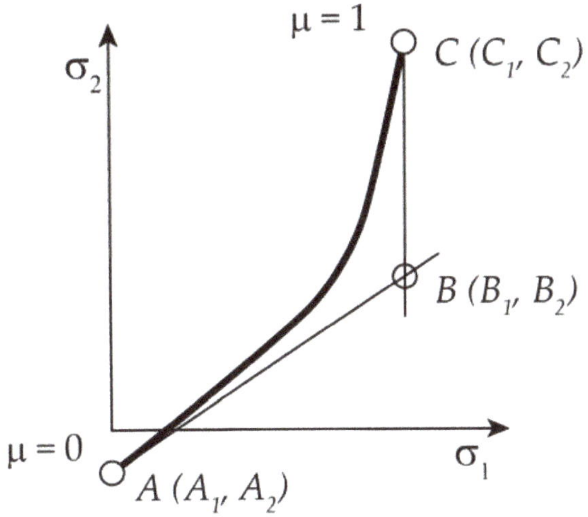

Figure 14. Bezier curve passing through reference points *A* and *C*.

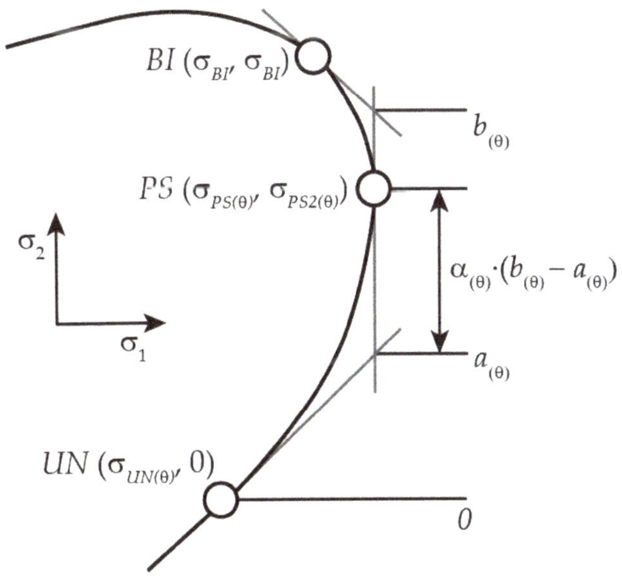

Figure 15. Definition of the plane strain point PS via extra input parameter α [50].

Using the Bezier curve to define the yield criterion, the functional relation can be subsequently expressed by the Equation (9):

$$\varphi + \sigma_f = \frac{\sigma_i}{f_i(\mu, c)} \tag{9}$$

where:

$$f_i(\mu, c) = A_i(c) + 2\mu \cdot (B_i(c) - A_i(c)) + \mu^2 \cdot (A_i(c) + C_i(c) - 2B_i(c)) \tag{10}$$

$$c = cos(2\theta) \tag{11}$$

μ = Bezier parameter ($0 \leq \mu \leq 1$)

A, B, C, are vectors formulating in a two-dimensional stress plane σ_1 and σ_2 the plane stress state condition ($\sigma_3 = 0$). The yield criterion is completely dependent on the three parameters mentioned above but is not uniquely determined due to the interdependence of the parabolic functions. The yield surface is described as the individual parts of the Bezier curve affected by the hinge-point B and passing through the reference points A and C. The total yield surface is constructed through the four quadrants expressing the different stress states. Continuity of the first derivative is required between any part of the yield function. The strain vector is derived as follows:

$$d\varepsilon_{ij} = d\lambda \frac{\partial \varphi}{\partial \sigma_{ij}} \tag{12}$$

$$\frac{\partial \varphi(\sigma_1, \sigma_2, \theta)}{\partial \sigma_{ij}} = \frac{\partial \varphi}{\partial \sigma_1} \cdot \frac{\partial \sigma_1}{\partial \sigma_{ij}} + \frac{\partial \varphi}{\partial \sigma_2} \cdot \frac{\partial \sigma_2}{\partial \sigma_{ij}} + \frac{\partial \varphi}{\partial cos(2\theta)} \frac{\partial cos(2\theta)}{\partial \sigma_{ij}} \tag{13}$$

$$\frac{\partial \varphi}{\partial \sigma_{ij}} = \begin{Bmatrix} \frac{\partial \varphi}{\partial \sigma_{xx}} \\ \frac{\partial \varphi}{\partial \sigma_{yy}} \\ \frac{\partial \varphi}{\partial \sigma_{xy}} \end{Bmatrix} = \begin{bmatrix} \frac{\partial \sigma_1}{\partial \sigma_{xx}} & \frac{\partial \sigma_2}{\partial \sigma_{xx}} & \frac{\partial cos(2\theta)}{\partial \sigma_{xx}} \\ \frac{\partial \sigma_1}{\partial \sigma_{yy}} & \frac{\partial \sigma_2}{\partial \sigma_{yy}} & \frac{\partial cos(2\theta)}{\partial \sigma_{yy}} \\ \frac{\partial \sigma_1}{\partial \sigma_{xy}} & \frac{\partial \sigma_2}{\partial \sigma_{xy}} & \frac{\partial cos(2\theta)}{\partial \sigma_{xy}} \end{bmatrix} \cdot \begin{Bmatrix} \frac{\partial \varphi}{\partial \sigma_1} \\ \frac{\partial \varphi}{\partial \sigma_2} \\ \frac{\partial \varphi}{\partial cos(2\theta)} \end{Bmatrix} = [D] \cdot \begin{Bmatrix} \frac{\partial \varphi}{\partial \sigma_1} \\ \frac{\partial \varphi}{\partial \sigma_2} \\ \frac{\partial \varphi}{\partial cos(2\theta)} \end{Bmatrix} \tag{14}$$

$$[D] = \begin{bmatrix} \frac{1}{2}(1 + cos(2\theta)) & \frac{1}{2}(1 - cos(2\theta)) & \frac{sin^2(2\theta)}{\sigma_1 - \sigma_2} \\ \frac{1}{2}(1 - cos(2\theta)) & \frac{1}{2}(1 + cos(2\theta)) & -\frac{sin^2(2\theta)}{\sigma_1 - \sigma_2} \\ sin(2\theta) & -sin(2\theta) & -\frac{2sin(2\theta)cos(2\theta)}{\sigma_1 - \sigma_2} \end{bmatrix} \tag{15}$$

$$[D] = \begin{bmatrix} \frac{1}{2}(1 + c) & \frac{1}{2}(1 - c) & \frac{s^2}{\sigma_1 - \sigma_2} \\ \frac{1}{2}(1 - c) & \frac{1}{2}(1 + c) & -\frac{s^2}{\sigma_1 - \sigma_2} \\ s & -s & -\frac{2sc}{\sigma_1 - \sigma_2} \end{bmatrix} \tag{16}$$

where:

$s = sin\ (2\theta)$
$c = cos\ (2\theta)$

The independent variables in the above relations are the two principal stresses (σ_1 and σ_2) and angle θ between the material axes and directions of the principal stresses.

3.2. Kinematic Hardening Law

The Yoshida–Uemori model is based on the principle of two yield surfaces, where the inner surface behaves as a purely kinematic model (i.e., its shape does not change and only the origin of its coordinate system shifts due to the strain hardening). The second surface (outer one-bounding surface) behaves as a purely isotropic model (i.e., the origin of its coordinate system does not change and this surface is increasing due to the strain hardening). The outer yield surface (bounding surface) is thus the limiting factor for the displacement of the inner surface [51].

The chosen model allows us to properly describe all the phenomena occurring during the forming of thin sheets. These are mainly the so-called transient Bauschinger effect during the change of stress state and eventually the sharp yield point of the material. This model also allows expressing the change of Young's modulus depending on the plastic strain. As the most important input for proper setting of the required parameters defining the Yoshida–Uemori model, the true stress–strain curve (hardening curve) measured from the static tensile test and cyclic test as well was used. The change of the mechanical properties due to the Bauschinger effect can be seen in Figure 16.

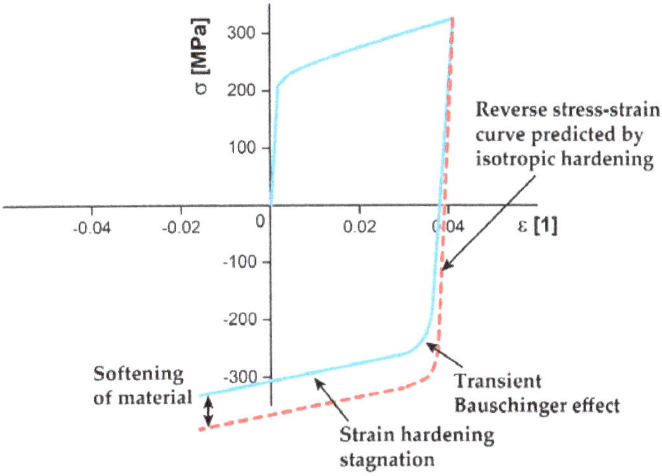

Figure 16. Change of mechanical properties due to the Bauschinger effect [51].

3.3. *Definition of the Material Model in the Software PAM-STAMP 2G*

Three material models were created in the software PAM-STAMP 2G for the numerical simulation of the bending process. In all cases, the Vegter anisotropic yield criterion based on the measured material characteristics from static tensile tests and HBT were used. Such yield criterion was subsequently combined with different types of hardening laws and the selection of surface or volume elements in the FEA. Finally, three combinations of the mathematical computational model were tested in this paper.

3.3.1. Vegter Yield Criterion in Combination with the Isotropic Hardening Law

Vegter anisotropic yield criterion as defined in the software PAM-STAMP 2G allows taking into account the yield strengths in the individual testing directions (in this case in directions 0°, 45° and 90° with respect to the rolling direction), the yield strength under the equi-biaxial loading (HBT) and plastic strain ratios (Lankford parameters) both from the static tensile test and HBT. In addition to these values, Young's modulus, Poisson's ratio and density of the material to be tested must be entered into the material card. In the case of isotropic hardening law, the stress–strain curve is defined as the average curve from the measured values in the individual tested directions. A weighted average of the relevant quantity (acc. to standard EN ISO 10 113/2020) is calculated from Equation (17). The material card with the Vegter yield criterion in combination with the isotropic hardening law can be seen in Figure 17. For the calculation, a surface element of the deformation mesh, thus "surface blank", was chosen.

$$x = \frac{1}{4}(x_0 + 2x_{45} + x_{90}) \tag{17}$$

where:

x—weighted average of the monitored quantity

x_0, x_{45}, x_{90}—measured values in the relevant directions

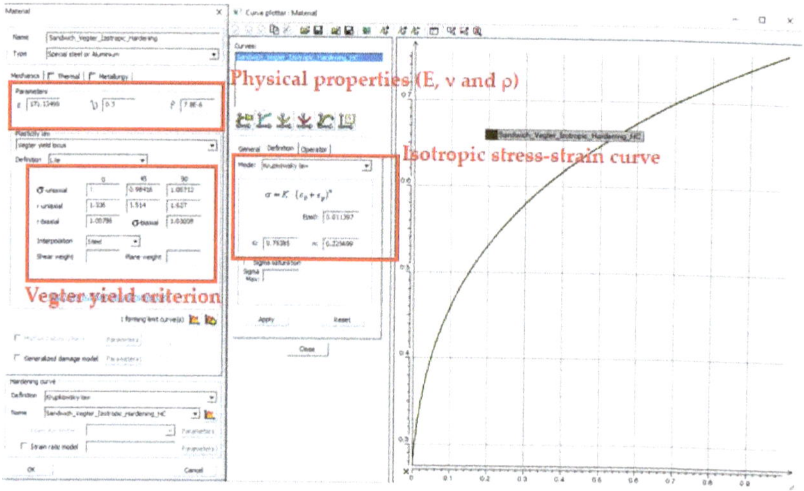

Figure 17. Material card (software PAM-STAMP 2G): Vegter yield criterion and isotropic hardening law.

3.3.2. Vegter Yield Criterion in Combination with the Kinematic Hardening Law

To determine the parameters for the model termed as the Yoshida–Uemori hardening law, it is necessary to calculate the parameters defining the kinematic hardening. For this purpose, the specialized software MatPara that has been developed and distributed by the authors of this model, was used. Input data for the calculation of the needed parameters are the data from tensile and cyclic tests. An example of fitting the measured curve to obtain material constants for the Yoshida–Uemori model can be seen in Figure 18. The material card for the Vegter yield criterion and kinematic hardening law is shown in Figure 19. For this material model, a surface element of mesh was chosen again.

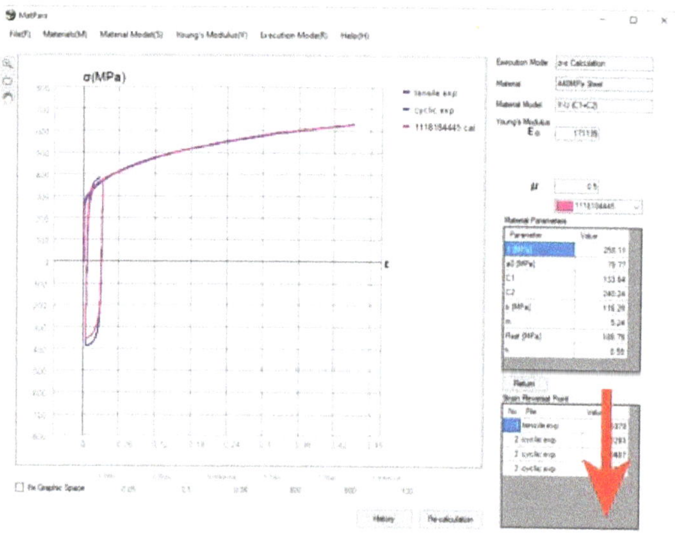

Figure 18. Software MatPara for calculation constants in the Yoshida–Uemori model.

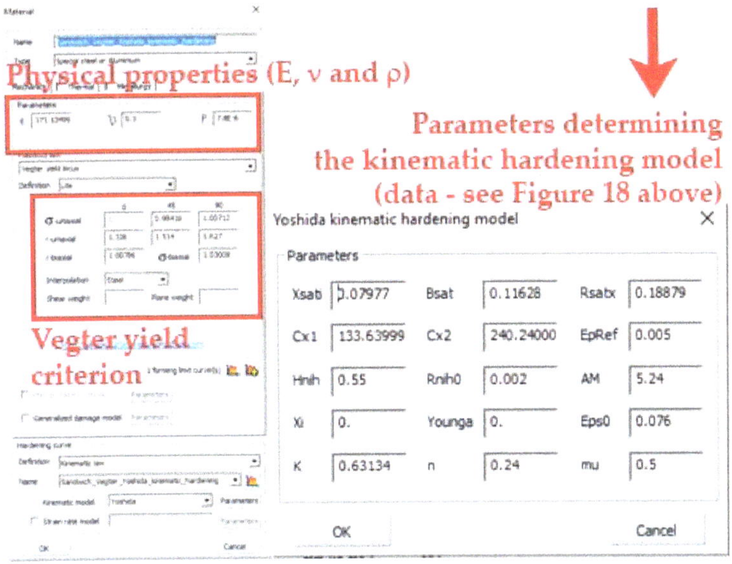

Figure 19. Material card: Vegter yield criterion and kinematic hardening law.

3.3.3. Vegter Yield Criterion in Combination with the Kinematic Hardening Law: Volume Element of the Deformation Mesh

Compared to the previous mathematical model, the volume element of the deformation mesh was chosen for this calculation. In the definition of the bend workpiece, a sheet is entered directly as a sandwich material and three layers are already taken into account for the calculation, as in the real process. Compared to the previous definition, the material definition must take into account the fact that the micro-alloyed steel layer itself has higher mechanical properties than the sandwich as a whole. In the previous cases, the sandwich was taken as a whole and (as is mentioned above) all measured values were related to the total sheet thickness. In this case, every layer was defined to directly correspond to its mechanical properties. In light of the micro-alloyed steel layer, there was only the need to recalculate the approximation constants to a smaller thickness (0.8 mm compared to the original 0.9 mm). The shape of the hardening curve did not change, the curve only shifts towards the higher values, which is reflected by an increase in the approximation constant K in Equation (1). On the other hand, it was quite difficult to determine the mechanical properties of the sandwich core. It was necessary to remove the layers of micro-alloyed steels and to test the mechanical properties of the thin central foil. From the measured values, the hardening curve was again constructed, see Figure 20.

When creating the model of the sheet for bending, the option "Volume Blank-sandwich" was selected in the PAM-STAMP 2G. This option allows (as was mentioned above) us to define the formed sheet by using existing volume elements with the possibility to define any number of layers and also any thickness of these layers. This is quite an advanced approach to specifying a sheet forming operation. However, when using volume elements, it is necessary to follow certain rules when selecting the suitable mesh parameters to keep the stability of the subsequent calculation. Dimensions of the volume element edges should not exceed a ratio of 1:4. Simultaneously, it is necessary to describe individual layers in the thickness direction. In this case, layers had thicknesses of 0.4–0.1–0.4 mm. The size of the element in the sheet thickness direction was thus more or less determined by the thinnest layer in the sandwich material. Furthermore, the minimal number of elements in the sheet thickness direction (which is usually five layers) must be observed. In this case, nine layers in the thickness direction and element having a size of 0.4 mm in the blank area were chosen so that both the minimal number of elements and the aspect ratio of 1:4 were fulfilled. The own definition of a blank with volume elements can be seen in Figure 21.

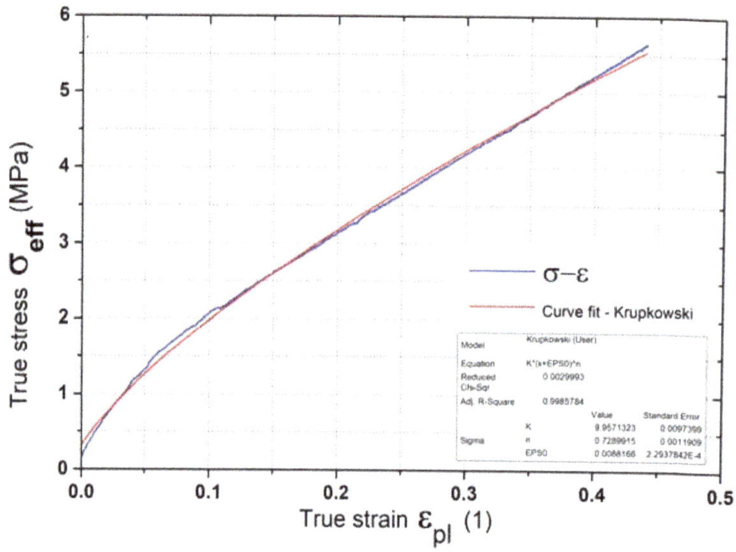

Figure 20. Stress–strain curve of the sandwich core from the acrylate copolymer.

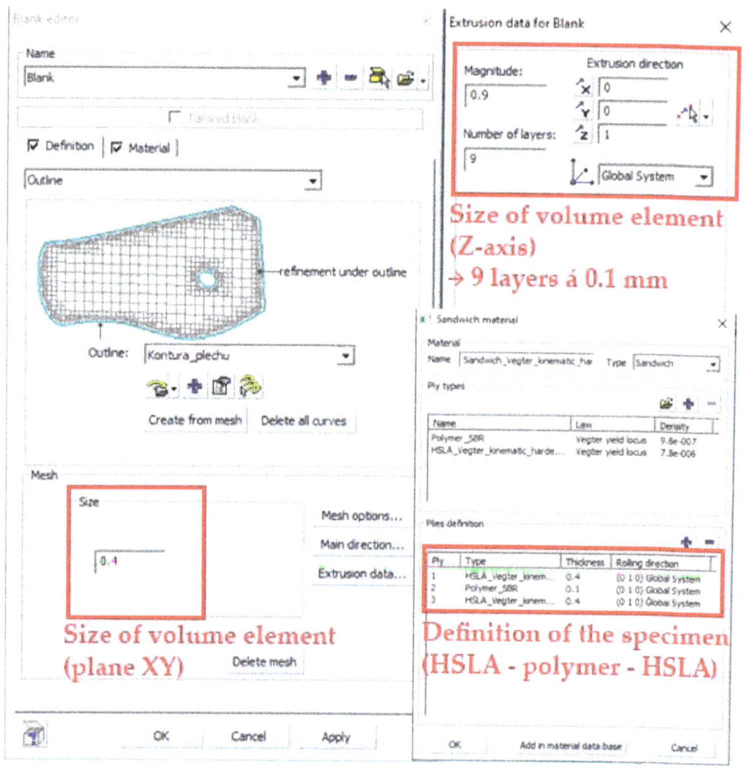

Figure 21. Definition of the blank using volume elements.

4. Results from the Finite Element Analysis (FEA)

Regarding the chosen material models, three calculations of the sheet metal forming process by bending technology were performed. In all cases, the same boundary conditions for the calculation were used (rigid tools, constant friction coefficient, constant bending punch feed rate, the position of the bottom dead centre, own course of the lower blank-holder force, etc.). For the evaluation of the numerical simulation accuracy, the criterion of matching the geometrical shape between the experimentally obtained shape and the shape obtained by the numerical simulations was chosen. When the springback computation was completed, the experimentally measured surface of the bent specimen was imported into the numerical simulation result and these two surfaces were subsequently compared. One of the post-processing options in the PAM-STAMP 2G is to express the dimensional deviations between two (or more) surfaces. This option was used, and the accuracy of the calculation was assessed based on this criterion. Due to the use of the volume element, the deviations were always related to the midplane of the sheet surfaces. Another possibility from the standard results evaluated in the sheet metal forming rested in the utilisation of the FLD diagram, which can be considered as a criterion for achieving the risk of rupture in the given part. For the chosen type of problem, any limit deformation was not achieved and the calculated deformations on the relevant bent part were well below the forming limit curve (FLC) for the tested sandwich material. The comparison of results between FEA and the real specimen can be seen in Figures 22–24.

Figure 22. FEA: Vegter–isotropic hardening law–surface elements.

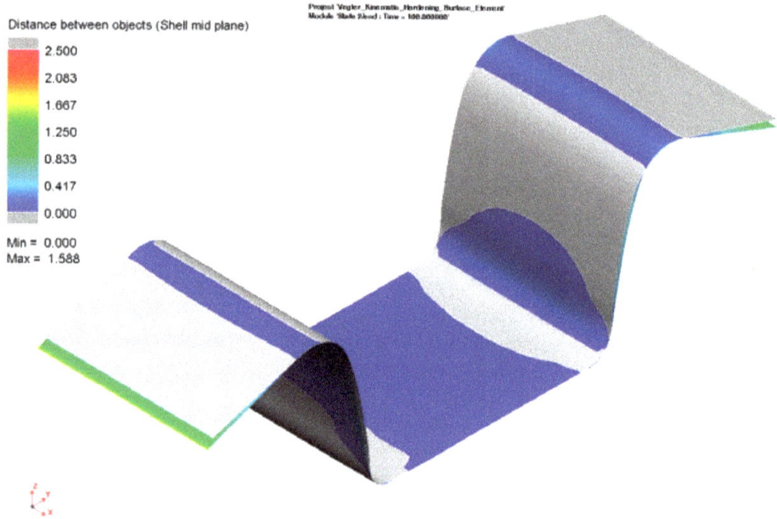

Figure 23. FEA: Vegter–kinematic hardening law–surface elements.

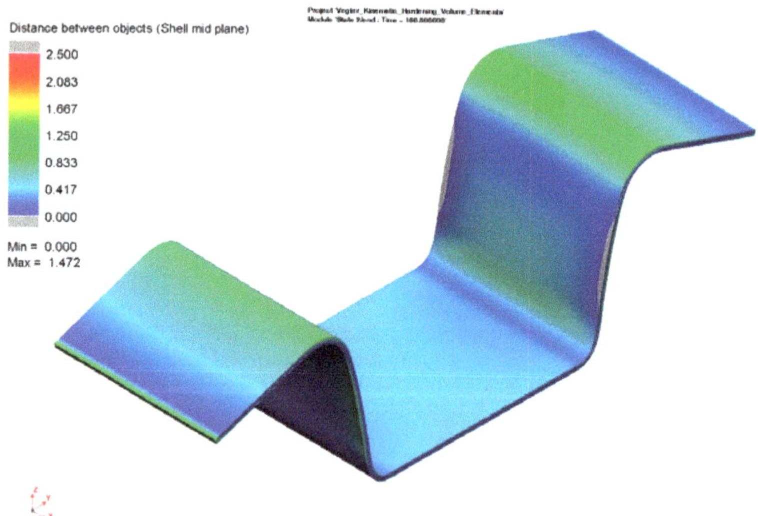

Figure 24. FEA: Vegter–kinematic hardening law–volume elements.

5. Discussion

Numerical simulations already represent a standard tool for predicting technological processes in industrial practice. In terms of practically achievable results, where high requirements are set on the accuracy and calculation speed, it is obvious that the proper choice of the mathematical model and the quality of input material data are some of the key factors influencing the final result of FEA. Previous studies have shown that the basic formability calculations can be performed just by using fundamental yield criterions in combination with a simple isotropic hardening law. The definitions of these models can be completed by performing the basic material tests; therefore, just using static tensile testing, where the time required to carry out these tests is relatively short. However, in cases where the technological operations involve the localisation of deformation (e.g., hemming) and there is also an increased requirement for predicting the final shape due to the springback effect, it is recommended to use the advanced computational models (e.g., Hill90, Barlat91, Barlat2000, Vegter or Yoshida 6th) in combination with a kinematic hardening law (e.g., Lemaitre and Chaboche or Yoshida–Uemori). The proper definition of these computational models means performing more time-consuming material tests. In addition to the standard static tensile test, there are tests performing under biaxial loading as well as shear, compression and cyclic tests. A frequently discussed issue is the selection of surface or volume elements of the computational mesh. From this point of view, for thin sheets up to a thickness of about 2 mm, it is generally recommended to use a surface element of the mesh. This approach significantly reduces the number of mesh elements, and the accuracy of calculation is considered to be sufficient compared to volume elements, where higher accuracy is assumed.

In this paper, the effect of the chosen calculation strategy on the FEA result was carried out. From the results presented in Figures 22–24, it is evident that the choice of the Vegter yield criterion in combination with the isotropic hardening law provides the worst results in light of the springback prediction. The maximum measured distance between the midplanes of the experimentally measured specimen and results obtained from the numerical simulation is about 2.5 mm (see Figure 22). To assess the deviations of the actual surface (experimentally obtained one) and the FEA surface, a longitudinal section was made along the axis of symmetry through both surfaces. Such a section can be seen in Figure 25, where all tested combinations with the experimentally measured shape (the real one), which is shown in blue colour, are compared. From the mutual comparison of these sections it can be seen that there is a fairly good matching in the bending punch radii (lower parts of the bending punch). However, the results of sections on the bending die drawing edge (upper parts of the specimen) already revealed considerable variations at the use of isotropic hardening law.

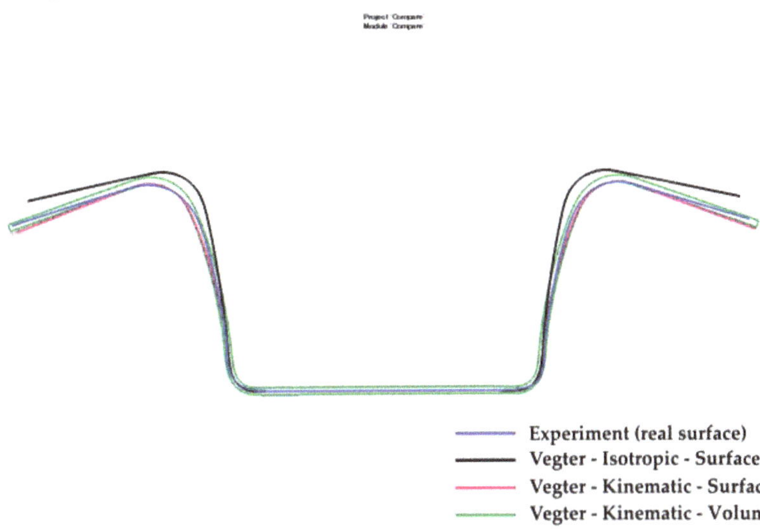

Figure 25. Final comparison along the section for results from FEA and experiments.

As in the previous case, a section through the resulting surfaces (see Figure 23) was performed when the Vegter yield criterion in combination with the kinematic law and surface elements was chosen. The final sections can be seen also in Figure 25. These results show quite a good shape matching both in the lower parts of the bending punch and in the upper parts of the specimen at bending on the drawing edge of the bending die. The maximum measured deviation between the midplanes in this case was 1.59 mm. It is quite important that this deviation occurs only in the peripheral parts of the specimen and is largely influenced by the flange length. For the proper assessment of the shape matching, the most important from a practical point of view is the shape matching in the areas of rounding (radii area) and the walls between these radii. From Figures 22–24 can be seen that minimal differences could be found in these zones (order of 0.1 mm).

The last evaluated computational model combined the Vegter yield criterion with kinematic hardening law and volume elements. The FEA result was assessed again based upon Figure 24 and also on the performed section, which can be seen in the green colour again in Figure 25. In addition, in this case, the red colour is the shape of the experimentally obtained specimen and the grey colour is from the FEA. The results again show a very good shape matching both in the lower parts of the bending punch (radii) as well as in the upper parts of the specimen on the drawing edge of the bending die. The maximum measured deviation between the midplanes in this case was 1.47 mm, and in light of the shape, similar conclusions can be made as in the previous case where surface elements were used.

In light of the quantitative comparison of achieved results, histograms of deviations between experimentally and numerically determined surfaces were used. In the following Figures 26–28, these histograms are shown (deviations between surfaces vs. relevant frequency) for all tested combinations. Concerning the obtained results and their comprehensible display, results were divided into five deviation regions with a step size of 0.5 mm (thus 0–0.5–1–1.5–2–2.5 mm).

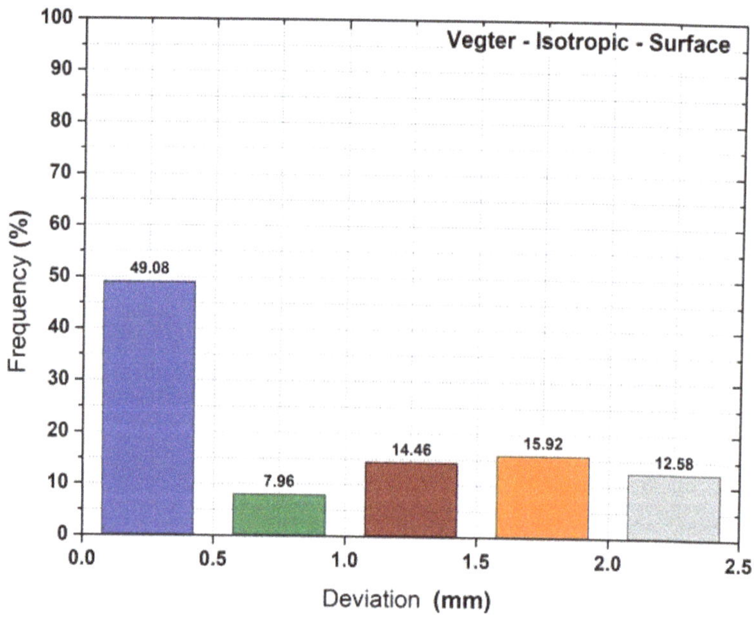

Figure 26. Distribution of deviations between determined surfaces: Vegter–isotropic hardening law–surface elements.

Figure 26 shows the distribution of deviations between real and computed surfaces by using the Vegter yield criterion, isotropic hardening law and surface elements.

The distribution of deviations between real and computed surfaces by using Vegter yield criterion, kinematic hardening law and surface elements is given in Figure 27.

Finally, Figure 28 illustrates almost the same situation as in the previous case; however, volume elements of the deformation mesh are used in this case.

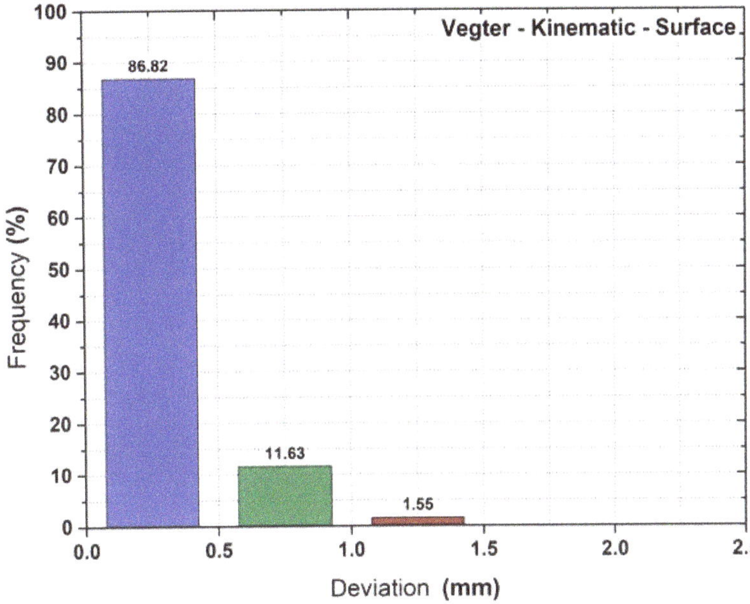

Figure 27. Distribution of deviations between determined surfaces: Vegter–kinematic hardening law–surface elements.

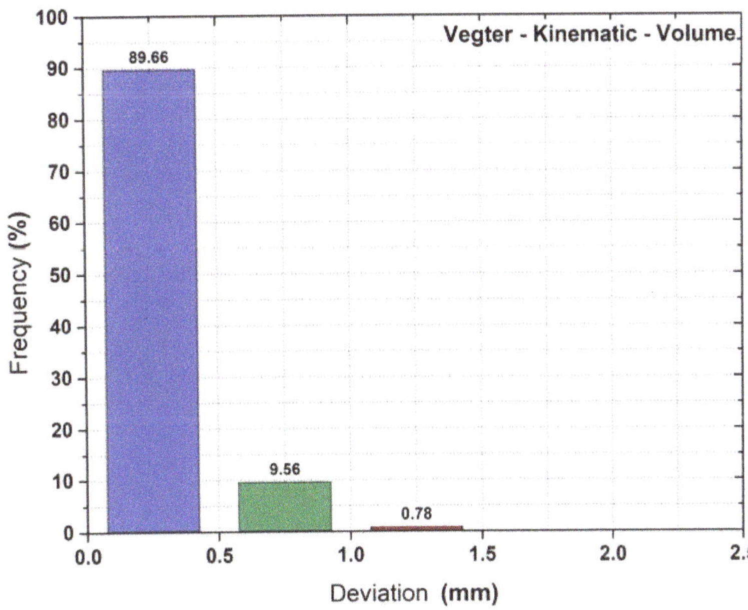

Figure 28. Distribution of deviations between determined surfaces: Vegter–kinematic hardening law–volume elements.

From the previous figures it is obvious that in light of the deviations between real and computed surfaces, the utilisation of the isotropic hardening law provides by far the worst results from all tested combinations. Only about 50% of such deviations could be found up to 0.5 mm. In addition to that, quite a high frequency of 12.58% was determined for the maximal range of deviations (2–2.5 mm). On the other hand, results from the kinematic hardening laws revealed maximal values up to 1.5 mm. More precisely, the maximal deviation was 1.588 mm for surface elements (see Figure 23), but its proportion was zero due to rounding. Moreover, the vast majority (about 90%) of these values could be found in the first range of deviations (0–0.5 mm). Slightly better results were determined by using the volume elements.

When an isotropic hardening law is chosen, it is evident that this model is not able to describe accurately enough the change in the stress state on the drawing edge, which significantly affects the stress and strain in the formed material and thus also the subsequent springback of the used material. From the results, it can be seen that in the lower parts of the specimen on the bending punch radii, where any change of stress state does not occur, the accuracy is quite sufficient for all chosen material models. In the upper parts on the bending die drawing edge, where a change of the stress state already occurs, the FEA results are greatly influenced by the choice of the calculation model. The kinematic hardening law provides significantly more accurate results than the isotropic hardening law. When comparing the FEA results, which use the same computational model (Vegter yield criterion and kinematic hardening law), but differ in the used meshing strategy (surface element vs. volume element), it can be seen that both models provide very similar results. Both used models show quite a high shape matching with the experimentally measured ones. The model with surface elements slightly underestimates the springback magnitude, while the model with volume elements slightly overestimates the springback magnitude compared to the experiment results. However, these differences are minimal, and the calculation accuracy can be considered satisfactory in both cases. On the other hand, the choice of a volume element in the case of sheets having lower thicknesses significantly increases the total number of elements used in the computational mesh. The model with surface elements had 4200 elements (square elements selected with the edge of 1 mm) compared to 236, 250 volume elements. Such a significant increase in the number of elements leads to a hundred-fold increase in the computational time (minutes for surface elements vs. hours for volume elements).

6. Conclusions

The major aim of the research presented in this paper was to determine the influence of the computational material model (generally meshing strategy) on the result of the FEA springback prediction for a sandwich material made from a micro-alloyed steel and acrylate copolymer. Different material models describing the deformation behaviour of the tested sandwich material in the plastic state (isotropic hardening law vs. kinematic hardening law) were tested in the experimental part. Furthermore, the effect of the selection of the dimensional type of mesh element (surface element vs. volume element) on the result of the numerical simulation were investigated as well. The following conclusions can be stated from the measured and computed results:

1. The isotropic hardening law cannot be used to correctly predict the springback of sandwich material in cases where the stress state changes during the forming process.
2. The kinematic hardening law provides a more accurate springback prediction compared to the isotropic hardening model regardless of the surface or volume element selection for the computational mesh.
3. The choice of the meshing strategy does not have any significant effect on the FEA result when the kinematic hardening law is used. The surface and volume elements give almost exactly comparable results for the springback prediction of the sandwich material.
4. From a quantitative point of view (using histograms of surfaces' deviations, see Figures 26–28), it was confirmed that the kinematic hardening law (regardless of the element type) has significantly higher accuracy in springback prediction than the isotropic hardening law. In addition to that, both kinematic hardening laws (surface

and volume type of mesh) have almost 90% of the surfaces' deviations up to 0.5 mm compared to the isotropic hardening law, where only 46% can be found up to 0.5 mm.
5. The definition of the sandwich material using layers of volume elements in the deformation mesh does not provide a significant improvement of the FEM result.
6. In the numerical simulation of forming the sandwich material, the measured values of the mechanical quantities can be related to the entire sheet (sandwich) thickness, and it is not necessary to distinguish the different deformation and stress behaviour of the individual layers.
7. From the calculation accuracy point of view, it does not make any sense to use volume elements of the deformation mesh for the thin sheets. Such an approach leads only to a significant increase in computational time.

Author Contributions: Conceptualisation, P.S. and J.S.; methodology, P.S.; investigation, J.S. and D.K.; resources, P.S. and J.S.; data curation, P.S. and J.S.; writing—original draft preparation, P.S.; writing—review and editing, P.S. and J.S.; visualisation, P.S. and J.S.; supervision, P.S.; funding acquisition, J.S. All authors have read and agreed to the published version of the manuscript.

Funding: This research received no external funding.

Institutional Review Board Statement: Not applicable.

Informed Consent Statement: Not applicable.

Data Availability Statement: Not applicable.

Conflicts of Interest: The authors declare no conflict of interest.

References

1. Fischer, S. Aluminium foldcores for sandwich structure application: Mechanical properties and FE-simulation. *Thin Walled Struct.* **2015**, *90*, 31–41. [CrossRef]
2. Grygorowicz, M.; Magnucki, K.; Malinowski, M. Elastic buckling of a sandwich beam with variable mechanical properties of the core. *Thin Walled Struct.* **2015**, *87*, 127–132. [CrossRef]
3. Sivaram, A.R.; Manikandan, N.; Krishnakumar, S.K.; Rajavel, R.; Krishnamohan, S.; Vijayaganth, G. Experimental study on aluminium based sandwich composite with polypropylene foam sheet. *Mater. Today Proc.* **2020**, *24*, 746–753. [CrossRef]
4. Arbaoui, J.; Schmitt, Y.; Pierrot, J.-L.; Royer, F.-X. Numerical simulation and experimental bending behaviour of multi-layer sandwich structures. *J. Theor. Appl. Mech.* **2014**, *52*, 431–442.
5. Li, X.; Lin, Y.; Lu, F. Numerical simulation on in-plane deformation characteristics of lightweight aluminum honeycomb under direct and indirect explosion. *Materials* **2019**, *12*, 2222. [CrossRef] [PubMed]
6. Wang, Z.; Li, Z.; Xiong, W. Numerical study on three-point bending behavior of honeycomb sandwich with ceramic tile. *Compos. Part B Eng.* **2019**, *167*, 63–70. [CrossRef]
7. Xie, S.; Feng, Z.; Zhou, H.; Wang, D. Three-point bending behavior of nomex honeycomb sandwich panels: Experiment and simulation. *Mech. Adv. Mater. Struct.* **2021**, *28*, 1917–1931. [CrossRef]
8. Bi, G.; Yin, J.; Wang, Z.; Jia, Z. Micro fracture behavior of composite honeycomb sandwich structure. *Materials* **2021**, *14*, 135. [CrossRef]
9. Gao, X.; Zhang, M.; Huang, Y.; Sang, L.; Hou, W. Experimental and numerical investigation of thermoplastic honeycomb sandwich structures under bending loading. *Thin Walled Struct.* **2020**, *155*, 106961. [CrossRef]
10. Wu, X.; Yu, H.; Guo, L.; Zhang, L.; Sun, X.; Chai, Z. Experimental and numerical investigation of static and fatigue behaviors of composites honeycomb sandwich structure. *Compos. Struct.* **2019**, *213*, 165–172. [CrossRef]
11. Yu, G.-C.; Feng, L.-J.; Wu, L.-Z. Thermal and mechanical properties of a multifunctional composite square honeycomb sandwich structure. *Mater. Des.* **2016**, *102*, 238–246. [CrossRef]
12. Liu, J.; Wang, Z.; Hui, D. Blast resistance and parametric study of sandwich structure consisting of honeycomb core filled with circular metallic tubes. *Compos. Part B Eng.* **2018**, *145*, 261–269. [CrossRef]
13. Gladkovsky, S.V.; Kuteneva, S.V.; Sergeev, S.N. Microstructure and mechanical properties of sandwich copper/steel composites produced by explosive welding. *Mater. Charact.* **2019**, *154*, 294–303. [CrossRef]
14. Nilsson, P.; Al-Emrani, M.; Atashipour, S.R. Fatigue-strength assessment of laser welds in corrugated core steel sandwich panels. *J. Constr. Steel Res.* **2020**, *164*, 105797. [CrossRef]
15. Nilsson, P.; Hedegård, J.; Al-Emrani, M.; Atashipour, S.R. The impact of production-dependent geometric properties on fatigue-relevant stresses in laser-welded corrugated core steel sandwich panels. *Weld World* **2019**, *63*, 1801–1818. [CrossRef]
16. Lee, H.S.; Yoon, J.H.; Yoo, J.T. Application of solid state joining technologies in aerospace parts. *Key Eng. Mater.* **2020**, *837*, 69–73. [CrossRef]

17. Pragana, J.P.; Contreiras, T.R.; Bragança, I.M.; Silva, C.M.; Alves, L.M.; Martins, P.A. Joining by forming of metal–polymer sandwich composite panels. *Proc. Inst. Mech. Eng. Part B J. Eng. Manuf.* **2019**, *233*, 2089–2098. [CrossRef]
18. Ablat, M.A.; Qattawi, A. Numerical simulation of sheet metal forming: A review. *Int. J. Adv. Manuf. Technol.* **2017**, *89*, 1235–1250. [CrossRef]
19. Zajkani, A.; Hajbarati, H. An analytical modeling for springback prediction during U-bending process of advanced high-strength steels based on anisotropic nonlinear kinematic hardening model. *Int. J. Adv. Manuf. Technol.* **2017**, *90*, 349–359. [CrossRef]
20. Li, Y.; Liang, Z.; Zhang, Z.; Zou, T.; Li, D.; Ding, S.; Xiao, H.; Shi, L. An analytical model for rapid prediction and compensation of springback for chain-die forming of an AHSS U-channel. *Int. J. Mech. Sci.* **2019**, *159*, 195–212. [CrossRef]
21. Hajbarati, H.; Zajkani, A. A novel analytical model to predict springback of DP780 steel based on modified Yoshida-Uemori two-surface hardening model. *Int. J. Mater.* **2019**, *12*, 441–455. [CrossRef]
22. Yang, X.; Choi, C.; Sever, N.K.; Altan, T. Prediction of springback in air-bending of advanced high strength steel (DP780) considering Young's modulus variation and with a piecewise hardening function. *Int. J. Mech. Sci.* **2016**, *105*, 266–272. [CrossRef]
23. Liu, J.; Xue, W. Unconstrained bending and springback behaviors of aluminum-polymer sandwich sheets. *Int. J. Adv. Manuf. Technol.* **2017**, *91*, 1517–1529. [CrossRef]
24. Gautam, V.; Raut, V.M.; Kumar, D.R. Analytical prediction of springback in bending of tailor-welded blanks incorporating effect of anisotropy and weld zone properties. *Proc. Inst. Mech. Eng. Part L J. Mater. Des. Appl.* **2018**, *232*, 294–306. [CrossRef]
25. Zhang, Z.K.; Wu, J.J.; Guo, R.C.; Wang, M.Z.; Li, F.F.; Guo, S.C.; Wang, Y.A.; Liu, W.P. A semi-analytical method for the springback prediction of thick-walled 3D tubes. *Mater. Des.* **2016**, *99*, 57–67. [CrossRef]
26. Jin, L.; Yang, Y.-F.; Li, R.-Z.; Cui, Y.-W.; Jamil, M.; Li, L. study on springback straightening after bending of the U-section of TC4 material under high-temperature conditions. *Materials* **2020**, *13*, 1895. [CrossRef]
27. Wasif, M.; Iqbal, S.A.; Tufail, M.; Karim, H. Experimental analysis and prediction of springback in V-bending process of high-tensile strength steels. *Trans. Indian Inst. Met.* **2020**, *73*, 285–300. [CrossRef]
28. Panthi, S.K.; Ramakrishnan, N. Semi analytical modeling of springback in arc bending and effect of forming load. *Trans. Nonferrous Met. Soc. China* **2011**, *21*, 2276–2284. [CrossRef]
29. Han, Y.S.; Yang, W.H.; Choi, K.Y.; Kim, B.H. A Study on the effect of input parameters on springback prediction accuracy. In *AIP Conference Proceedings*; American Institute of Physics: College Park, MD, USA, 2011; Volume 1383, pp. 1137–1142. [CrossRef]
30. Mertin, C.; Stellmacher, T.; Schmitz, R.; Hirt, G. Enhanced springback prediction for bending of high-strength spring steel using material data from an inverse modelling approach. *Procedia Manuf.* **2019**, *29*, 153–160. [CrossRef]
31. Gu, B.; He, J.; Li, S.; Chen, Y.; Li, Y. Cyclic sheet metal test comparison and parameter calibration for springback prediction of dual-phase steel sheets. *J. Manuf. Sci. Eng.* **2017**, *139*, 91010. [CrossRef]
32. Li, Y.; Li, A.; Yue, Z.; Qiu, L.; Badreddine, H.; Gao, J.; Wang, Y. Springback prediction of AL6061 pipe in free bending process based on finite element and analytic methods. *Int. J. Adv. Manuf. Technol.* **2020**, *109*, 1789–1799. [CrossRef]
33. Seo, K.-Y.; Kim, J.-H.; Lee, H.-S.; Kim, J.H.; Kim, B.-M. Effect of constitutive equations on springback prediction accuracy in the TRIP1180 cold stamping. *Metals* **2018**, *8*, 18. [CrossRef]
34. Liu, X.; Cao, J.; Chai, X.; Liu, J.; Zhao, R.; Kong, N. Investigation of forming parameters on springback for ultra high strength steel considering Young's modulus variation in cold roll forming. *J. Manuf. Process.* **2017**, *29*, 289–297. [CrossRef]
35. Pouraliakbar, H.; Khalaj, G.; Jandaghi, M.R.; Khalaj, M.J. Study on the correlation of toughness with chemical composition and tensile test results in microalloyed API pipeline steels. *J. Min. Metall. B Metall.* **2015**, *51*, 173–178. [CrossRef]
36. Jamli, M.R.; Farid, N.M. The sustainability of neural network applications within finite element analysis in sheet metal forming: A review. *Measurement* **2019**, *138*, 446–460. [CrossRef]
37. Angsuseranee, N.; Pluphrach, G.; Watcharasresomroeng, B.; Songkroh, A. Springback and sidewall curl prediction in U-bending process of AHSS through finite element method and artificial neural network approach. *Songklanakarin J. Sci. Technol.* **2018**, *40*, 534–539.
38. Miranda, S.S.; Barbosa, M.R.; Santos, A.D.; Pacheco, J.B.; Amaral, R.L. Forming and springback prediction in press brake air bending combining finite element analysis and neural networks. *J. Strain Anal. Eng. Des.* **2018**, *53*, 584–601. [CrossRef]
39. Zhu, Y.X.; Liu, Y.L.; Yang, H.; Li, H.P. Development and application of the material constitutive model in springback prediction of cold-bending. *Mater. Des.* **2012**, *42*, 245–258. [CrossRef]
40. Chatti, S.; Fathallah, R. A study of the variations in elastic modulus and its effect on springback prediction. *Int. J. Mater.* **2014**, *7*, 19–29. [CrossRef]
41. Jung, J.; Jun, S.; Lee, H.-S.; Kim, B.-M.; Lee, M.-G.; Kim, J.H. Anisotropic hardening behaviour and springback of advanced high-Strength steels. *Metals* **2017**, *7*, 480. [CrossRef]
42. Baara, W.A.B.; Baharudin, B.T.H.T.B.; Anuar, M.K.; Ismail, M.I.S. Effect of elastic module degradation measurement in different sizes of the nonlinear isotropic–kinematic yield surface on springback prediction. *Metals* **2019**, *9*, 511. [CrossRef]
43. Mulidrán, P.; Spišák, E.; Tomáš, M.; Slota, J.; Majerníková, J. Numerical prediction and reduction of hat-shaped part springback made of dual-phase AHSS steel. *Metals* **2020**, *10*, 1119. [CrossRef]
44. Mulidrán, P.; Šiser, M.; Slota, J.; Spišák, E.; Sleziak, T. Numerical prediction of forming car body parts with emphasis on springback. *Metals* **2018**, *8*, 435. [CrossRef]
45. Trzepiecinski, T.; Lemu, H.G. Effect of computational parameters on springback prediction by numerical simulation. *Metals* **2017**, *7*, 380. [CrossRef]

46. Naofal, J.; Naeini, H.M.; Mazdak, S. Effects of hardening model and variation of elastic modulus on springback prediction in roll forming. *Metals* **2019**, *9*, 1005. [CrossRef]
47. Sumikawa, S.; Ishiwatari, A.; Hiramoto, J.; Urabe, T. Improvement of springback prediction accuracy using material model considering elastoplastic anisotropy and Bauschinger effect. *J. Mater. Process. Technol.* **2016**, *230*, 1–7. [CrossRef]
48. Slota, J.; Šiser, M.; Dvorák, M. Experimental and numerical analysis of springback behavior of aluminum alloys. *Strength Mater.* **2017**, *49*, 565–574. [CrossRef]
49. Vegter, H.; van den Boogaard, A.H. A plane stress yield function for anisotropic sheet material by interpolation of biaxial stress states. *Int. J. Plast.* **2006**, *22*, 557–580. [CrossRef]
50. ESI Group. *Pam-Stamp 2020.5 User's Guide*; ESI Group: Paris, France, 2020.
51. Yoshida, F.; Uemori, T. A model of large-strain cyclic plasticity and its application to springback simulation. *Int. J. Mech. Sci.* **2003**, *45*, 1687–1702. [CrossRef]

machines

MDPI

Article

An Analytical Method for Tensor Visualization in a Plane

Tomáš Stejskal [1], Jozef Svetlík [1,*] and Jozef Dobránsky [2]

1 Department of Manufacturing Machinery, Faculty of Mechanical Engineering, The Technical University of Kosice, Letná 9, 04001 Košice, Slovakia; tomas.stejskal@tuke.sk
2 Department of Automobile and Manufacturing Technologies, Faculty of Manufacturing Technologies, The Technical University of Kosice, Letná 9, 04001 Košice, Slovakia; jozef.dobransky@tuke.sk
* Correspondence: jozef.svetlik@tuke.sk; Tel.: +421-55-603-2195

Abstract: The rendering of tensor glyphs is a progressive process of visualizing the vector space both in fluid dynamics and the latest medical scanning. Nowadays, the rendering accuracy is ensured by numerical methods based on interpolation of tensor functions. The tensor glyph functions to visualize significant properties of the vector space. Not all these properties are visualized at all times. The number of properties and their unambiguity depend on the method chosen. This work presents a direct analytical expression covering rank two tensors in a plane. Unlike the methods used so far, this method is accurate and unambiguous one for tensor visualization. The method was applied to the simplest tensor type, which presented an advantage for the method's analytical approach. The analytical approach to the planar case is significant also because it provides instruction on how to expand analytical calculations to cover higher spatial dimensions. In this way, numerical methods for tensor rendering can be replaced with an accurate analytical method.

Keywords: tensor glyph; golden section; vector space

Citation: Stejskal, T.; Svetlík, J.; Dobránsky, J. An Analytical Method for Tensor Visualization in a Plane. *Machines* **2022**, *10*, 89. https://doi.org/10.3390/machines10020089

Academic Editor: Benoit Eynard

Received: 17 December 2021
Accepted: 20 January 2022
Published: 25 January 2022

1. Introduction

The rendering of tensor glyphs is a progressive process of visualization in vector spaces. Tensor glyphs are objects that represent multidimensional data using several admissible rules [1]. The main uses are in visualizing tensors of fluid dynamics, stress tensors, or a Jacobian matrix of a velocity field [2], as well as diffuse tensors of magnetic resonance in medicine [3,4].

A rank two tensor glyph in 2D or 3D space must satisfy several requirements. The following requirements are fully satisfied in some glyph structures and satisfied only to a limited degree in others. The conditions to be satisfied by general tensors are invariance under isometric domain transformation, scaling invariance, direct encoding of real eigenvalues and eigenvectors, glyph uniqueness with respect to the relevant tensor, and continual glyph changes mapping continuous changes in the tensor.

Kratz et al. team focused on combining statistical and analytical methods, using especially Mohr diagrams in the area of stress tensors [5]. Some rendering techniques stemmed from knowledge of the tensor eigenvectors. With hyperstreamlining, these data were enhanced, thus improving the complexity of the visualization technology. However, the analysis was based on visualizing two-dimensional, scalar-derived charts, thus reducing information on tensor properties [6–8]. To proceed with the analysis of the entire tensor invariant part, Zobel and Scheuermann introduced the term extreme points [9].

Kindlmann and Schultz introduced superquadratic tensor glyphs that satisfy all of the requirements, but only where symmetric tensors are concerned [10,11].

In the work of Globus et al. [12], the condition of uniqueness was not satisfied, because only the real eigenvalues were represented in the form of an ellipsoid, and it did not hold for complex values. Similarly, Theisel et al. [13] showed glyphs that failed to satisfy the condition of uniqueness, as well as that of change continuity. Technical mechanics approaches the visualization of tautness tensors in several ways. Mohr circles [14] visualize only eigenvalues, and therefore lack constancy under rotation, which fails to satisfy the condition of invariance under isometric domain transformation. Lately, the research into tensor visualization has been exploring different metrics such as Frobenius, Euclidean, Wasserstein, and Fisher Rao [15].

All of the papers mentioned above address individual tensor properties through numerical methods, in which the results are made more exact through interpolation, subject to the preservation of the usual tensor invariants [16,17]. However, this condition is not satisfied at all times. Technical mechanics operate with, for example, the ellipse of inertia, [18], which violates the first tensor invariant, as pointed out below.

The ingenuity of the proposed method compared to the methods used so far lies in the accurate analytical calculation of tensor rendering. The proposed analytical calculation applied to rank two tensors in the plane is capable of precise rendering of all tensor properties in the form of an analytical tensor glyph.

We are not aware of the existence of a direct analytical calculation of a tensor shape in the form of graphic visualization. The analytical calculation proposed for rank two tensors in a plane makes it possible to accurately represent all tensor properties in the form of an analytical tensor glyph.

Analytical calculation of tensor glyphs was applied to the planar case, as it is the simplest one to compute. Section 2 describes decomposition of the general matrix into its symmetrical and antisymmetrical component. This facilitates further simplification of the analytical calculation with the sole focus on the symmetrical tensor. Section 3 presents how tensor invariants are determined, pointing out, among other things, the manner in which complex results in quadratic form can be represented by real numbers. Analytical calculations require the third invariant be defined, in addition to the two known tensor invariants, as it is directly related to the symmetrical matrix norm. The complete analytical procedure for tensor glyph computing is presented in Section 4. This section also features tensor glyph illustrations. Section 6 introduces a special tensor glyph shape, enabling measurement of the golden section proportions. The significance of this lies in the fact that such proportions will materialize under exactly the 45-degree angle of the reference vector viewed from the aspect of basic tensor glyph symmetry. This finding leads to new useful considerations, for example in terms of the optimization of topological structures in the field of materials strength [19].

2. Theoretical Background

All mathematical objects have underwent development that started with simpler objects. Before a tensor could be created, the vector needed to be defined. The vector definition was preceded by a scalar definition.

A multidimensional array can also be precisely described by an object of a tensor type. Tensor rank is determined from the number of simultaneous directions representing a physical value. It then follows that the number of tensor components is derived from the following formula:

$$C = N^R \tag{1}$$

where C is the number of components, N is the number of dimensions, and R is the tensor rank.

For example, the definition of tensor rank for a 3D space is as follows. A tensor of rank R is an array of 3^R values (in 3D space) called tensor components, which combine with multiple directional implicators (basis vectors) to form a quantity that does not vary with changes in the coordinate system [20].

Based on the above, a table of precisely defined tensor types and numbers of tensor components can be constructed (Table 1).

Higher dimensions are described in metric spaces by the defined spatial Pythagorean theorem. Spatial structures where the space dimension is smaller than the tensor rank are marked in grey. The position of planar tensors analyzed in the present paper is marked in green.

Table 1. Number of spatial object components.

Tensor Rank R	0	1	2	3	4
Common Name	Scalar	Vector	Second-Order Tensor	Third-Order Tensor	Fourth-Order Tensor
Dimensions N					
0	0^0	1	1	1	1
1	1	1	1	1	1
2	1	2	4	8	16
3	1	3	9	27	81
4	1	4	16	64	256

Most often, a tensor is defined as a multidimensional array. Formally, it is defined through indices that, subject to the observation of certain rules, enable transformation operations of matrix calculation [21]. The definition of a tensor as a multidimensional array satisfying the transformation law dates back to the work of Ricci [22]. Another tensor definition takes the form of a multilinear map. This definition better shows the independence of the tensor from its base in the geometric object sense [23,24]. In some mathematical applications, the tensor is defined based on the tensor product of vectors in a vector space. This definition leads to a tensor being an object without components, thus enabling the extension of the linear algebra concept to multilinear algebra [25,26].

According to Table 1, a rank two tensor in a plane has four components, which can be expressed in the form of a square matrix. This view leads to a simplified tensor definition, drawing on the planar case. The tensor definition derives from tensor components and the transformation prescription. The tensor components alone do not create a spatial object. A spatial object can emerge only if the tensor transformation is applied in every direction.

Building on this consideration, a tensor can be derived from the procedure used for rotating the vector or the coordinate system. The vector rotation involves a transformation element in the form of a special square matrix, with a functional relation existing between the individual values. This rotational matrix is not a tensor, but the calculation principle can be applied to tensors as well. A general tensor in a plane can be defined by four different real numbers, without a functional relation between the individual values. However, a mathematical decomposition of general matrices into their symmetric and antisymmetric parts is available.

Each general matrix, and hence the matrix of a general tensor, can be decomposed through a simple modification of the addition of the symmetric and antisymmetric tensors:

$$\mathbf{T} = \mathbf{T_S} + \mathbf{T_A} \tag{2}$$

$$\begin{bmatrix} b_{11} & b_{12} \\ b_{21} & b_{22} \end{bmatrix} = \begin{bmatrix} b_{11} & \frac{1}{2}(b_{12}+b_{21}) \\ \frac{1}{2}(b_{21}+b_{12}) & b_{22} \end{bmatrix} + \begin{bmatrix} 0 & \frac{1}{2}(b_{12}-b_{21}) \\ \frac{1}{2}(b_{21}-b_{12}) & 0 \end{bmatrix} \tag{3}$$

where $\mathbf{T_S}$ is the symmetric tensor and $\mathbf{T_A}$ is the antisymmetric tensor.

In terms of its size, the antisymmetric part does not depend on the general tensor's symmetric part. The general matrix expression of the general tensor is as follows:

$$\begin{bmatrix} b_{11} & b_{12} \\ b_{21} & b_{22} \end{bmatrix} = \begin{bmatrix} a_{11} & a_{12} \\ a_{21} & a_{22} \end{bmatrix} + \begin{bmatrix} c_{11} & c_{12} \\ c_{21} & c_{22} \end{bmatrix} \tag{4}$$

$$\mathbf{B} = \mathbf{A} + \mathbf{C} \tag{5}$$

where b_{ij} represents the general tensor components, a_{ij} represents the symmetric tensor components, and c_{ij} represents the antisymmetric tensor components.

3. Invariants of the Tensor in a Plane and Tensor Eigenvalues

Rotating the coordinate system with respect to the initial state yields a square matrix, but the information about the initial state prior to rotation is lost. In the initial state, the initial vector $\boldsymbol{r}_0$ is defined by angle α. If the rotation angle φ is zero, then the components of the initial vector $\boldsymbol{r}_0$ lie along the main coordinate axis. Unknown components of the initial vector, λ_1, λ_2, are the eigenvalues of the transformation matrix that emerges from rotating the coordinate system, and this matrix is diagonal.

Despite varying angle φ, a rotational matrix always preserves the determinant (6). In this case, the matrix eigenvalues equal one, so the initial angle must equal 45°. However, as demonstrated below, a rotational matrix cannot be a tensor:

$$\det\begin{vmatrix} \cos\varphi & -\sin\varphi \\ \sin\varphi & \cos\varphi \end{vmatrix} = 1 \tag{6}$$

Let us generalize this requirement also for the tensor matrix, whose initial state can be expressed as the product of the eigenvalues and the identity matrix:

$$\mathbf{A} = \lambda\mathbf{E} \tag{7}$$

where **A** is a square matrix and **E** is an identity matrix.

Let the determinant of any tensor matrix remain constant, too. Then the difference between the matrices results in the characteristic equation, and the respective determinant equals zero.

The λ parameter in this equation is an unknown variable. Adjustments lead to a quadratic equation:

$$\lambda^2 = \lambda(a_{11} + a_{22}) + (a_{11}a_{22} - a_{12}a_{21}) = 0 \tag{8}$$

$$\lambda^2 - \lambda m + n = 0 \tag{9}$$

Solving the quadratic equation, we arrive at the unknown values of λ_1 and λ_2. To represent the shape, it is necessary to determine the domain of values for real roots only. The validity lies in the value interval of the initial angle α, as shown in Figure 1.

$$\alpha = \arctan\frac{\lambda_2}{\lambda_1} \quad for\ \alpha \in \left(-\frac{3\pi}{4}, \frac{\pi}{4}\right) \tag{10}$$

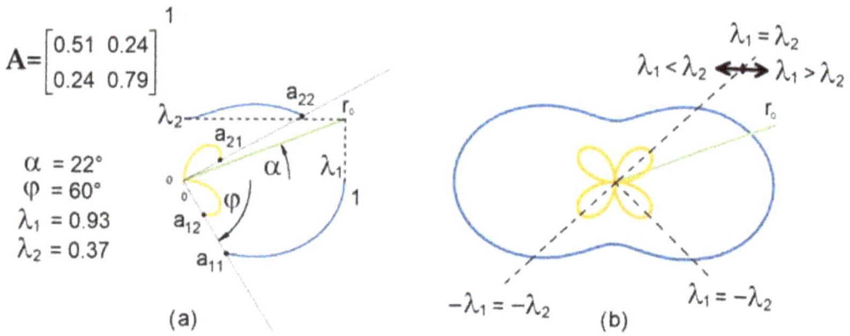

Figure 1. Symmetric tensor and its components rotated by (**a**) $\varphi = 60°$ and (**b**) $\varphi = 360°$.

The asymmetry of the domain of angle α is due to its approximation to the domain of the quadratic equation. When the angle is larger than any enabled by the domain, the quadratic equation has complex roots.

Each tensor can be expressed as a polynomial, which is called the quadratic form. For a planar tensor, it is enough to solve a polynomial of the second order, which corresponds to the quadratic equation. The equation roots can be real or complex. Complex roots emerge when the eigenvalues satisfy $\lambda_1 < \lambda_2$. This exchange is the result of the parameters

λ_1, λ_2 swapping their positions in the main diagonal square matrix. In such a case, the discriminant is a negative number and the corresponding matrix is negative-definite, because the eigenvalues do not change. Each complex solution matches exactly one real solution, which is arrived at by swapping the order of eigenvalues to satisfy $\lambda_1 > \lambda_2$. The eigenvalues' order does not result in a loss of information from the swap of their position. Thus, visualization can have the same form in both real and complex solutions. By reshuffling the order of eigenvalues, a solution in the range of real numbers is arrived at, because the tensor matrix is positive-definite. Such a result can be represented graphically.

If we rotate the coordinate system with respect to the tensor's initial state, the tensor's transformation matrix components $(a_{11}, a_{22}, a_{12}, a_{21})$ must duly change. The quadratic equation coefficients m, n cannot change, because the initial state does not change either. These quadratic equation coefficients are called tensor invariants. Therefore, the following relations automatically hold for invariants:

$$m = \mathrm{tr}\mathbf{A} = (a_{11} + a_{22}) = constant \tag{11}$$

Trace is the first invariant (sum of values of the main diagonal):

$$n = \det\mathbf{A} = (a_{11}a_{11} - a_{12}a_{21}) = constant \tag{12}$$

The determinant is the second invariant.

The above relations are commonly known. However, yet another tensor invariant exists, which has the form of the initial vector's square length. This corresponds to the sum of the squares of the eigenvalues (15). The size of this vector stays unchanged during transformation.

The invariant in the form of the initial vector's length can be expressed in absolute value:

$$|\mathbf{r}| = \sqrt{\lambda_1^2 + \lambda_2^2} \tag{13}$$

or by its square, which does not change either:

$$l = \lambda_1^2 + \lambda_2^2 = |\mathbf{r}|^2 = konst \tag{14}$$

Inserting it in the quadratic equations (Appendix A) and making adjustments, we arrive at the following result:

$$l = m^2 - 2n = \lambda_1^2 + \lambda_2^2 = a_{11}^2 + a_{22}^2 + 2a_{12}a_{21} \tag{15}$$

Further analysis uncovers the fact that the initial vector length corresponds to the symmetric part of the general matrix norm (Frobenius or Hilbert–Schmidt norm). A square symmetric matrix norm is as follows:

$$||\mathbf{T}_S||_F = \sqrt{\mathrm{tr}^2\mathbf{T}_S - 2\det\mathbf{T}_S} \tag{16}$$

Inserting the components into the equation below (17) yields the same result:

$$||\mathbf{T}_S||_F = \sqrt{(a_{11} + a_{22})^2 + 2(a_{11}a_{22} - a_{12}a_{21})} = \sqrt{a_{11}^2 + a_{22}^2 + 2a_{12}a_{21}} = \sqrt{l} \tag{17}$$

In the literature, a matrix norm is not considered to be an additional tensor invariant. Nevertheless, the additional invariant is necessary for the analytical calculation of the tensor's graphical representation.

The above formulas show that the rotational matrix maintains the determinant and the matrix norm, yet cannot be a tensor, as it does not satisfy the first invariant. This case is special because it must satisfy two conditions. Based on Equation (15) and the double root discriminant, we have:

$$m^2 - l - 2n = 0 \tag{18}$$

$$m^2 - 4n = 0 \tag{19}$$

It follows from these conditions that the following must hold for a rotational matrix:

$$l = 2n \tag{20}$$

Inserting and adjusting the components, we get:

$$a_{11} = a_{22} \tag{21}$$

which is satisfied only in the case of the rotational matrix; however, the first invariant requires a condition (11), which is not satisfied.

4. Analytical Procedure for Solving the Reverse Task

This section highlights the key to the tensor's graphical representation. The previous task can be reversed in the sense that the rotation angle φ can be the unknown value and the parameters $\lambda_1,\ \lambda_2$ can be known.

This task is not addressed in practice in the way outlined here, likely due to the fact that to solve it, two invariants in the plane are not enough, and it is suitable to derive the solution for a symmetric tensor only. A general asymmetric tensor can be addressed only when the symmetric tensor has been solved.

Since this task is not commonly addressed in the literature, we outline the manner of deriving transformational relations. For unambiguous solving of the reverse task, the above-mentioned relations are used simultaneously with the condition of tensor symmetry:

$$a_{12} = a_{21} \tag{22}$$

The rotational angle of the coordinate system is determined with the help of the angle formed by the matrix eigenvector $\mathbf{u}(u_1, u_2)$ and the initial coordinate system. The eigenvector can be derived from relation (7). After having been multiplied by the eigenvector, the following holds true:

$$\mathbf{Au} = \lambda \mathbf{Eu} \tag{23}$$

Only one eigenvector satisfies the above equality notation. Its orientation is determined by angle φ with respect to the initial tensor state. Making use of commonly known adjustments, we arrive at the following relation:

$$\tan\varphi = \frac{u_1}{u_2} = \frac{-(a_{11} - \lambda_1)}{a_{12}} \tag{24}$$

A symmetric tensor in a plane is unequivocally determined by three independent values. Rotating the coordinate system from its initial state will change the tensor components, but the tensor will remain the same; it will just be represented in a different direction. Thus, a tensor is an omnidirectional object. Based on this notion, the tensor shape can be represented in a plane through components under a given angle of the coordinate system's rotation. Gradually changing the angle of rotation all the way to 360°, some shape characteristics of a symmetrical tensor should emerge in the plane.

Based on the above-mentioned formulas, the $a_{22}(\lambda_1, \lambda_2, \varphi)$ component can be expressed, resulting in a quadratic form after derivation (Appendix B):

$$a_{22_{1,2}} = \frac{2\lambda_2 \cot^2\varphi + \lambda_1 + \lambda_2 \mp \sqrt{(2\lambda_2 \cot^2\varphi + \lambda_1 + \lambda_2)^2 - 4(1 + \cot^2\varphi)(\lambda_1\lambda_2 + \lambda_2^2 \cot^2\varphi)}}{2(1 + \cot^2\varphi)} \tag{25}$$

One root of the quadratic equation corresponds to the initial state and the other to the state rotated by angle φ. In terms of solving the task, only the rotated state is of interest. Subsequently, the remaining components of the symmetric tensor are expressed as follows:

$$a_{11} = \lambda_1 + \lambda_2 - a_{22} \tag{26}$$

$$a_{21} = \cot\varphi(\lambda_1 - a_{11}) \tag{27}$$

$$a_{12} = a_{21} \tag{28}$$

The graphical representation creates an omnidirectional shape of the symmetric tensor for the initial vector $\mathbf{r_0}(\lambda_1, \lambda_2)$, as shown in Figure 1. Figure 1b shows the point of the mutual swap of the values on the main diagonal, so that the solutions of the equation be real numbers.

Figures 2–4 show symmetric tensor glyphs under various angles α.

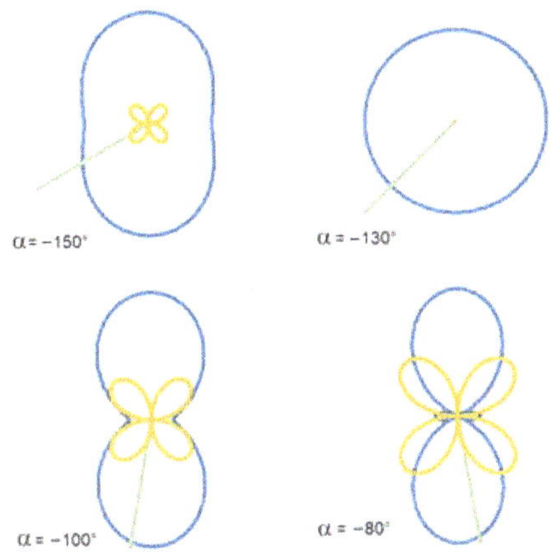

Figure 2. Symmetric tensor glyphs: $\alpha = -150°, -135°, -100°,$ and $-80°$.

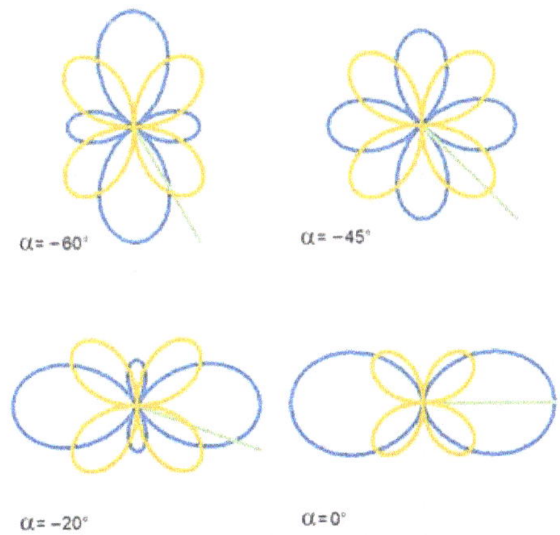

Figure 3. Symmetric tensor glyphs: $\alpha = -60°, -45°, -20°,$ and $0°$.

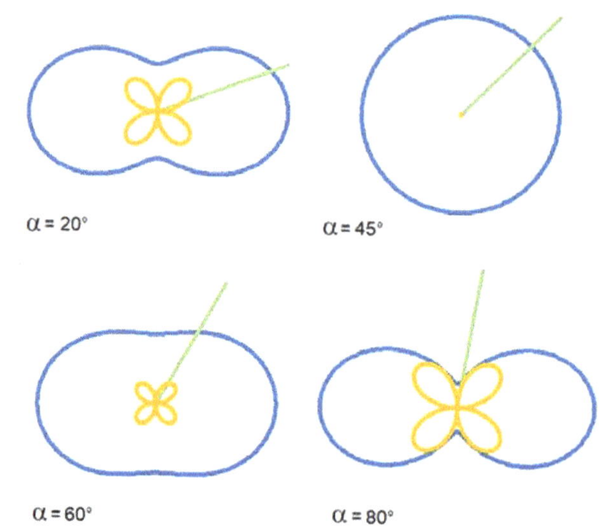

Figure 4. Symmetric tensor glyphs: $\alpha = 20^\circ, 45^\circ, 60^\circ$, and 80°.

As shown, symmetric tensor glyphs repeat their shapes with changes in the initial angle. Different shapes are limited to a single quadrant.

5. General Tensor

The previous section introduced the symmetric tensor with three freely exchangeable components. The antisymmetric tensor features only one freely exchangeable component. Therefore, the only possible geometric shape corresponding to the antisymmetric tensor is a circle. If it should be a different shape, the tensor determinant would not be constant when rotated by angle φ, which would violate the tensor invariant condition.

When a general tensor is expressed in a matrix form, the antisymmetric tensor components do not change with changes in rotation angle φ. The absolute value of these constant components represents the circle's radius:

$$c_{12} = \frac{1}{2}(b_{12} - b_{21}) \tag{29}$$

$$c_{21} = -c_{12} \tag{30}$$

The general tensor maintains the product of symmetric and antisymmetric tensor invariants:

$$n_G = n_S + n_A \tag{31}$$

$$m_G = m_S + m_A \tag{32}$$

$$l_G = l_S + l_A \tag{33}$$

with index G representing the general tensor invariants, and S and A being the invariant indices of the symmetric and antisymmetric tensors, respectively. Applying the above formulas, the general tensor can be represented graphically, as shown in Figures 5 and 6.

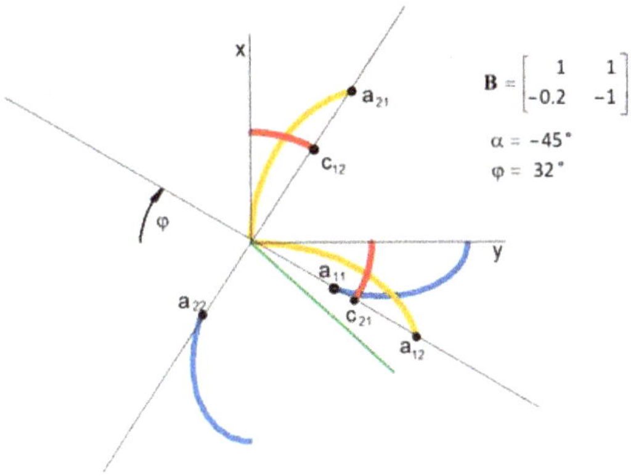

Figure 5. General tensor components rotated by $\varphi = 32°$.

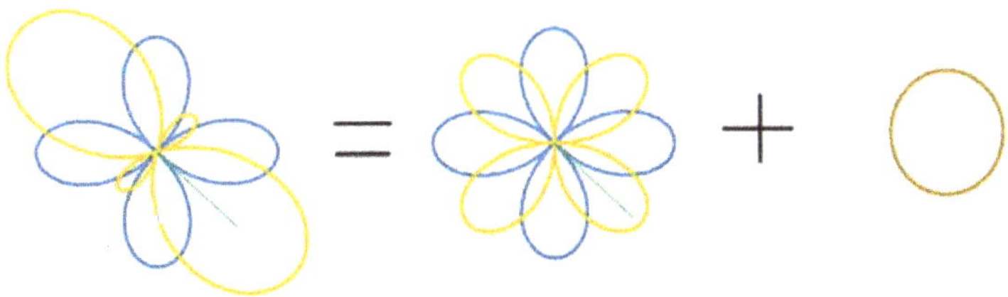

Figure 6. General tensor is a product of symmetric and antisymmetric tensors, using parameters in Figure 7.

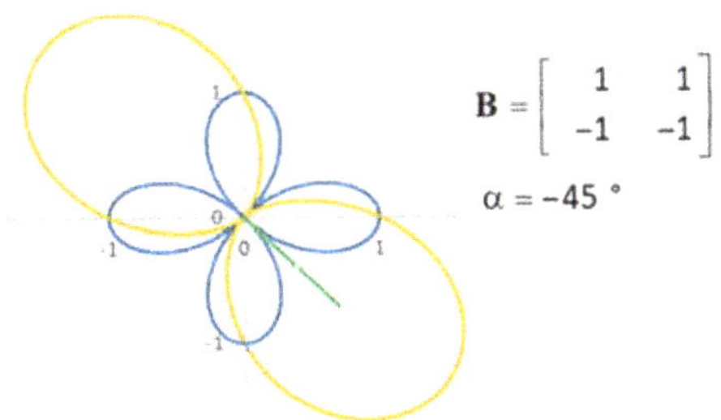

Figure 7. General tensor with $\alpha = -45°$.

Figures 7–10 show glyphs representing various general tensor values.

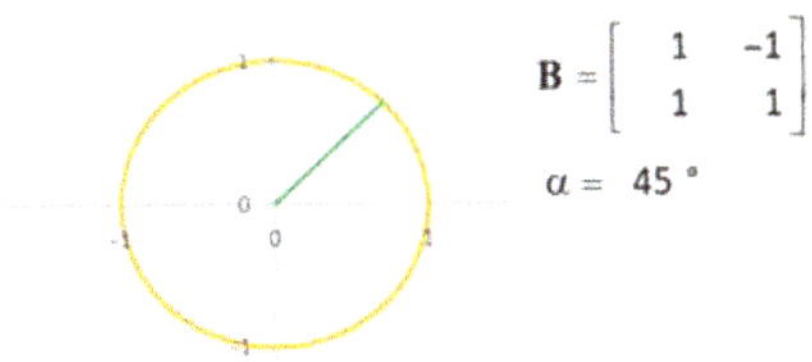

Figure 8. General tensor with $\alpha = 45°$.

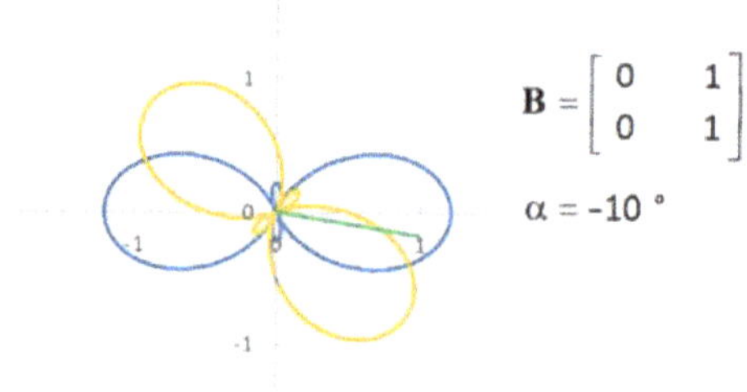

Figure 9. General tensor with $\alpha = -10°$.

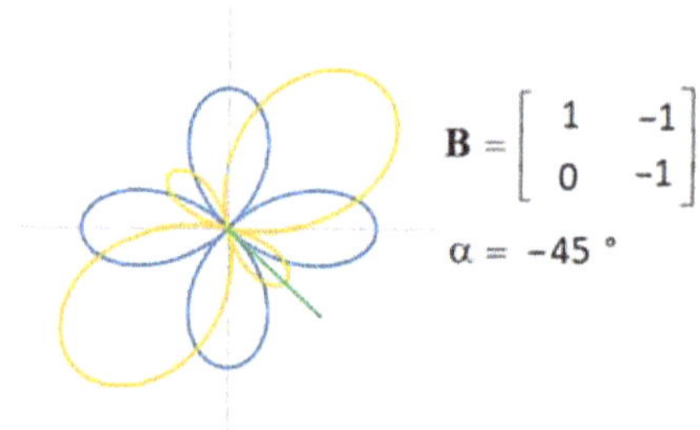

Figure 10. General tensor with $\alpha = -45°$.

Components of this rotational matrix can be a part of some general tensor. Upon defining the square matrix components, the matrix can be decomposed into symmetric and antisymmetric tensors. Typically, for this tensor, the initial angle always equals $\alpha = 45°$. However, with every rotation of the rotational matrix by angle φ, a new general tensor emerges. Figures 11–14 show examples of different tensors corresponding to some of the rotational matrix states.

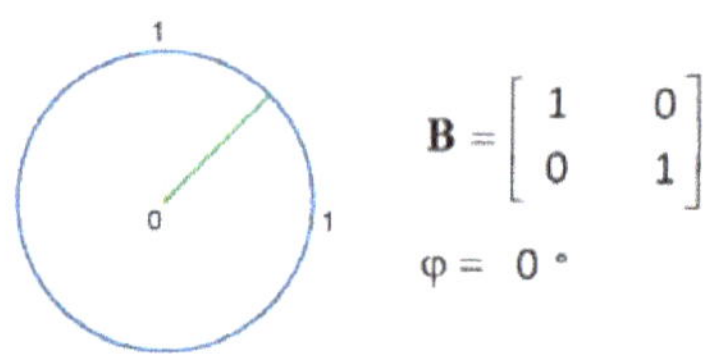

Figure 11. General tensor for rotational matrix with angle $\varphi = 0°$.

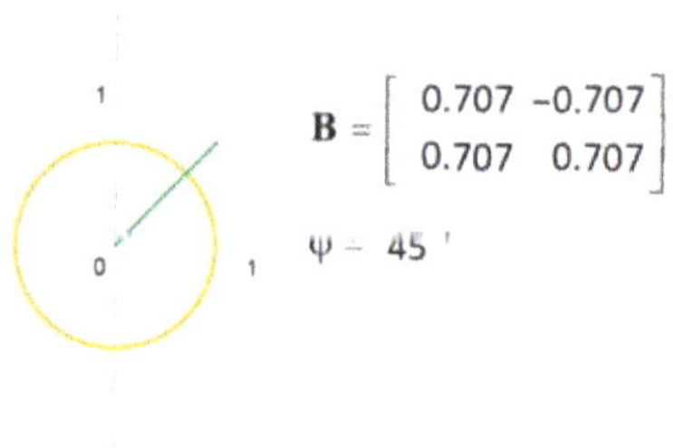

Figure 12. General tensor for rotational matrix with angle $\varphi = 45°$.

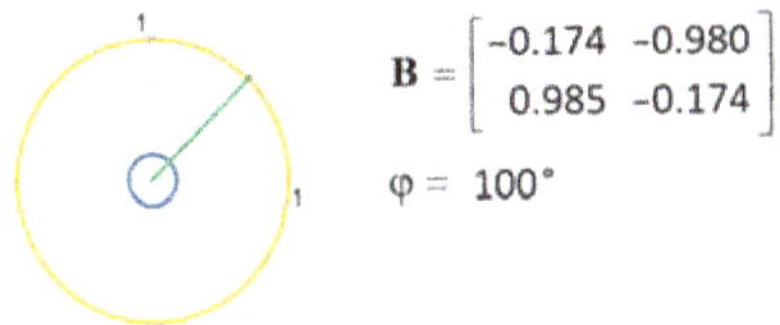

Figure 13. General tensor for rotational matrix with angle $\varphi = 100°$.

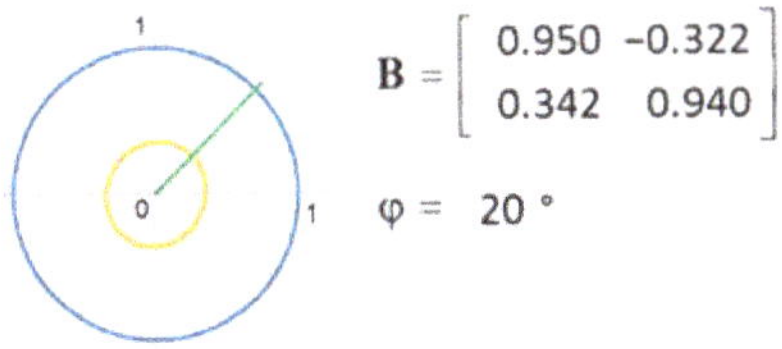

Figure 14. General tensor for rotational matrix with angle $\varphi = 20°$.

As can be seen, in all cases, the tensor shape is made of the diagonal matrix components (a_{11}, a_{22}) and (a_{12}, a_{21}), which give rise to two concentric circles.

6. Golden Section in Tensor Visualization

Shape proportions created by diagonal symmetric tensor components depend solely on the angle of the initial vector. With the initial state $\alpha = 0°$, the maximum values of tensor component parts in the x-axis direction display the golden section ratio φ (Figure 15).

$$\frac{a_{11_{x\ \max}} - a_{12_{x\ \max}}}{a_{11_{x\ \max}}} = \varphi = 0.618\ldots \tag{34}$$

This fundamental relation is introduced without being derived mathematically. It was obtained through measurement. If we view the symmetric tensor as a glyph, the shape of which changes symmetrically, in terms of glyph symmetry, this condition is achieved under a 45° angle.

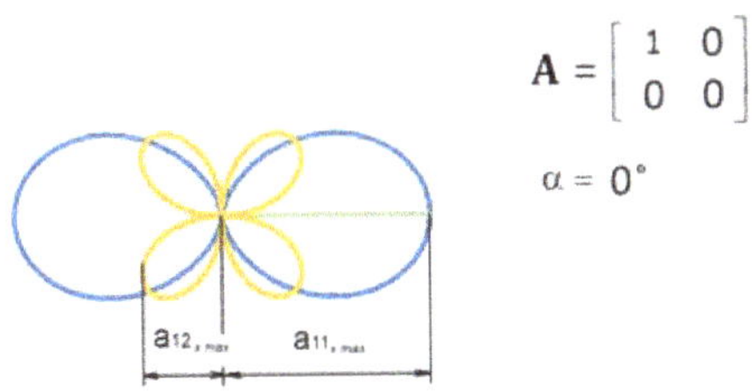

Figure 15. Symmetric tensor with $\alpha = 0°$.

7. Discussion and Conclusions

Tensors are fundamental building blocks used to describe physical phenomena. In the orthogonal coordinate system, their shape diversity is limited, as shown in Figures 2–14. It follows from the analysis that these analytical glyphs are unambiguous in terms of their shape. The analytical calculations presented in Section 4 are also highly likely to hold by analogy for 3D space. A substantial barrier to creating analytical calculations has been the absence of the third invariant in 2D space. In a similar fashion, the fourth tensor invariant for 3D space is also missing. It is necessary to derive the calculations for the symmetrical tensor first, followed by the asymetrical tensor. In addition, the change in the eigenvalues order on the main tensor diagonal is necessary to ensure that solving the equation always yields a real number instead of a complex one. The changed order does not lead to loss of information on tensor properties and the real numbers can be visualized in 3D rendering.

An interesting observation is that the symmetric tensor shape under zero angle of the initial state, as shown in Figure 15, has a proportional relation to the golden section ratio. This is significant because this commonly known principle, abundantly present in organic nature, was found at the heart of the tensor calculation. By means of the constant φ, the symmetric tensor is connected to the complex plane in the same way as the significant constants e and π. This is, therefore, likely to be a fundamental principle. The golden section ratio can also be encountered in relation to the strength of materials theory [19], as mentioned in the conclusion of an important paper by Michell [27]. These references to the golden section ratio are not limited to the plane only; they can also be found in attempts to describe higher dimensions. A paper dealing with transformations of a Pythagorean-like formula for surfaces immersed in three-dimensional space forming constant sectional curvature in a Riemann sphere examined the first, second, and third fundamental forms of the surface, proving that the immersed surfaces are totally round spheres with Gauss-like curvature $\varphi + c$, where φ is the golden ratio [28]. This, too, is a fundamental principle.

This paper's main contributions are as follows:

(a) A new analytical approach to deriving rank two tensor components in a plane;
(b) Analytic tensor glyph visualization of the rank two tensor in a plane by representing all components in every direction;
(c) Definition of a new tensor invariant that can be added to all dimensions;
(d) Pointing out the golden ratio law present in the representation of a planar tensor with a zero initial angle (as shown in Figure 15).

Tensor glyphs imaging is a progressive and useful method when applied to in-depth analysis of dynamic processes. Currently, only numerical methods for imaging exist, resting on different principles mentioned in the introductory section. Due to computation constraints, some principles are not capable of determining all tensor glyph properties. In analytical calculation, this problem ceases to exist. However, a special transfer system for transforming complex results into the real space had to be used; details of the technique used to that end were given in Section 3.

Author Contributions: Conceptualization, T.S.; methodology, T.S.; resources, T.S.; writing—original draft preparation, T.S.; writing—review and editing, J.S. and J.D.; project administration, J.S. All authors have read and agreed to the published version of the manuscript.

Funding: This work was supported by the Slovak Research and Development Agency under contract no. APVV-18-0413, KEGA 016TUKE-4/2021 and KEGA 063TUKE-4/2021.

Institutional Review Board Statement: Not applicable.

Informed Consent Statement: Not applicable.

Data Availability Statement: Not applicable.

Conflicts of Interest: The authors declare no conflict of interest.

Appendix A

Determining the third invariant l.

Let there be the following input equations:

$$\lambda^2 - \lambda m + n = 0 \tag{A1}$$

Let us insert the following components (A1) λ_1, λ_2 into the equation:

$$\begin{aligned} \lambda_1^2 - \lambda_1 m + n = 0 \\ \lambda_2^2 - \lambda_2 m + n = 0 \end{aligned} \quad \text{(A2)}$$

Let us add up the equations:

$$\lambda_1^2 + \lambda_2^2 - m(\lambda_1 + \lambda_2) + 2n = 0 \quad \text{(A3)}$$

Let us insert the parameters m, l:

$$l - m^2 + 2n = 0 \quad \text{(A4)}$$

$$l = m^2 - 2n \quad \text{(A5)}$$

As shown, the third invariant is also linearly independent of other invariants. Nonetheless, a certain form of quadratic dependence does exist between the invariants. We insert the tensor components (A5) into the equation:

$$l = (a_{11} + a_{22})^2 - 2(a_{11}a_{22} - a_{12}a_{21}) \quad \text{(A6)}$$

$$l = a_{11}^2 + 2a_{11}a_{22} + a_{22}^2 - 2a_{11}a_{22} + 2a_{12}a_{21} \quad \text{(A7)}$$

$$l = a_{11}^2 + a_{22}^2 + 2a_{12}a_{21} \quad \text{(A8)}$$

Appendix B

Calculating the components of a_{22}.

Let there be the following input equations:

$$m = \lambda_1 + \lambda_2 = a_{11} + a_{22} \quad \text{(A9)}$$

$$n = \lambda_1\lambda_2 = a_{11}a_{22} - a_{12}a_{21} \quad \text{(A10)}$$

$$l = \lambda_1\lambda_1 + \lambda_2\lambda_2 = a_{11} + a_{22} = m^2 - 2n = a_{11}^2 + a_{22}^2 + 2a_{12}a_{21} \quad \text{(A11)}$$

$$a_{12} = a_{21} \quad \text{(A12)}$$

$$\tan\varphi = \frac{-(a_{11} - \lambda_1)}{a_{12}} \quad \text{(A13)}$$

The input equations shall be adjusted as follows:

$$2n = m^2 - l \quad \text{(A14)}$$

$$a_{12} = a_{21} = \cot\varphi(\lambda_1 - a_{11}) \quad \text{(A15)}$$

Let us determine the product of the components a_{12}, a_{21}:

$$a_{11} = m - a_{22} = \lambda_1 + \lambda_2 - a_{22} \quad \text{(A16)}$$

$$a_{12}a_{21} = \cot^2\varphi(\lambda_1 - a_{11})^2 \quad \text{(A17)}$$

$$a_{12}a_{21} = \cot^2\varphi(\lambda_1 - \lambda_1 - \lambda_2 + a_{22})^2 \quad \text{(A18)}$$

$$a_{12}a_{21} = \cot^2\varphi(a_{22} - \lambda_2)^2 \tag{A19}$$

Let us express the component l:

$$l = a_{11}^2 + a_{22}^2 + 2a_{12}a_{21} \tag{A20}$$

$$l = (m - a_{22})^2 + a_{22}^2 + 2\cot^2\varphi(a_{22} - \lambda_2)^2 \tag{A21}$$

$$0 = -l + m^2 - 2ma_{22} + a_{22}^2 + a_{22}^2 + 2\cot^2\varphi(a_{22} - \lambda_2)^2 \tag{A22}$$

$$0 = 2n - 2ma_{22} + 2a_{22}^2 + 2\cot^2\varphi(a_{22} - \lambda_2)^2 \tag{A23}$$

$$0 = n - ma_{22} + a_{22}^2 + \cot^2\varphi(a_{22} - \lambda_2)^2 \tag{A24}$$

$$0 = n - ma_{22} + a_{22}^2 + \cot^2\varphi\left(a_{22}^2 - 2a_{22}\lambda_2 - \lambda_2^2\right) \tag{A25}$$

$$0 = n - ma_{22} + a_{22}^2 + a_{22}^2\cot^2\varphi - 2a_{22}\lambda_2 a_{22}^2\cot^2\varphi + \lambda_2^2\cot^2\varphi \tag{A26}$$

$$0 = a_{22}^2\left(1 + \cot^2\varphi\right) - a_{22}\left(2\lambda_2\cot^2\varphi + m\right) + n + \lambda_2^2\cot^2\varphi \tag{A27}$$

Component a_{22} is calculated from the quadratic equation (A27):

$$a_{22_{1,2}} = \frac{(2\lambda_2\cot^2\varphi + m) \mp \sqrt{(2\lambda_2\cot^2\varphi + m)^2 - 4(1 + \cot^2\varphi)(n + \lambda_2^2\cot^2\varphi)}}{2(1 + \cot^2\varphi)} \tag{A28}$$

$$a_{22_{1,2}} = \frac{2\lambda_2\cot^2\varphi + \lambda_1 + \lambda_2 \mp \sqrt{(2\lambda_2\cot^2\varphi + \lambda_1 + \lambda_2)^2 - 4(1 + \cot^2\varphi)(\lambda_1\lambda_2 + \lambda_2^2\cot^2\varphi)}}{2(1 + \cot^2\varphi)} \tag{A29}$$

References

1. Borgo, R.; Kehrer, J.; Chung, D.H.; Maguire, E.; Laramee, R.S.; Hauser, H.; Ward, M.; Chen, M. Glyph-based visualization: Foundations, design guidelines, techniques and applications. In *Eurographics State of the Art Reports*; The Eurographics Association: Hannover, Germany, 2013; pp. 39–63.
2. Gerrits, T.; Rössl, C.; Holger Theisel, H. Glyphs for general second-order 2d and 3d tensors. *IEEE Trans. Vis. Comput. Graph.* **2016**, *23*, 980–989. [CrossRef] [PubMed]
3. Sarwar, T.; Ramamohanarao, K.; Zalesky, A. Mapping connectomes with diffusion MRI: Deterministic or probabilistic tractography? *Magn. Reson. Med.* **2019**, *81*, 1368–1384. [CrossRef] [PubMed]
4. Muhammed, A.M.; Aswathi, V. Analysis of Visualization Techniques in Diffusion Tensor Imaging (DTI). In Proceedings of the 2018 Second International Conference on Advances in Electronics, Computers and Communications (ICAECC), Bengaluru, India, 9–10 February 2018.
5. Kratz, A.; Björn, M.; Hotz, I. A visual approach to analysis of stress tensor fields. In *Scientific Visualization: Interactions, Features, Metaphors*, 1st ed.; Hagen, H., Ed.; Schloss Dagstuhl—Leibniz-Zentrum für Informatik GmbH: Kaiserslautern, Germany, 2011; Volume 2, pp. 188–211.
6. Hashash, Y.; Yao, J.; Wotring, D. Glyph and hyperstreamline representation of stress and strain tensors and material constitutive response. *Int. J. Numer. Anal. Methods Geomech.* **2003**, *27*, 603–626. [CrossRef]
7. Schroeder, W.; Martin, K.M.; Lorensen, W. The design and implementation of an object-oriented toolkit for 3D graphics and visualization. In Proceedings of the 7th Annual IEEE Conference on Visualization (Visualization 96), San Francisco, CA, USA, 27 October–1 November 1996.
8. Delmarcelle, T.; Hesselink, L. Visualizing second-order tensor fields with hyperstreamlines. *IEEE Comput. Graph. Appl.* **1993**, *13*, 25–33. [CrossRef]

9. Zobel, V.; Scheuermann, G. Extremal curves and surfaces in symmetric tensor fields. *Visual Comput.* **2018**, *34*, 1427–1442. [CrossRef]
10. Kindlmann, G. Superquadric tensor glyphs. In Proceedings of the Sixth Joint Eurographics-IEEE TCVG Conference on Visualization, Konstanz, Germany, 19–24 May 2004; pp. 147–154.
11. Schultz, T.; Kindlmann, G.L. Superquadric glyphs for symmetric second-order tensors. *IEEE TVCG* **2010**, *16*, 1595–1604. [CrossRef] [PubMed]
12. Globus, A.; Levit, C.; Lasinski, T. A tool for visualizing the topology of three-dimensional vector fields. In Proceedings of the the second annual IEEE conference on Visualization, San Diego, CA, USA, 22–25 October 1991; pp. 33–40.
13. Theisel, H.; Weinkauf, T.; Hege, H.C.; Seidel, H.P. Saddle connectors—An approach to visualizing the topological skeleton of complex 3d vector fields. In Proceedings of the 14th IEEE Visualization, Washington, DC, USA, 22–24 October 2003; pp. 225–232.
14. Crossno, P.; Rogers, D.H.; Brannon, R.M.; Coblentz, D. Visualization of salt-induced stress perturbations. In Proceedings of the 15th IEEE Visualization Conference, Austin, TX, USA, 10–15 October 2004; pp. 369–376.
15. Feragen, A.; Fuster, A. Geometries and interpolations for symmetric positive definite matrices. In *Modeling, Analysis, and Visualization of Anisotropy*, 1st ed.; Schultz, T., Özarslan, E., Hotz, I., Eds.; Springer: Dordrecht, The Netherlands, 2017; pp. 85–113.
16. Hlawitschka, M.; Scheuermann, G. HOT-lines: Tracking lines in higher order tensor fields. In Proceedings of the 16th IEEE Visualization Conference, Minneapolis, MN, USA, 23–28 October 2005; pp. 27–34.
17. Healy, D.; Timms, N.E.; Pearce, M.A. The variation and visualisation of elastic anisotropy in rock-forming minerals. *Solid Earth* **2020**, *11*, 259–286. [CrossRef]
18. Lord, N. The moment of inertia of an elliptical wire. *Math. Gaz.* **2014**, *98*, 121–125. [CrossRef]
19. Fleisch, D.A. *A Student's Guide to Vectors and Tensors*; Cambridge University Press: Cambridge, UK, 2011; p. 134.
20. Sharpe, R.W. *Differential Geometry: Cartan's Generalization of Klein's Erlangen Program*; Springer Science & Business Media: New York, NY, USA, 2000; pp. 194–200.
21. Kline, M. *Mathematical Thought from Ancient to Modern Times: Volume 3*; Oxford University Press: New York, NY, USA, 1990; pp. 1127–1128.
22. Lee, J.M. Smooth manifolds. In *Introduction to Smooth Manifolds*; Springer: New York, NY, USA, 2013; pp. 304–316.
23. Lang, S. The Tensor Product. In *Algebra. Graduate Texts in Mathematics*; Springer: New York, NY, USA, 2002; Volume 211, pp. 601–603.
24. Jeevanjee, N. *An Introduction to Tensors and Group Theory for Physicists*; Birkhäuser: New York, NY, USA, 2011; pp. 39–54.
25. Dodson, C.T.J.; Poston, T. Tensor Geometry Springer-Verlag. *Grad. Texts Math.* **1991**, *120*, 105–106.
26. Stejskal, T.; Dovica, M.; Svetlík, J.; Demeč, P.; Hrivniak, L.; Šašala, M. Establishing the Optimal Density of the Michell Truss Members. *Materials* **2020**, *13*, 3867. [CrossRef] [PubMed]
27. Michell, A.G.M. The limits of economy of material in frame-structures. *Philos. Mag.* **1904**, *8*, 589–597.
28. Aydin, M.E.; Mihai, A.A. Note on Surfaces in Space Forms with Pythagorean Fundamental Forms. *Mathematics* **2020**, *8*, 444. [CrossRef]

MDPI
St. Alban-Anlage 66
4052 Basel
Switzerland
Tel. +41 61 683 77 34
Fax +41 61 302 89 18
www.mdpi.com

Machines Editorial Office
E-mail: machines@mdpi.com
www.mdpi.com/journal/machines

www.ingramcontent.com/pod-product-compliance
Lightning Source LLC
Chambersburg PA
CBHW040820250926
51850CB00007BA/1223
* 9 7 8 3 0 3 6 5 6 7 4 6 4 *